GOR ■ Publications

Springer

Berlin
Heidelberg
New York
Hong Kong
London
Milan
Paris
Tokyo

Hans-Otto Günther
Paul van Beek
Editors

Advanced Planning and Scheduling Solutions in Process Industry

Mit 79 Abbildungen
und 82 Tabellen

Springer

Professor Dr. Hans-Otto Günther
Technical University of Berlin
Department of Production Management
Wilmersdorfer Straße 148
10585 Berlin
Germany

Professor Dr. Paul van Beek
Wageningen University
Operations Research & Logistics Group
and Management Studies Group
Hollandseweg 1
6706 KN Wageningen
The Netherlands

ISBN 3-540-00222-7 Springer-Verlag Berlin Heidelberg New York

Bibliographic information published by Die Deutsche Bibliothek
Die Deutsche Bibliothek lists this publication in the Deutsche Nationalbibliografie; detailed bibliographic data available in the internet at *http.//dnb.ddb.de*

Springer-Verlag Berlin Heidelberg New York
a member of BertelsmannSpringer Science + Business Media GmbH

http://www.springer.de

Printed in Germany

Cover design: design & production, Heidelberg

SPIN 10904022 42/3130 – 5 4 3 2 1 0 – Printed on acid-free paper

Table of Contents

Advanced planning and scheduling in process industry

Hans-Otto Günther[1] **and Paul van Beek**[2]

[1] Department of Production Management, Technical University Berlin, Wilmersdorfer Str. 148, 10585 Berlin, Germany (e-mail: ho.guenther@pm-berlin.net)

[2] Operations Research & Logistics Group, Wageningen University, Hollandseweg 1, 6706 KN Wageningen, The Netherlands (e-mail: Paul.vanBeek@wur.nl)

The past decade has shown an increasing level of interest, research and application of quantitative models and computer based planning systems in process industry. Mathematical models and related algorithms form the basis of so-called Advanced Planning Systems (APS) which have gained considerable attention in practice. APS are primarily concerned with supporting decision making activities in supply chain management at the strategic, tactical and operational decision level (cf. Knolmayer et al., 2002). In contrast to classical MRP-systems, APS take into account the limited availability of resources and employ true optimization techniques (cf. Fleischmann et al., 2003). In particular, OR methodology has been applied to analyze and support the design of supply networks, the production and distribution of final products to customers, the planning and scheduling of operations and control issues arising in the production of food and beverages, chemicals and pharmaceuticals, for instance (cf. Miller, 2001; Stadtler and Kilger, 2003).

This book provides successful solutions to problems of production planning and scheduling, logistics and supply chain management. Fifteen papers previously published in *"OR Spectrum – Quantitative Approaches in Management"*, the journal of the German OR society GOR, have been selected for publication in this volume. All papers have been peer-reviewed according to the standard of the journal. The book comprises reports on the state of the art, applications of quantitative models and methods, as well as case studies and success stories from industry. Moreover, it reflects recent developments in co-operative decision support, such as multi-agent systems, as well as environmental issues. The book is divided into three parts. Part I presents contributions to the methodologies on which APS are based. Part II is focussed on industrial applications and case studies. Finally, Part III covers environmental aspects which are of increasing importance in many types of process industries. All contributions in this book deal with production and logistics problems arising in process industry.

Process industries

Among the various industries which have adopted advanced planning techniques, the so-called process industries play an outstanding role. These industries are characterized by repetitive production operations carrying out specific physical (e.g. blending or milling) and chemical reactions. Products are gained in bulk, often liquid or gaseous, rather than in discrete units. Processing can take place with batches or with continuous flow of material. Examples of process industries include production of food and beverages, the chemical, pharmaceutical, steel and semiconductor industries, as well as the generation of energy.

In particular, the chemical industry faces an increasing cost burden implied by a greater internationalisation of the markets. Compared to the discrete parts manufacturing industries, however, achieving competitive advantages through efficient reorganisation of the supply chains proves much more difficult. This is due, on the one hand, to the complex processing conditions in chemical production. On the other hand, the potential to rationalise is often limited by improvements in chemical process engineering. Hence, optimising the supply chain by use of advanced planning systems provides new opportunities to tackle the ever-rising cost burden and to strengthen one's own competitive position.

Another also important process industry concerns the food industry. Also this type of industry has adopted advanced planning techniques during the last decade. Successful and competitive food supply systems require that crucial elements like costs, quality, technology and safety be taken into account in a multidisciplinary way. Furthermore, the environmental impact of (agricultural) production, processing and distribution is an increasing priority both in industry and in academia. The same applies to the chemical industry. This has stimulated the development of so-called "green supply chains" (see Jongen and Meulenberg, 1998).

Process industries show a considerably increased complexity compared to discrete parts manufacturing (cf. Reklaitis et al., 1996). For instance, the complexity of scheduling chemical processes is determined by such factors as batch size constraints, shared intermediates, flexible proportions of input and output goods, production of by-products, limited predictability of processing times and yields, mixing and blending processes, carrying out processes without interruption, use of multi-purpose equipment, sequence and usage dependent cleaning operations, finite intermediate storage and use of product specific storage devices, cyclical material flows, no-wait production for certain types of products, usage of secondary resources, such as energy, steam, or cooling water, and complex packaging and filling operations. Very often, time and cost intensive cleaning procedures as well as the necessity of detailed quality control after each batch are the major motivations for operating the production system in campaign mode, i.e. the equipment units required by a particular type of product or process are set up according to the corresponding recipe. When set-up and cleaning efforts are significant, e.g. in the production of specialty chemicals, only a few campaigns of a specific product, possibly, merely a single campaign are carried out each year. In the case of pharmaceutical ingredients, for instance, the processing mode is subject to strict legal regulations. For the sake of pureness, exhaustive cleaning operations are prescribed, which can take as long as two weeks. As a consequence, most of the scheduling approaches developed for discrete parts manufacturing are hardly applicable. Hence, despite the analogy of the fundamental planning and scheduling problems, there are a number of major issues which have to be reflected by the mathematical models and solution algorithms employed.

In recent years, multi-site production planning has emerged as one of the most challenging problems in the process industry. As a consequence, the focus in production planning and scheduling is shifting from the management of individual chemical processes to a holistic view of the various logistics and production stages, i.e. an approach in which suppliers, production plants and customers are considered as constituents of an integrated network.

Methodology

Due to the great advances in information technology and in the development of optimization algorithms and modelling tools, significant progress in the application of OR methodology has been achieved, particularly in the field of supply chain management (cf., Zijm, 2000). Many of the successful implementations of OR methodology in industry represent solutions tailored to the needs of the specific application environment and are developed without the aid of comprehensive software systems designed for the support of enterprise-wide planning functions. Others are intended to interact with the available APS or ERP systems employed by the particular organization. As a key element, advanced planning solutions utilize models and algorithms to determine optimal or at least reasonable values of the decision variables, e.g. lot sizes or locations, while considering all the constraints arising from the particular planning problem. In some cases, simple heuristics can be used which are based on rules of thumb and common understanding. Especially, when a quick response to unpredictable variations in supply, demand or capacity is needed, efficient heuristics are often preferred to exact optimization methods which usually require considerable computational effort. Large scale optimization problems, for instance, strategic design of logistics networks or multi-site supply network planning, require sophisticated optimization techniques. To support these decisions, APS also provide exact optimization techniques mainly based on mathematical programming (LP, MILP).

Despite the limitations of general purpose optimization algorithms, mathematical programming is considered as a standard method to solve a variety of production and logistics problems in the process industries, since this approach offers the advantage of being able to incorporate very easily the particular operating mode of the production system and the network of material flows between the various elements of a supply chain or between different units within a plant. Another advantage of the general framework of mathematical programming is the availability of powerful standard optimization software. Therefore, this approach is adopted in many industrial applications as well as research projects for developing an application-oriented solution methodology. Undoubtedly, special purpose algorithms are more efficient in a number of applications. This is true for solution procedures based on project scheduling methods, for instance, as well as problem-specific heuristics which, in many cases, allow solutions to be determined in real time. The major disadvantage of special purpose algorithms, however, is that they have to be redesigned, as soon as certain application features change, or even have to be abandoned.

There is a tremendous body of literature dealing with mathematical methodology applied in planning and scheduling. The first paper by *Kallrath "Planning and scheduling in the process industry"* provides an overview of the state-of-the-art of planning and scheduling problems with special emphasis on the chemical process industries. The approaches laid out particularly consider the difficulty of production planning and scheduling within complex multi-stage, capacitated production systems. In the appendix, a benchmark problem is summarized which has already been used extensively for research on planning and scheduling in chemical batch production. The contribution of the paper by *Neumann, Schwindt and Trautmann "Advanced production scheduling for batch plants in process industries"* has two issues. At first, it outlines the architecture of an APS for process industries and illustrates the various modules of which such systems consist. At second, it presents a powerful approach for batching and batch scheduling in chemical batch plants. The following paper by *Grunow, Günther and Lehmann "Campaign planning for multi-stage batch processes in the chemical industry"* addresses the problem of co-ordinating campaigns within a multi-site production network in the chemical industry. Aggregation techniques and hierarchical model formulations based on mathematical and constraint programming are suggested which allow the user to determine campaign schedules and corresponding equipment configurations with only reasonable computational effort. The subsequent paper by *Timpe "Solving planning and scheduling problems with combined integer and constraint programming"* presents a very efficient solution strategy based on combined integer and constraint programming. Its applicability is shown through a real-world planning and scheduling problem from the chemical industry. The paper by *Ivanescu, Fransoo and Bertrand "Makespan estimation and order acceptance in batch process industries when processing times are uncertain"* examines the performance of aggregate makespan estimation models. The authors consider uncertain processing times and provide a theoretical basis for order acceptance policies which can be used instead of detailed scheduling models. The final paper of the methodology section of this volume focuses on collaborative supply chain decision making. In their paper *"A multi-agent system for chemical supply chain simulation and management support"*, *García-Flores and Wang* describe the architecture of a multi-agent system that can be used to simulate the dynamic behaviour and support the management of chemical supply chains over the Internet.

Industrial applications and case studies

Meanwhile, developers of commercially available software for production planning and control have recognized that considerable cost savings and faster service in supply chains may be achieved by making use of mathematical methods. This development produced numerous applications of APS in the process industries. At the same time, many industrial companies have strengthened their efforts to employ mathematical methods for solving different problems arising in the field of production and logistics. A major driving force behind this development lies in the globalization of the world economy which has facilitated the co-operation between different partners working together in world-wide logistics networks. Especially, in the chemical industry, huge amounts of goods and values are transferred between the various economic regions. Figures 1 and 2 exhibit the world network of major chemical and food trade flows, respectively. The corresponding manufacture and distribution of goods has to be managed from a global economic point of view as well as from a company perspective. Hence, considerable cost savings can be gained from optimizing the structure and the operations of complex supply networks linking plants, suppliers, distribution centres and customers. Consequently, in order to improve the performance of the entire logistic chain, more sophisticated quantitative approaches and more effective decision support are needed. Indeed, OR provides a wealth of suitable tools for supporting decisions at the strategic, tactical and operational level. OR can also be used for finding integrated tools combining these three decisions levels.

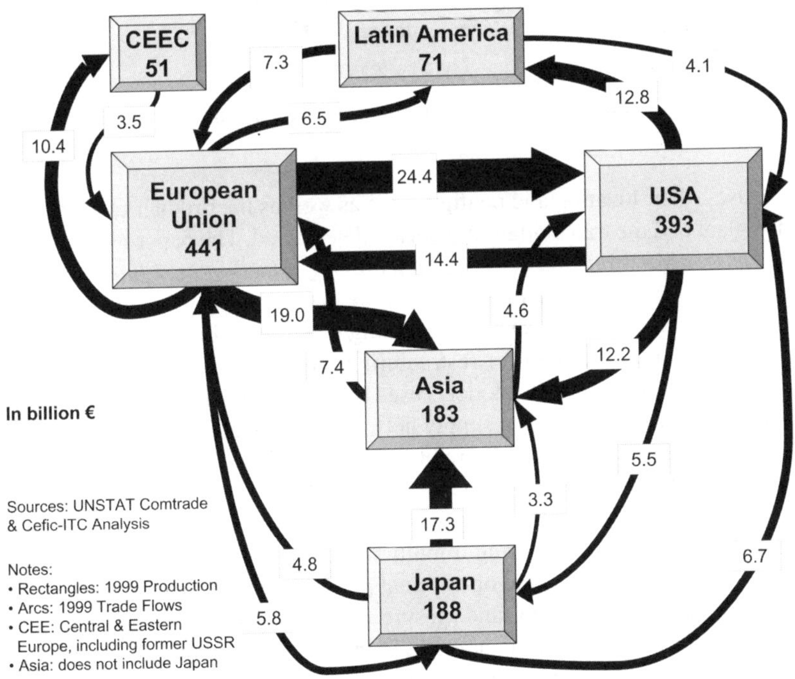

Fig. 1. World network of major chemical trade flows

Today, supply chain management is recognized as a core competitive weapon. To support and integrate activities along the supply chain, APS are being offered by different software vendors. Despite the different orientation and industry focus of APS, a common structure underlying most of the APS can be identified (cf. Meyr et al., 2003). At the *strategic network design phase*, the number of plants and

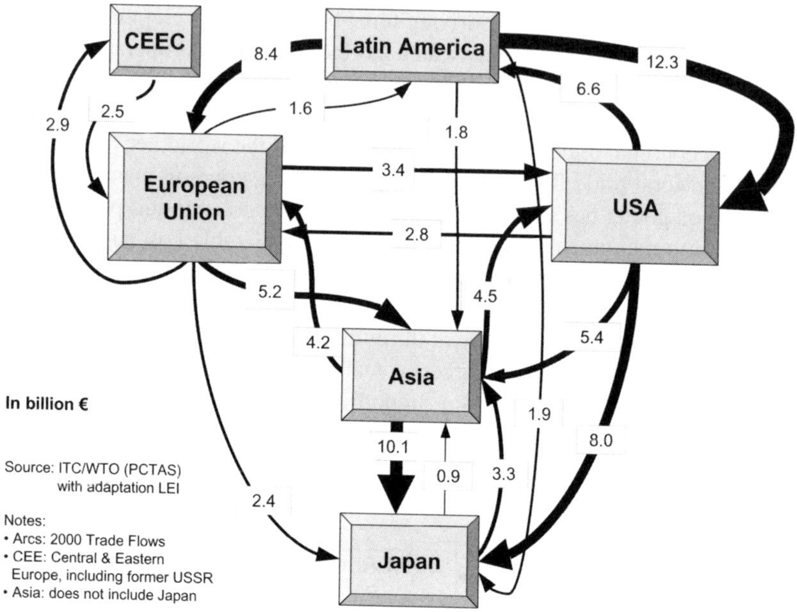

Fig. 2. World network of major food trade flows

warehouses, their location and configuration as well as the product types assigned to each plant and the transportation links are determined. The paper by *Wouda, van Beek, van der Vorst and Tacke "An application of mixed-integer linear programming models on the redesign of the supply network of Nutricia Dairy & Drinks Group in Hungary"* is a fine example showing how mixed-integer linear programming (MILP) can be applied for restructuring the supply network of a company producing dairy products. The model is used by the Nutricia Dairy & Drinks Group in Hungary to evaluate different supply network scenarios. As a result, production and transportation costs could be reduced considerably. Another MILP success story is presented by *Kallrath* in his second paper *"Combined strategic and operational planning - An MILP success story in chemical industry"*. He develops an MILP model for reconfiguring a multi-site production network in the chemical industry. In his strategic and operational planning approach, decision variables refer to the purchase, opening and closure of production units and to aggregate production plans for a long-term horizon. In the real world application considered, significant financial benefits could be gained.

The core module at the mid-term level of APS consists of *supply network planning*. In the chemical industry, production stages are often assigned to different plants at geographically distributed locations. Production activities within such multi-site logistics networks have to be well co-ordinated in order to avoid excessive inventories, inefficient capacity utilization and poor customer service. *"An integrated system solution for supply chain optimization in the chemical process industry"* is provided by *Berning, Brandenburg, Gürsoy, Mehta and Tölle*. They consider a complex supply network optimization problem at Bayer AG and apply a tailor-made optimization algorithm in order to determine production paths and facilities for each production campaign. Their approach is embedded into a collaborative planning concept which allows all plants working together as partners in a supply chain. In their paper *"Plant co-ordination in pharmaceutics supply networks", Grunow, Günther and Yang* consider the case of supply-network planning for the production of active ingredients in the chemical-pharmaceutical industry, where enormous cumulative lead-times of up to two years and complex networks of material flows between the different locations occur. They introduce several aggregation schemes and a novel MILP model formulation which is based on a continuous representation of time. Moreover, they propose an iterative near-optimal solution procedure which can be successfully applied to even exceptionally large real life problem instances.

In APS the short-term level includes *production planning and detailed scheduling* (PPDS) as the major module. Material requirements planning, transport planning, demand fulfilment and available-to-promise may be used to supplement PPDS in a specific application environment. In the process industries PPDS is often complicated by the diversity of processing tasks, the necessity of cleaning operations and constraints arising from individual process plans. Thus, no general solution approach exists. For a make-and-pack production system, consisting of a central production and a subsequent packaging stage, *Méndez and Cerdá* propose *"An MILP-based approach to the short-term scheduling of make-and-pack continuous production plants"*. Their approach is based on a continuous representation of time which allows a more compact model formulation compared to conventional model formulations based on a dense grid of discrete time periods. Examples from industry are used in order to illustrate the computational performance of the approach presented.

The final paper *"Long term planning of inter-company energy supply concepts"* in the application section of this volume by *Fichtner, Wietschel and Rentz* presents a detailed techno-economical analysis of energy supply networks. An optimization model is proposed in order to assess investment and long-term operational strategies for a network of five industrial companies. Computational results show that a combined cycle power plant would be the most promising option to fulfil the future energy demand of the companies involved.

Environmental issues

The entire logistic process, from the supply of raw materials, components and subassemblies to the distribution of end-products, the disposal of waste and non-usable by-products and the remanufacture of goods after usage, makes up a complex logistic chain which raises the need for integrating the design, management and control of production and logistics operations. Meanwhile, great efforts have been undertaken in many industries to develop and utilize environmentally friendly technologies. In addition, concepts for recycling and remanufacture of goods after their useful life have been developed and successfully applied. The subject of reverse logistics has also attracted increasing attention from researchers in the field of Management as well as Operations Research.

In their paper *"Business Case Océ: Reverse logistic network redesign for copiers" Krikke, van Harten and Schuur* address the strategic design of reverse logistic networks. The authors report on a business case study from Océ, a Dutch producer of copiers. By means of an MILP model, they investigate the economical consequences of different options to design the reverse logistics network. In addition, logistics performance indicators and issues of mathematical modelling are discussed. The two final papers in this volume deal with electronic scrap recovery. *"Models for recycling electronics end-of-life products"* are presented by *Sodhi and Reimer*. They develop a variety of mathematical programming models reflecting the different viewpoints of the generators, recyclers and material processors in the electronics industry. Their models aim at optimizing recycling operations and can also be used to evaluate the economic sustainability of electronics recycling. In their case study paper *"Integrated planning of acquisition, disassembly and bulk recycling: A case study on electronic scrap recovery"*, *Spengler, Ploog and Schröter* investigate the acquisition, disassembly and bulk recycling planning of a major electronic scrap recycling enterprise which recovers industrial and consumer electronics. They develop an integrated MILP model and demonstrate its application to determine the daily recovery program for a major electronic scrap recovery centre that faces limited processing capacities. Moreover, they show that considerable economic benefits can be gained compared with basic short-term recovery programs used in the examined recycling enterprise.

Final remarks

To summarize, this volume aims to provide insights into recent research efforts which employ quantitative methods for problem solving in different areas of production management and logistics occurring in process industry. It shows how OR can be applied successfully to both theoretical analysis and practical decision support. This is demonstrated not only for traditional areas of OR applications, but also for novel problem instances that emerge from new developments in the world of production and logistics. This book has greatly benefited from the cooperation among the authors, reviewers and editors. We would like to express our sincere thanks to the reviewers for their excellent and timely refereeing. Last, but not least, we thank all 36 authors for their contributions which made this book possible.

References

Fleischmann B, Meyr H, Wagner M (2003) Advanced planning. In: Stadtler H, Kilger C (eds) (2003) Supply chain management and advanced planning, 2nd edn, pp 71–96. Springer, Berlin Heidelberg New York

Jongen WMF, Meulenberg MTG (eds) (1998) Innovation of food production systems. Wageningen Press, Wageningen, The Netherlands

Knolmayer G, Mertens P, Zeier A (2002) Supply chain management based on SAP systems – Order management in manufacturing companies. Springer, Berlin Heidelberg New York

Meyr H, Wagner M, Rohde J (2003) Structure of advanced planning systems. In: Stadtler H, Kilger C (eds) (2003) Supply chain management and advanced planning, 2nd edn, pp 99–104. Springer, Berlin Heidelberg New York

Miller T (2001) Hierarchical operations and supply chain planning. Springer, Berlin Heidelberg New York

Reklaitis V, Sunol AK, Rippin DWT, Hortacsu O (eds) (1996) Batch processing systems engineering. Springer, Berlin Heidelberg New York

Stadtler H, Kilger C (eds) (2003) Supply chain management and advanced planning, 2nd edn. Springer, Berlin Heidelberg New York

Zijm WHM (2000) Towards intelligent manufacturing planning and control systems. OR Spectrum 22: 313–345

Planning and scheduling in the process industry

Josef Kallrath[1,2]

[1] BASF-AG, GVC/S (Scientific Computing) - C13, 67056 Ludwigshafen, Germany
(e-mail: josef.kallrath@basf-ag.de)
[2] Astronomy Department, University of Florida, Gainesville, FL 32661, USA
(e-mail: kallrath@astro.ufl.edu)

Abstract. Since there has been tremendous progress in planning and scheduling in the process industry during the last 20 years, it might be worthwhile to give an overview of the current state-of-the-art of planning and scheduling problems in the chemical process industry. This is the purpose of the current review which has the following structure: we start with some conceptional thoughts and some comments on special features of planning and scheduling problems in the process industry. In Section 2 the focus is on planning problems while in Section 3 different types of scheduling problems are discussed. Section 4 presents some solution approaches especially those applied to a benchmark problem which has received considerable interest during the last years. Section 5 allows a short view into the future of planning and scheduling. In the appendix we describe the Westenberger-Kallrath problem which has already been used extensively as a benchmark problem for planning and scheduling in the process industry.

Key words: Mixed integer programming – Supply chain optimization – Process industry – Planning – Scheduling

1 Introduction

1.1 Special features in the process industry

In the process industry *continuous* and *batch production systems* can be distinguished. There exists also semi-batch production which combines features from both. Plants producing only a limited number of products each in relatively high volume typically use special purpose equipment allowing a continuous flow of materials in *long campaigns*, i.e., there is a continuous stream of input and output products with no clearly defined start or end time. Alternatively, small quantities of a large number of products are preferably produced using multi-purpose equipment which are operated in *batch mode*, i.e., there is a *well-defined start-up*, e.g., filling in some products, well-defined follow-up steps defined by specific recipes, e.g., heating the product, adding other products and let them react, and a *clearly defined end*, e.g., extracting the finished product. Batch production involves an integer number of batches where a batch is the smallest quantity to be produced, e.g., 500 kg. Several batches of the same product following each other immediately establish a *campaign*. Production may be subject to certain constraints, e.g., campaigns are built up by a discrete number of batches, or a minimal campaign length (or production quantity) has to be observed. Within a fixed planning horizon, a certain product can be produced in several campaigns; this implies that campaigns have to be modelled as individual entities.

Another special feature in the refinery or petrochemical industry or process industry in general is the *pooling* problem (see, for instance, [28], or Chapter 11 in [42]), an almost classical problem in nonlinear optimization. It is also known as the *fuel mixture problem* in the refinery industry but it also occurs in blending problems in the food industry. The pooling problem refers to the intrinsic nonlinear problem of forcing the same (unknown) fractional composition of multi-component streams emerging from a pool, e.g., a tank or a splitter in a mass flow network. Structurally, this problem contains indefinite bilinear terms (products of variables) appearing in equality constraints, e.g., mass balances. The pooling problem occurs in all multi-component network flow problems in which the conservation of both mass flow and composition is required and both the flow and composition quantities are variable.

Non-linear programming (NLP) models have been used by the refining, chemical and other process industries for many years. These nonlinear problems are non-convex and either approximated by linear ones and solved by linear programming (LP) or approximated by a sequence of linear models. This sequential linear programming (SLP) technique is well established in the refinery industry but suffers from the drawback of yielding only locally optimum solutions. Although many users may identify obviously sub-optimal solutions from experience, there is no validation of those which are not obviously so, as this would require truly globally optimal solutions. From an end-user point of view, the problems of existing technology are becoming ever more acute. Since the market for products such as gasoline and chemicals are becoming increasingly amalgamated, many planning problems now necessarily involve multiple production facilities in geographically separate sites, with concomitant interactions and interconnections. These are hard to solve and much more prone to giving sub-optimal local solutions, particularly if they stretch over many time periods. However, recent advances in optimization algorithms have yielded experimental academic codes which do find truly globally optimal solutions to these NLP models. Non-convex nonlinear models are not restricted to the oil refining and petrochemical sector, but arise in logistics, network design, energy, environment, and waste management as well as finance and their solution asks for global optimization.

In the chemical process industry, the proper description of the reaction kinetics leads to exponential terms. If, in addition, plants operate in discrete modes or connections between various units, e.g., tanks and crackers or vacuum columns have to be chosen selectively, then mixed-integer nonlinear optimization problems need to be solved. Process network flow or process synthesis problems [30] usually fall into this category, too. Examples are heat exchanger or mass exchange networks.

Planning and scheduling is part of company-wide logistics and supply chain management. However, to distinguish between those topics, or even to distinguish between planning and scheduling is often a rather artificial approach. In reality, the border lines between all those areas are diffuse. There are strong overlaps between scheduling and planning in production, distribution or supply chain management and strategic planning.

1.2 Some comments on planning and scheduling in the process industry

Although the boundary between planning and scheduling is diffuse let us try to work out a few structural elements of planning and scheduling, which may include the following features:

- multi-purpose (multi-product, multi-mode) reactors,
- sequence-dependent set-up times and cleaning cost,
- combined divergent, convergent and cyclic material flows,
- non-preemptive processes (no-interruption), buffer times,
- multi-stage, batch & campaign production using shared intermediates,
- multi-component flow and nonlinear blending,
- finite intermediate storage, dedicated and variable tanks.

Structurally, these features often lead to allocation and sequencing problems, knapsack structures, or to the pooling problem. As there is no clear definition of the border line between planning and scheduling problems, we try to illuminate the subject from different angles by summarizing a few aspects and objectives of planning and scheduling and try to develop a kind of an informal definition serving as a platform. In production or supply chain planning, we usually consider material flow and balance equations connecting sources and sinks of a supply network. Time-indexed models using a relative coarse discretization of time, *e.g.*, a year, quarters, months or weeks are usually accurate enough. LP, MILP and MINLP technologies are often appropriate and successful for problems with a clear quantitative objective function as outlined in Section 2, or quantitative multi-criteria objectives.

In scheduling problems the focus on time is more detailed and may require even continuous time formulations. Furthermore, one faces rather (conflicting) goals than objectives: the optimal use of resources, minimal makespan, minimal operating cost or maximum profit versus more qualitative goals such as reliability (meet demand in time, proper quality, etc.) and robustness; such qualitative goals are often hard to quantify. The short-term operational aspects of operating a set of chemical reactors, food producing machines or distillation columns in a refinery are of primary interest. Users are mostly interested in feasible, acceptable and robust schedules, the objectives are usually somewhat vague, but it is common that the possibility to interact and to re-schedule, as well as the stability of solutions in cases of re-scheduling are highly appreciated. Scheduling problems are usually NP-hard, no standard solution techniques are available and, actually, in many cases we are facing feasibility problems rather than optimization problems. The solution approaches found in the literature are:

- *exact* and *deterministic methods* such as mathematical optimization including MILP and MINLP, graph theory (GT) or constraint programming (CP), or *hybrid approaches* in which MILP and CP are integrated,
- *meta-heuristics* (evolutionary strategies, tabu search, simulated annealing, ...) as described briefly in Section 4.2.4 or [48].

In addition to these remarks it is worthwhile to comment on the difference between *offline* and *online scheduling* [64]. Offline scheduling as mostly discussed in this article, in the ideal case, assumes that all data of a problem are given, i.e., full knowledge of the future (of course, this is also an approximation since our knowledge of future demand or orders is uncertain), and is close to planning except for the length of the time horizon and the resolution of time. Online scheduling as a special case of (combinatorial) on-line optimization [11] makes decisions based on past events and current data without information about future events relevant for the current decision problem; many decisions have to be made before all data are available and decisions once made cannot be changed. It may involve current process control data, updated demand data and orders, but misses orders which may enter the system in the near future and within the horizon of the current schedule to be determined. The goal is to exploit uncertain (w.r.t. the future) and incomplete information in such a way to improve the final quality of its overall performance, i.e., the quality of schedules over rolling time horizons. Unlike in stochastic optimization, where known data are subject to stochastistic uncertainties, the uncertainty in on-line scheduling only arises from the uncertainty of future data.

2 Model features in planning problems

Planning in the process industry is used to create production, distribution, sales and inventory plans based on customer and market information while observing all relevant constraints. In particular, operational plans have to be determined which are aimed to structure future production, distribution and other related activities according to business objectives. It is common practice that, based on these operational plans, detailed schedules are worked out which define the precise timing and sequencing of individual operations as well as the assignment of the required resources over time. Planning tools and software packages from various vendors are designed to incorporate new market and operational information quickly and help business users to keep their operations performing at their optimum. Especially, nowadays it is possible to find the optimal way to meet business objectives and to fulfill all production, logistics, marketing, financial and customer constraints and especially

- to accurately model *single site* and *multi-site networks*;
- to perform *capital planning* and *acquisition or divestiture analysis*, i.e., to have the possibility to change the structure of a manufacturing production network through investment and to determine the best investment type, size and location

based on user defined rules relating to business objectives and available resources, e.g., Kallrath [40]; the results of such analysis can lead to non-intuitive solutions that provide management with scenarios that could dramatically increase profits;
- to produce *integrated enterprise solutions* and to enable a cross-functional view of the planning process involving production, distribution and transport, sales, marketing and finance functions.

Planning as part of the supply chain management may focus on short and mid-term *sales and operations planning* or long-term *acquisition, consolidation, and capacity analysis* with a strategic focus. In the literature and in available software packages we usually find *time-indexed models* supporting multi-period analysis, i.e., nearly all the data may vary over time and allow to evaluate scenarios that involve time dependent aspects such as seasonal demand patterns, new product introductions, shutdown of production facilities for maintenance periods. These models include the following main structural objects:

- **Locations** can be production or storage sites, hosting plants and tanks, or demand points hosting tanks.
- **Facilities** typical are production, wrapping or inventory units that are characterized by their functional properties. Especially, in the process industry we find multi-stage production systems involving units with general product-mode relationships. Their functional properties are attributes such as capacity, throughput rates, product recipes, yields, minimum production utilization rates, fixed and variable costs, or storage limitations. Facilities can be existing or potential (for design studies). Production facilities may be subject to batch and campaign constraints across periods.
- **Demand Points** may represent customers, regional warehouse locations or distributors who specify the quantity of a product they request. A demand point can be also seen as a sink of the planning model, i.e., a point where a product leaves the system and is not further traced. Demand may be subject to certain constraints, e.g., satisfying a minimum quantity of demand, observing origins of production or supplying a customer always from the same origin.
- **Inventories** may be physically fixed entities such as tanks or warehouses but also moveable entities (e.g., drums, containers, boxes, etc.). They can be defined as dedicated
  1. to a single product from one production source,
  2. dedicated to a specific product, or
  3. free to accept any product from any source or origin.

  We may encounter tank farms, and especially *multi-purpose storage entities*, i.e., variable and multi-product tanks.
- **Products** may be classified as raw materials, intermediates, finished and salable products. A product may have several of these attributes, and it can be purchased from suppliers, produced or sold. Products are produced according to the capabilities at the facilities and the recipes assigned. Products may establish a product group, e.g., additives. Product requirements are based on market demand which is characterized by volume, selling price, package type, time,

origin and location or by other products in which they are used as intermediate products.
- **Suppliers** or **vendors** may provide products for purchase under different offering schemes. This includes the ability to link the product supply to locations and describe contractual pricing mechanisms or availability. The solver may choose the optimal supplier.

Regarding the overall business and strategic objectives the model needs to incorporate data describing the

- costs, i.e., certain fixed costs, variable costs (production, transportation, inventory, external product purchase, energy, resources and utilities), and further
- commercial aspects: financial aspects such as depreciation plans, discount rates, investment plans, foreign currency exchange rates, duties and tariffs, as well as site dependent taxes.

Maximize operating cash flow and *maximize net present value* (NPV) objective functions are used to determine the financial and operating impacts of mergers, acquisitions, consolidation initiatives, and capital spending programs effecting business. In detail this may include:

1. maximize the net profit (free design reactors; open and close facilities),
2. maximize the contribution margin for a fixed system of production units,
3. maximize the contribution margin while satisfying a minimum percentage of demand,
4. minimize the cost while satisfying full demand (allow external purchase of products),
5. maximize total sales neglecting cost,
6. maximize total production for a fixed system of production reactors,
7. maximize total production of products for which demand exists,
8. minimize energy consumption or the usage of other utilities,
9. minimize the deviation of the usage of resources from their average usage,
10. multi-criteria objectives, e.g., maximize contribution margin and minimize total transportion volume.

Planning involves the determination of operational plans that support different short- or mid-term objectives for the current business. By using different objective functions it is possible to create operational plans that support strategies such as market penetration, top-line growth, or maximization of cash flow to support other business initiatives.

If, besides this broad structure, the focus is on a more detailed representation of physical entities, we find that planning models and their constraints may involve the following features (in alphabetic order):

- **Batch production**: The quantity of a specific product being produced in a campaign possibly over several periods must be an integer multiple of some pre-defined batch size.
- **Buy, build, close or sell specific production assets**: This feature is used for closing, or selling acquisition, consolidation and capacity planning to determine

the NPV and operational impacts of adding or removing specific assets or groups of assets to the network.
- **Campaign production**: This allows to impose a lower and/or an upper bound on a contiguous production run (campaign) possibly across periods; this feature is also known under the name *minimal runs*.
- **Delay cost**: Penalty cost apply if customer orders are delivered after the requested delivery date.
- **Minimum production requirements**: Minimum utilization rates modelled as semi-continuous variables have to be observed for specific production units and/or entire production locations for each production time period.
- **Multi-locations**: This can be production sites, storage sites, and demand points.
- **Multi-purpose production units**: If a unit is fixed to a certain mode, several products are produced (with different mode-dependent daily production rates), and vice-versa, a product can be produced in different modes. Daily production can be less than the capacity rates. A detailed mode-changing production scheme may be used to describe the cost and time required for sequence-dependent mode-changes.
- **Multi-stage production**: Free and fixed recipe structures allow for the production of multiple intermediate products before the manufacture of the final product with convergent and divergent product flows. The recipes may depend on the mode of the multi-purpose production unit.
- **Multi-time periods**: Non-equidistant time period scales are possible for commercial and production needs. For instance, demand may be forecast weekly for the first quarter of the year and then quarterly for the remainder of the year.
- **Nonlinear pricing** for the purchase of products or utilities (energy, water, etc.) or nonlinear cost for inventory or transportation may lead to convex and concave structures in order to model volume and price discount schemes for the products or services purchased, while in addition, contract start-up and cancellation fees may lead to additional binary variables.
- **Order lost cost**: Penalty cost are incurred if products are not delivered as requested and promised.
- **Packaging** machines are optimized to increase machine throughput and assure that priority is given to the most profitable products.
- **Product swaps**: With the objective of saving transportation and other cost companies often arrange joint supply agreements called *swaps*. For example: Company 1 based in Europe as well in the US has a production shortage of product A in the US and thus purchases a defined quantity of product A in the US from company 2. Company 2 (also located in the US and Europe) has a customer in Europe requesting product A and thus purchases a defined quantity product A from company 1 in Europe. Both companies get product A where they need it and avoid the cost of shipping the product. Without this type of supply agreement company 1 would have to ship product A from its European plant to the US, and company 2 would have to ship product A from its US manufacturing plant to Europe.
- **Production origin tracing**: It is possible to define fixed, free or unique origins for specific demands. For example, a customer may require that his demand is

satisfied only from a specific plant in the network, or it may not be supplied from a set of plants, or the customer only requests that he is supplied from one unique plant during the whole planning horizon.
- **Shelf-life time**: Product aging time can be traced. This allows for the application of constraints such as: maximum shelf-life time, disposal costs for time expired products, and the setting of selling prices as a function of product life.
- **Transportation** and **logistics**: Transportation quantities are appropriately modelled by the use of semi-continuous variables. This allows minimum and maximum shipment quantities to be defined for each source location, destination location, product, and transport mean combination. The logistics involves the costs and lead times and constraints (minimum shipment quantities) associated with moving intermediate and finished products between facilities and demand points. The mean of transport may be chosen by the optimizer and nonlinear cost functions have to be considered as well.

This list covers many features but may be incomplete.

3 Types of scheduling problems in the process industry

In this review, we present some typical structures and categories of scheduling problems. Their treatment differs by the mathematical techniques applied:

- batch and campaign planning,
- scheduling problems in the chemical process industry including lot-sizing and sequencing,
- time-precedence and aggregate resource constraints,
- nonlinear scheduling problems including blending.

3.1 Batch and campaign planning

For a given planning problem formulated as a time-indexed model, it may be desirable to include certain constraints in the model which allow to incorporate batch and campaign features. To be as general as possible we might want to consider batch reactors which can be, for example, operated in different modes producing, at a given time, several products in each mode with different free or fixed recipes leading to a general mode-product relation ([42], pp.153–155, 320–324). Thus, in a certain mode several products are produced (with different daily production rates), and vice-versa, a product can be produced in different modes. Daily production can be less than the capacity rates. This is an important feature in demand-driven joint production in which several products are produced in fixed or variable ratios to each other.

In time-indexed model formulations where variables p_{pt} define the production quantity [e.g., in tons] of a product p in period (time-interval) t it is not easy to model batch or campaign restrictions if the batch or minimal campaign size is larger than the capacity per period. Assume that production is performed in batches of 200 tons, and that the time intervals have a length of ten days with a daily production rate of 10 tons/day. The minimum time to produce the batch would cover 20 days, or exactly two time intervals. A plan looking like $p_{p4} = 45$ tons, $p_{p5} = 100$ tons, and $p_{p6} = 55$ tons covers three periods (the first and third only partial) to produce exactly 200 tons, and thus provides more degrees of freedom.

Brockmüller and Wolsey [12] solved the problem for a special case (daily production equals the capacity rates). Their approach, which falls into the class of discrete lotsizing and scheduling problems (DLSP, [20, p.146]), uses explicitly the feature that production equals the capacity rates in order to determine a priori the number of periods to produce a campaign of specified minimal size. However, if daily production can take any value between a lower bound, e.g., zero, and the capacity rate per day, or if a product is produced, for example, according to general mode-product relations, then this a priori information is not available. Kallrath [38] overcomes this restriction and has developed an extension which can be added to any production planning model formulated as a proportional lotsizing and scheduling problem (PLSP, [24]), i.e., based on time-indexed formulations with at most one setup- or mode-change per period. It allows to add constraints involving accumulated quantities over several time-slices thus implementing the concept of contiguity into the model. This feature is relevant to any kind of process industry. It allows to model batch and campaign production or to require that a certain time-lag between successive mode-changes is observed. The key idea used in the technical approach is to identify which time-indexed production contributions belong to certain contiguous components, e.g., campaigns, over several time slices and to replace products of continuous and binary variables, or absolute value terms by linear relations involving additional binary variables. This approach has successfully been applied to production planning problems in the chemical industry. Other approaches are described by [31], [7] or [47].

3.2 A typical scheduling problem in the chemical industry

Westenberger and Kallrath (1994) in a cooperational work of Bayer AG and BASF Aktiengesellschaft formulated a typical but generic scheduling problem with the objective to push the development of algorithms for scheduling problems in process industry. Their proposal to establish a working group to develop standardized benchmark problems for plannig and scheduling in the chemical industry initiated many research projects and activities. The *Westenberger-Kallrath problem* has been understood as a typical scheduling problem occurring in process industry including the major characteristics of a real batch production process (involving multi-product facilities, multi-stage production, combined divergent and convergent product flows, variable batch sizes, non-preemptive processes, shared intermediates, alternative recipes, flexible proportions of output products, blending processes, sequence and usage dependent cleaning operations, finite intermediate storage, cyclic material flows, re-usage of carrier substances, and no-wait production for certain types of products) so as to encourage researchers and engineers to test their algorithms and software tools by applying them to this test case. Solutions based on mathematical programming techniques, heuristics, simulation, genetic algorithms, evolutionary strategies etc. may be sent to the author. Contributions and results are collected under http://www.math.tu-berlin.de/chemical-benchmarks/; the problem is summarized in the appendix of this review paper. A mathematical description has been given in order to guarantee a unique interpretation of the test problem. The problem description contains all data necessary to perform own test calculations; no special chemical knowledge is needed. Everybody should feel free to transform it to his needs as long as equivalency is guaranteed. Mathematical methods and typical results for related problems can be found in literature [21], [41], [60], [17].

The test and benchmark problem includes all features described above and should be solved for a set of different objective functions among them:

1. minimization of makespan,
2. maximization of profit by optimizing the product mix,
3. minimization of investment cost by optimization of stock and production capacitics.

Over the last 6 years many publications ([3], [8], [10], [14], [15], [43], [54], [58], [63], [69], [9], [52]) appeared in which different techniques have been applied to produce solutions to the problem; some results are summarized in [52].

3.3 Time-precedence and aggregate resource constraints

For a general discussion of such problems see, for instance, [56], [53] and [13]. A complex scheduling problem in the process industry [42] containing precedence and aggregate resource constraints has been tackled by many authors and is briefly summarized below.

The client uses a set of machines and employs a number of workers. He receives orders from his customers. Each order demands a certain quantity of a product which can be produced on the client's machines. The machines are operated and supervised by the workers. Orders are often split up into several identical jobs, which are necessary in order to produce the required quantity of the product, because, typically, orders demand a larger quantity of a specific product than the machine capacity allows to produce in one job. A job for a given order is processed on a machine according to a specific *procedure* or *process plan*. It consists of a deterministic sequence of tasks defining how to produce a specific product. The size of a job is limited by the capacity of the machine. Each task has a pre-defined demand for labour and a certain duration. The workers are allocated to different tasks in order to keep the jobs running. Allocation of the workers has to comply with working regulation rules, e.g., taking breaks, washing, equally spread labour among the workers, limits on labour intensive work, over-occupation rate and over-time. The objective is to minimize the makespan and/or to minimize the (variation in the) number of workers. In [42] we find a model including assignment and sequencing decisions, and a time-indexed formulation to describe the detailed personnel requirements. Another solution technique to solve the problem is described in Section 4.2.3.

3.4 Nonlinear scheduling problems in refineries

Planning and scheduling has a long history in the refinery industry and dates back to the 1950s. If treated in the context of mathematical optimization, refinery scheduling leads to MINLP problems due to the presence of the pooling problem. It is common practice in the refinery industry that the scheduling problem usually comes up after a production planning problem based on mass balances has been solved for a medium time horizon, consisting of monthly or quarterly periods. The data generated by the production planning problem are input data to the scheduling problem which determines a detailed crude oil processing schedule, process unit schedule and blending and shipping schedules. The purpose of the scheduling problem is to transform the production plan into a schedule useful for all operations within a time horizon of a few days. In that sense the scheduling problem is rather a feasibility problem than an optimization problem. While in typical production scheduling problems degeneracy and symmetry cause large problems, the nonlinear features in the refinery scheduling problems could destroy some symmetry and may lead to useful relaxations.

A typical example is a medium-sized refinery which schedules the production and the storage of oil products for the next two to four weeks. The input data for the scheduling problem are provided by production plans generated by solving an LP problem. The model describes the typical processes in a refinery, e.g., the flows of crude oil from tankers to crude oil tanks, from the tanks to the initial production units, i.e., top distillation units, and the distribution to intermediate tanks or to further production units, e.g., vacuum distiller or thermal cracker, and finally, filling the product tanks and the delivery of the final products. The model takes into account that there are several types of oils and blending components (intermediate products) with different chemical or physical properties and that they are treated differently by the production units. In addition the model considers minimum and maximum capacity restrictions for tanks and production units. Besides this the units require a minimal run time period once they are activated.

The model includes logical constraints determining the daily production schedule and the assignment of raw material and intermediates to tanks and production units. The objective of the optimization problem is to minimize deviations from targets, e.g., deviations from the pre-determined production plan, or, alternatively, to minimize the use of blending components instead of crude oil products. For a case study within a confidential research project, the model included 16 crude oils, 4 blending components, 17 crude oil tanks, 5 production units and 8 final products.

The features and the constraints of the refinery model described above can be summarized and classified into the following groups:

- flow of oils and blending components (linear mass balance equations),
- quality constraints and capacity limits of production units and tanks (inequalities),
- proportional composition of streams (nonlinear equations),
- assignment of tanks and production units (equation & inequalities involving binary variables).

For a 30-days scenario the model contains 80,000 constraints, 70,000 continuous variables, 2,500 binary variables and 130,000 nonzero entries in the matrix. The problem is very difficult to solve not only because of its size but, primarily, because of the combination of *nonlinear constraints* and *binary variables*. The nonlinear character of the mathematical model is caused by the fact that different types of crude oils or blending components of different properties are mixed, whereas the decisions on the daily production schedule and the assignment of crude oils and intermediates to tank and production units require the use of binary variables.

The solution of this MINLP problem has been derived using the software package `XPRESS-MP` by Dash Optimization. For this problem, Dash had extended their Branch&Bound-facilities by a special recursion algorithm for the nonlinear terms. Solutions have been obtained on a PC with Pentium I processor within acceptable time (approximately one hour for a 17-days scenario).

4 Solution approaches

4.1 Solution approaches used in planning

Most of the planning problems in the process industry lead to MILP or MINLP models and contain the following building blocks: *tracing the states of plants*, *modeling production*, *balance equations for material flows*, *transportation terms*, *consumption of utilities*, *cost terms*, and *special model features*. Mode-changes, start-up and cancellation features, and nonlinear cost structures require many binary variables. Minimum utilization rates and transportation often require semi-continuous variables. Special features such as batch and campaign constraints across periods require special constraints to implement the concept of contiguity. The model, however, remains linear in all variables. Only if the pooling problem occurs, *e.g.*, in the refinery industry or the food industry, we are really facing a MINLP problem. For a review on algorithms used in LP, MILP, NLP, and MINLP the reader is referred to [39].

Using state-of-the art commercial solvers, e.g., `XPRESS-MP` [`XPRESS-MP` is by Dash Optimization, http://www.dashoptimization.com] or `CPLEX` [`CPLEX` is by ILOG, http://www.ilog.com], MILP problems can be solved quite efficiently. In the case of MINLP, the solution efficiency depends strongly on the individual problem and the model formulation. However, as stressed in [39] for both problem types, MILP and MINLP, it is recommended that the full mathematical structure of a problem is exploited, that appropriate reformulations of models are made and that problem specific valid inequalities or cuts are used. Software packages may also differ with respect to the ability of *pre-solving techniques*, *default-strategies* for the Branch&Bound algorithm, *cut generation* within the Branch&Cut algorithm, and last but not least *diagnosing and tracing infeasibilities* which is an important issue in practice.

Current activities to solve planning problems more efficiently are focused on the construction of useful valid inequalities for certain substructures of planning problems. Those inequalities may a priori be added to a model, and in the extreme case they would describe the complete convex hull. As an example we consider the mixed-integer inequality

$$x \leq C\lambda \quad , \quad 0 \leq x \leq X \quad ; \quad x \in \mathbb{R}_0^+ \quad , \quad \lambda \in \mathbb{N} \tag{1}$$

which has the valid inequality

$$x \leq X - G(K - \lambda) \quad \text{where } K := \left\lceil \frac{X}{C} \right\rceil \quad \text{and} \quad G := X - C\,(K - 1) \tag{2}$$

This valid inequality (2) is the more useful, the more K and X/C deviate. A special case arising often is the situation $\lambda \in \{0, 1\}$. Another example, taken from ([70], p. 129) is

$$A_1\alpha_1 + A_2\alpha_2 \leq B + x \qquad x \in \mathbb{R}_0^+ \qquad \alpha_1, \alpha_2 \in \mathbb{N} \tag{3}$$

which for $B \notin \mathbb{N}$ leads to the valid inequality

$$\lfloor A_1 \rfloor \alpha_1 + \left(\lfloor A_2 \rfloor \alpha_2 + \frac{f_2 - f}{1 - f} \right) \leq \lfloor B \rfloor + \frac{x}{1 - f} \tag{4}$$

where the following abbreviations are used:

$$f := B - \lfloor B \rfloor \quad , \quad f_1 := A_1 - \lfloor A_1 \rfloor \quad , \quad f_2 := A_2 - \lfloor A_2 \rfloor \tag{5}$$

The dynamic counterpart of valid inequalities added a priori to a model leads to cutting plane algorithms which avoid adding a large number of inequalities a priori to the model (note, this can be equivalent to finding the complete convex hull). Instead, only those useful in the vicinity of the optimal solution are added dynamically. For the topics of valid inequalities and cutting plane algorithms the reader is referred to the well written book by Wolsey [70].

Using these techniques, for some BASF planning problems including up to 100,000 constraints and up to 150,000 variables with several thousand binary variables, good solution with integrality gaps below 2% have been achieved within 30 minutes on standard Pentium machines [39].

4.2 Solution approaches used in scheduling

The complexity of scheduling problems can easily exceed today's hardware and algorithmic capabilities. Nevertheless, there are numerous promising contributions (see, for instance, [67], [45], [55], [4], [65], [63], [52], [69]) for problems in the process industry. What makes scheduling problems so difficult? Using exact methods such as MILP, in some cases it is not even possible to find feasible integer solutions because feasible integer solutions exist often only very deep in the B&B tree. In many cases it is very difficult to derive useful upper and/or lower bounds. Scheduling problems usually suffer from poor LP relaxations. Resource constraints can easily be fulfilled with fractional values of the binary variables used in time-indexed formulations and thus lower bounds are very weak. Even using parallel algorithms and powerful hardware, scheduling problems might be too complex, and often cannot be solved with MILP methods, at least not yet. If we meet such cases, it is also worthwhile to apply another exact method: constraint programming (CP). Heipcke [33] investigated successfully a very difficult scheduling problem (Section 10.5 in [42]), which became a benchmark problem in both the MIP and CP community as well as amongst scientist using graph theory or meta-heuristics. Heipcke originally applied both methods, CP versus MILP, to this problem and later contributed to the combination of both techniques [34]. Timpe [68] reports a successful application of a combined MIP-CP approach to a scheduling problem in the chemical industry. CP [46] has been developed in the 1980s out of Logic Programming and Constraint Solving and has been applied successfully to a large range of industrial applications, especially to discrete (optimization) problems. CP is a technique for discrete optimization that uses a tree search and performs domain reduction at each node. CP models typically include a wide range of constraint types, e.g., special global constraint operators such as *all different* or *cumulative*. Unfortunately, in the CP community the motivation is low to develop a common language which would allow to formulate a problem in a very compact way as is the case for mathematical programming languages. Instead, most and generic CP applications require the users to program in C or C++; this makes it very difficult to port the model to different hardware platforms and to maintain the software over a longer period.

If CP also fails, the last resort might be to use heuristic approaches [48], e.g., simulated annealing or tabu search. Heuristics exploiting the structure of scheduling chemical batch processes can also lead to good results as the two-stage solution procedure by Blömer and Günther [10]) demonstrates. In the first stage, an LP-based heuristic produces an initial solution. The proposed *time grid heuristic* defines a time grid that includes only a limited number of feasible periods in which a processing task is allowed to start. Thus, the size of the original multi-period MILP model is reduced in a controlled manner and optimal solutions of the relaxed model are obtained within reasonable computational time.

4.2.1 Processes and the state-task network (STN) representation

The approaches developed at Imperial College have at their heart novel process representations which allow to apply several exact optimization methods and decomposition techniques. The first approach was the mathematical programming approach of Kondili et al. [44] based on a discrete representation of time and the newly introduced *state-task network* (STN) representation of the process. The STN representation has three main advantages:

- it distinguishes the process operations from the resources that may be used to execute them, and therefore provides a conceptual platform in which the unique assignment assumption is relaxed and unit-to-task allocation is optimized;
- it avoids the use of task precedence relations which become very complicated in multipurpose plants (a task can be scheduled to start if its input materials are available in the correct quantities and other resources, *e.g.*, processing equipment and utilities, are also available, regardless of the plant history);

- it provides a means of describing very general process recipes, involving batch splitting and mixing and material recycles, as well as storage policies including zero-wait, no-intermediate storage, multipurpose storage tanks and so on.

The formulation of Kondili et al. [44] (described in more detail in Kondili *et al.* [45] is based on the definition of binary variables that indicate whether tasks start in specific units of equipment at the beginning of each time period, together with associated variable batch sizes. Other key variables are the quantity of material in each state held in dedicated storage devices over each time interval, and the quantity of each utility required for processing tasks. Their key constraints are related to equipment and utility usage, material balances and capacity constraints. The common, discrete time grid captures all the plant resource utilizations in a straightforward manner; discontinuities in these are forced to occur at the predefined interval boundaries. Their approach was hindered in its ability to handle large problems by the weakness of the allocation constraints and the general limitations of discrete-time approaches such as the need for relatively large numbers of grid points to represent activities with significantly different durations. Shah et al. [66] modified the model to improve its relaxation properties significantly and therefore increase the scope of applicability considerably.

Pantelides [55] presented a critique of the STN and associated scheduling formulations and argued that despite its advantages, it suffers from a number of drawbacks:

- the model of plant operation is somewhat restricted (each operation is assumed to use exactly one major item of equipment throughout its operation);
- tasks are always assumed to be processing activities which change material states (changeovers or transportation activities have to be treated as special cases);
- each item of equipment is treated as a distinct entity (this introduces solution degeneracy if multiple equivalent items exist);
- different resources (materials, units, utilities) are treated differently, giving rise to many different types of constraints, each of which must be formulated carefully to avoid unnecessarily increasing the integrality gap.

Pantelides [55] then proposed an alternative representation, the *resource-task network* (RTN), based on a uniform description of all resources. In contrast to the STN approach, where a task consumes and produces materials while using equipment and utilities during its execution, in this representation, a task is assumed only to consume and produce resources. Processing items are treated as though consumed at the start of a task and produced at the end. Furthermore, processing equipment in different conditions (e.g., "clean" or "dirty") can be treated as different resources, with different activities (e.g., "processing" or "cleaning") consuming and generating them – this enables a simple representation of changeover activities. He also proposed a discrete-time scheduling formulation based on the RTN which, due to the uniform treatment of resources, only requires the description of three types of constraints, and does not distinguish between identical equipment items.

4.2.2 Decomposition: batching and batch scheduling

The scheduling problem described in Appendix A has also been tackled by the decomposition approach developed by Schwindt and Trautmann [63] and Trautmann [69]. The basic idea of this approach is to decompose the problem into *batching* and *batch scheduling*. In the first step, the number and the size of the batches to be produced is determined. The second step generates a feasible schedule and computes the start and end times of the batches.

The approach considers multi-stage production using multi-purpose equipment. Final products are produced according to a sequence of *tasks*. Production requires reactor time, utilities (such as energy, water, etc.), storage capacities, and possibly personnel. The connections between reactors and storage devices are described by a fixed topology which allows divergent, convergent and cyclic material flows. For further details and underlying assumptions see Neumann et al. [52].

The batching step decomposes the demand into feasible and appropriate batches. The decision variables associated with this step are the number of batches per task, the size of the batches, and the production or conversion rates between input and output product flows. In this step, the objective is to minimize the number of batches weighted by the process times of those batches subject to the constraints:

- the remaining quantity of pre-products not consumed in the production process has to observe the lower and upper storage limits (safety stock, storage capacity);
- the size of the batches has to fulfill the *just-in-time constraints* for those intermediate products for which no storage capacity is available (reactors charge directly to reactors of the next production stage);
- the size of batches and the production or conversion rates have to be within the technological limits.

In [69] and Neumann et al. [52] we find a mixed-integer formulation of the batching problem with nonlinear constraints assuming that for all batches of a task the same size is chosen. The batching problem is solved approximately within seconds.

During the second step, batch scheduling, a feasible schedule of minimal makespan is generated providing the start and end times of all tasks as well as the assignment of resources to the task. The Branch&Bound method presented in Trautmann [69] and Schwindt and Trautmann [63] for solving the batch-scheduling problem, is based on models and methods of resource-constrained project planning (see, for instance, [13] or [62]). The key idea of this method is a decomposition of the problem into *temporal scheduling* and *generation of additional constraints*. Time planning corresponds to the solution of a problem with relaxed resource constraints, i.e., initially it is assumed that all resources are available without any limit. The remaining constraints are time-window constraints originating from production breaks, earliest time for delivering a product, due dates, expiration dates of stored products etc. This temporal scheduling problem corresponds to the determination of the longest path in a graph and can (when appropriately modified to account for forced production breaks and calendar conditions, e.g., respecting holidays) be solved by an extended *label–correcting-algorithm* [69], i.e., computing a schedule whose activities start as early as possible while observing all lower and upper bounds on the differences between termination and starting times is equivalent to compute the longest path from the source to all nodes in a graph [51]. If the plan generated is feasible for the whole problem, the algorithm stops. Otherwise, a point in time is determined at which, now considering the limited resources, a resource constraint would be violated or a batch is started without having uniquely fixed all resources. In the first case, the violation of the resource constraints is eliminated by adding precedence relations between activities; in the second case, the resources are selected. In both cases, the alternative precedence relations or the alternative resources are enumerated by adding them to the search tree. The new subproblem, differing from the previous one by the constraints added, is solved as a temporal scheduling problem. If this problem is not feasible, another subproblem of the tree is chosen. If it is solvable, the scheme of adding additional constraints is continued until a feasible plan is found. The first feasible plan already imposes an upper bound but the search tree might be further explored. In order to restrict the computing time, only a partial search is applied namely to that part of the tree in which one might expect to find good solutions (*filtered beam search*).

Although Trautmann and Schwindt only consider the makespan objective, their approach is open to account for other objective functions, e.g., to minimize the delay with respect to some due dates. Another advantage of their approach is that re-scheduling is supported, for instance, if some resources suddenly fail.

4.2.3 Special algorithms for problems with time-precedence and aggregate resource constraints

Some algorithms applied successfully to the problem described in Section 3.3 are based on the original work by Bartusch et al. [5]. The basic idea is to permit temporal constraints in the form of arbitrary, context-sensitive time lags between start and/or completion times of activities and time-dependent resource requirements and availabilities in the form of piece-wise constant step functions. The B&B algorithm constructs feasible schedules along the time axis, starting at time 0 and, at every decision time t, branches into feasible decisions (about which set of jobs to start at t) with respect to the resource constraints, while always observing the temporal constraints. It thus always maintains a set of best possible feasible partial schedules for the different subproblems created by these decisions and seeks to either complete them into a full schedule, or to discard a branch because the best value that can be obtained by this branch exceeds the current best objective value.

This approach is combined with heuristics to generate good feasible solutions in order to start with good upper bounds, different strategies for exploring the subproblems created (best-first-search, depth-first-search, user-defined search), and several methods for generating good lower bounds for best possible value of the subproblems. In addition, it uses a powerful domination rule which permits discarding a subproblem if the currently "active" jobs in the partial schedule have been encountered already in a different branch with a better lower bound. This domination rule is based on an efficient search tree implementation of all sets of active jobs encountered so far. The most recent progress and results have been achieved by the group around R. Möhring at TU Berlin and are based on Lagrangian Relaxation and Branch&Bound techniques.

Lagrangian relaxation: In order to compute lower bounds on the optimal objective function value for resource-constrained project scheduling problems, an approach via Lagrangian relaxation has been suggested [49]. The Lagrangian relaxation is based on a well-known time-indexed integer linear programming formulation of the problem [57]. The same Lagrangian relaxation has been used before [18]. The basic idea is to relax the resource-constraints, and to penalize their violation in the usual Lagrangian fashion. This results in a Lagrangian subproblem which is a so-called project scheduling problem with start-time dependent costs. In this problem, each job incurs a cost which depends on its start time, the jobs are subject to precedence constraints (or arbitrary time lags), and the objective is to find a schedule which has minimal costs. This problem, or special cases thereof, has been addressed frequently in the literature. [50] give an overview of these results. As it turns out, the project scheduling problem with start-time dependent costs can be efficiently solved as a minimum cut problem in a directed graph. This insight is the key to the practical efficiency of the Lagrangian approach, which uses a sub-gradient method to iterate the Lagrangian multipliers and provides reasonable strong lower bounds within very moderate computation times. Note that the lower bounds are theoretically the same as those obtained by the LP-relaxation of the problem, but in the current case the Lagrangian relaxation method works much faster than Simplex type algorithms or interior point methods. In addition, [49] propose to exploit the dual information from solutions of the Lagrangian relaxation in order to compute also feasible solutions to resource-constrained project scheduling problems. The basic idea is borrowed from previous work on approximation algorithms for machine scheduling problems. In this context, it was shown by various authors that LP-based list-scheduling, combined with the concept of so-called α-completion times, can lead to schedules with constant worst-case performance guarantees. The rationale behind this approach is that a solution of a relaxed problem also holds valuable information to compute feasible solutions to the original, resource-constrained problem. [49] use simple list scheduling algorithms which are based on priority lists according to α-completion times of jobs in the solutions of the Lagrangian subproblems. Feasible solutions are computed in each iteration of the subgradient optimization. The computational experiments with this approach are very promising in terms of both computation time and solution quality. In fact, on a set of well established benchmark instances, the Lagrangian based approach provides solutions which are comparable to those of state-of-the-art algorithms from the literature. Moreover, for instances close to real-world scenarios, the results are clearly favorable to those obtained with a constraint propagation approach by [34], or a branch-and-bound algorithm by [27]. Note that the Lagrangian approach can handle many regular, and also non-regular objective functions.

Branch-and-Bound: For resource-constrained problems with arbitrary minimal and maximal time lags, also called *time windows* or *generalized precedence constraints*, it is already NP-hard to compute a *feasible* schedule. Many order theoretic insights into the structure of optimal solutions have been already obtained by [5]. Partially based on ideas of their work, branch-and-bound algorithms have been proposed and evaluated more recently by [19], [62], and [27]. The underlying idea of the algorithms is that *time-feasible schedules* are enumerated by systematically resolving *resource conflicts*. Here, a time-feasible schedule denotes a schedule which does not violate the time lag constraints, and a resource conflict is a time phase during which the schedule violates the resource constraints. The resource conflicts are resolved by introducing additional precedence relations between jobs, or sets of jobs ([19], [62]), a concept which is based on an order theoretic representation theorem of optimal schedules (see Theorem 3.8 in [5]). In contrast, [27] resolve resource conflicts by a dynamic update of release dates instead of introducing precedence relations. Thus, their algorithm is not based on the order theoretic concept described in [5], but on a very simple dominance property instead. At a first glance, this technique has the drawback that resource conflicts are resolved only locally. Nevertheless, subject to several additional features which help to truncate large parts of the enumeration tree, the computational results show that the algorithm performs better than previous algorithms which are based on the idea to resolve resource conflicts. Compared to the previous branch-and-bound approaches, the efficiency is partly due to the efficient update of the time-feasible schedules in each node of the enumeration tree. Other branch-and-bound algorithms by [34], [22], and [23] are based on constraint propagation. These algorithms rely on the idea to reduce the possible start times of jobs as much as possible by propagating corresponding lower and upper bounds in every node of the enumeration tree. The results with these algorithms are also very good. As a rule of thumb, however, for large-scale instances, the computation times of all available branch-and-bound implementations are prohibitive. In these cases the algorithms are generally used as heuristics by only evaluating parts of the enumeration tree.

4.2.4 Heuristics and meta-heuristics

Besides the exact methods described so far, a variety of heuristics, according to Glover [29] better called *meta-heuristics*, are used to solve scheduling problems [21] by simulating a given system and evaluating its *function of merit* (the objective function in exact optimization). Some well known meta-heuristics, i.e., techniques which are not problem specific and are based on generic principles and schemes which can be used to construct problem-specific heuristics are: *genetic algorithms* (GA), *simulated annealing* (SA) and *tabu search* (TS). All meta-heuristics have in common that they usually lack the proof of convergence and the proof of optimality. However, they can be effectively used to improve a given solution by performing a local search coupled with some exchange mechanisms based on appropriate neighborhood relations. SA and TS have the advantage that they can leave *poor* local optima and move to *better* solutions.

SA (see, for instance, [1], [25] or [2]) links the probability of accepting a solution which is worse than the reference solution to a temperature-like parameter which describes the cooling of metals. This approach which introduces a non-deterministic argument for accepting an inferior solution is the key-element to leave *poor* local optima.

TS [29] is a meta-strategy for guiding known heuristics past the traps of local optimality. It exploits knowledge from previous solutions and thus uses an abstract memory. Popularized by Glover in the early 90s, TS has been applied to integer programming problems involving scheduling, routing, traveling salesman and related problems.

GA ([35], [36]) – an algorithm in the class of evolutionary algorithms – uses population of solutions subject to survival of the fittest criteria, mutation and recombination of positive properties in very solutions.

5 Conclusions and summary

We have provided an overview of planning and scheduling in the process industry. The state-of-the-art technology based on mathematical, especially mixed-integer optimization for planning is quite advanced and appropriate for solving real world planning problems. Mixed integer optimization can provide a quantitative basis for decisions and allow to cope most successfully with complex problems and it has proven itself as a useful technique to reduce costs and to support other objectives. Despite that, this technology has not yet found its way into many commercial software packages. For scheduling problems, there is not yet a commonly accepted state-of-the-art technology although some promising approaches have been developed, especially for job shop problems. Nevertheless, the majority of software packages is still based on pure heuristics.

What will the future hold for us in planning and scheduling? There is a growing number of software packages available which support – at least to some extent – the application of exact methods for a variety of planning problems. That way, planning based on mathematical, especially mixed-integer optimization becomes more and more the state-of-the art in the chemical, food and pharmaceutical industry and as well in refineries. For scheduling this is not yet the case. Most approaches are still based on heuristics. But there is light at the end of the tunnel: hybrid approaches in which MILP, constraint programming and graph theory are used together are in the process of being developed, and may be a common language for scheduling problems.

There is a trend over the last few years bringing mathematical programming and the constraint programming community closer to each other. This results in hybrid approaches (see, for instance, [32] and [37]), i.e., in a language and algorithms combining elements from both communities. This may have a great impact on supply chain problems and scheduling. In 1999, the European Commission awarded the project LISCOS (Large Integrated Supply Chain Optimization Software) with several million Euros. The technical core of this project initiated by BASF's mathematical consultant group and 8 other partners is the development of MIP-CP hybrid techniques (http://www.liscos.fc.ul.pt). Timpe [68] describes a successful application of this techniques to a real world planning and scheduling problem in the process industry.

Another focus of modeling which is possible now due to increased computer power available is the opportunity to solve design and operational planning problems, or strategic and operational planning problems simultaneously in one model. The motivation, ideas how to to this and successful examples are provided by [40].

Finally, it is now also possible to give up the assumption that all data have to be treated as determinstic data – note that in this article we focussed only on deterministic models. However, as some data, e.g., demand forecast in planning models, or production data in scheduling may be subject to uncertainties, is seems to be advantegous if we could give up the assumption that planning and modeling is exclusively based on deterministic data. In that case, stochastic optimization is the mean of choice. Nowadays, there exist powerful solution techniques to combine mixed-integer programming and stochastic optimization (see, for instance, Schultz [61] or Carøe and Schultz [16]) when data are subject to uncertainties. Successful applications of this techniques to scheduling problems in the chemical process industry are reported, for instance, by Sand et al. [59] or Engell et al. [26]. Alternatively, to these techniques there exists an approach [6] to find robust solutions to LP-problems including uncertain data.

While MIP has already well established itself in planning, further quantum leaps in scheduling are to be expected from the combination of several techniques such as mathematical optimization, graph theory and constraint programming, and to exploit problem specific structures. Exact, or at least much better scheduling techniques knock at our doors and might, say, within 5 to 10 years, play a similar role as does MIP in planning nowadays.

Appendix: A benchmark problem (Westenberger and Kallrath)

Due to the increasing interest in the benchmark scheduling problem formulated by Westenberger and Kallrath (1994) in a joint internal publication by Bayer AG (Leverkusen) and BASF Aktiengesellschaft (Ludwigshafen), this appendix summarizes it and thus makes it available to a broader audience. The case study covers most of the features that contribute to the complexity of batch process scheduling in industry. The problem is to some extent presented in the language of an MILP model. This is to guarantee uniqueness in the interpretation. Nevertheless, people interested in the problem should be encouraged to try other approaches, e.g., time-continuous formulations. A state-task-network representation of the production process and the possible assignment of tasks to production units are given in Figure 3 in [10].

1 Description of the problem

The production process considered consists of a network of 17 processing tasks, 19 states, 9 production units, and 37 divergent, convergent and cyclic material flows. Moreover, the production process includes flexible proportions of output goods (see Task 2/2), cyclical material flows (recycling of output from Task 3/3 into State 1), and several intermediates (see state nodes 5, 9, 10, and 12), which do not allow storage between processing steps. All processing tasks are performed in batch mode with lower and upper bounds on the batch sizes. These bounds are pre-determined from properties of the processes and the capacities of the reactors. Batch sizes are treated as decision variables and may be different even for the same type of product or the same production unit.

Table 1. Description of production units

Production unit	Number of processing lines	Number of products	maximum batch size	minimum batch size
R_u^P	L_u	P_u / unit	A_u^+ / kg	A_u^- / kg
R_1	1	1	10	3
R_2	1	2	20	5
R_3	1	2	10	4
R_4	1	4	10	4
R_5	1	2	10	4
R_6	2	3	7	3
R_7	2	5	12	4

In this production planning problem a variety of products are produced by a process plant consisting of a number U of production units $R_u^P, u = 1, \ldots, U$. We suppose an one-to-one relation of process steps and production units in the plant. Every production unit R_u^P contains L_u *production lines* $R_{ul}^L, l = 1, \ldots, L_u$. In the current benchmark example we have $U = 7$ and the vector L_u specified in Table 1. Each production unit R_u^L generates a number P_u of *products* $P_{up}, p = 1, \ldots, P_u$. The number of products per production unit is given in Table 1 as well.

Remark. The production is realized in batch mode, i.e., certain quantities of ingredients are processed by a given *process time per batch process* T_{ulp}. Process times depend on the production line R_{ul}^L used for processing and the product to be made (see Table 2), but in our case, not on actual batch sizes.

In our example we have at most two different ingredients per batch (see product P_{73}) and at most 2 generated products per batch. For instance, productions of P_{21} and P_{22} are coupled. The same holds for P_{31} and P_{32}.

Remark. Note that F_{211n}^{o1} is restricted by $0.2 \leq F_{211n}^{o1} \leq 0.7$ and that F_{311n}^{o1} is explained in Section 5 of this appendix. Only those processing lines are allowed to make product P_{up} which have a process time T_{ulp} for this product declared in Table 2. For example, production of P_{73} is allowed to run on R_{71}^L, but not to run on R_{72}^L.

We refer to batch processes as B_{ulpn}, where B_{ulpn} is a pointer to the nth batch process P_{up} running on production R_{ul}^L. The integer variable $\nu_{ulp} \in \mathbb{N}_0$ describes the total number of batch processes to be performed for product P_{up} in production line R_{ul}^L. In case of two products we declare the first one as the main product and the second one as a side product. The side product can be interpreted as a less attractive product or as a waste product. Each batch has a *specific batch size* described by the variable a_{ulpn} subject to a maximum size A_u^+ and a minimum size A_u^-. Values for A_u^+ and A_u^- are given in Table 1.

The fraction of the kth ingredient of a batch B_{ulpn} is specified by F_{ulpn}^{ik}. The fraction of kth output is given by F_{ulpn}^{ok}. Because of a maximum of two ingredients and a maximum of two output components, it is sufficient to specify the fraction of the first ingredient and the fraction of the main product only.

Table 2. Description of batch processes

P_{up}	$P_{up'}$	F^{o1}_{ulpn}	r^P_u	R^L_{ul}	T_{ulp}	P_{vr}	$P_{v'r'}$	F^{i1}_{ulpn}
P_{11}	–	1.	R_1	R^L_{11}	0.05 d	P_0	–	1.
P_{21}	P_{22}	F^{o1*}_{211n}	R_2	R^L_{21}	0.1 d	P_{11}	–	1.
P_{31}	P_{32}	F^{o1**}_{311n}	R_3	R^L_{31}	0.05 d	P_{22}	–	1.
P_{41}	–	1.	R_4	R^L_{41}	0.1 d	P_{21}	–	1.
P_{42}	–	1.	R_4	R^L_{41}	0.1 d	P_{21}	–	1.
P_{43}	–	1.	R_4	R^L_{41}	0.1 d	P_{31}	–	1.
P_{44}	–	1.	R_4	R^L_{41}	0.1 d	P_{31}	–	1.
P_{51}	–	1.	R_5	R^L_{51}	0.15 d	P_{21}	–	1.
P_{52}	–	1.	R_5	R^L_{51}	0.15 d	P_{31}	–	1.
P_{61}	–	1.	R_6	R^L_{61}	0.100 d	P_{42}	–	1.
P_{62}	–	1.	R_6	R^L_{61}	0.125 d	P_{43}	–	1.
P_{63}	–	1.	R_6	R^L_{61}	0.150 d	P_{44}	–	1.
P_{61}	–	1.	R_6	R^L_{62}	0.125 d	P_{42}	–	1.
P_{62}	–	1.	R_6	R^L_{62}	0.150 d	P_{43}	–	1.
P_{63}	–	1.	R_6	R^L_{62}	0.150 d	P_{44}	–	1.
P_{71}	–	1.	R_7	R^L_{71}	0.100 d	P_{51}	–	1.
P_{72}	–	1.	R_7	R^L_{71}	0.100 d	P_{52}	–	1.
P_{73}	–	1.	R_7	R^L_{71}	0.100 d	P_{41}	P_{61}	1/2
P_{74}	–	1.	R_7	R^L_{71}	0.150 d	P_{62}	–	1.
P_{75}	–	1.	R_7	R^L_{71}	0.150 d	P_{63}	–	1.
P_{71}	–	1.	R_7	R^L_{72}	0.150.d	P_{51}	–	1.
P_{72}	–	1.	R_7	R^L_{72}	0.150 d	P_{52}	–	1.
P_{74}	–	1.	R_7	R^L_{72}	0.150 d	P_{62}	–	1.
P_{75}	–	1.	R_7	R^L_{72}	0.150 d	P_{63}	–	1.

The columns in this table have the following meaning:
main and side product, fraction of main product, production unit
process time per batch, ingredient 1 and 2, fraction of ingredient 1

If we introduce the variables m^{ik}_{ulpn} and m^{ok}_{ulpn} with

$$m^{i1}_{ulpn} = F^{i1}_{ulpn} a_{ulpn} \quad , \quad m^{i2}_{ulpn} = (1 - F^{i1}_{ulpn}) a_{ulpn} \tag{6}$$

$$m^{o1}_{ulpn} = F^{o1}_{ulpn} a_{ulpn} \quad , \quad m^{o2}_{ulpn} = (1 - F^{o1}_{ulpn}) a_{ulpn} \tag{7}$$

to describe the quantities of ingredients and output of products in a batch process B_{ulpn}, the mass conservation reads

$$m^{i1}_{ulpn} + m^{i2}_{ulpn} = a_{ulpn} = m^{o1}_{ulpn} + m^{o2}_{ulpn} \tag{8}$$

Each batch B_{ulpn} has a unique *start time* t^S_{ulpn} and a unique *end time* t^E_{ulpn} related by

$$t^E_{ulpn} = t^S_{ulpn} + T_{ulp} \quad , \quad \forall \{ulpn\} \tag{9}$$

Remark. For the sake of simplicity we assume that process time T_{ulp} does not depend on the actual batch size of any batch job B_{ulpn}.

Call a batch $B_{u'l'p'n'}$ the *successor* to batch B_{ulpn} if both batches have to run on the same production line and if there is no other job in between them running on the same processing line. We can express this condition mathematically.

A batch job $B_{u'l'p'n'}$ is called *successor* to batch B_{ulpn} if $u = u'$, $l = l'$ and $t^S_{ulpn} \subset t^S_{ulp'n'}$ is true and for each $B_{ulp''n''}$ with $p'' \subset P_u + 1$ and $n'' \subset N_{ulp''} + 1$ one of two relations (10) or (11)

$$\{(t^S_{ulp''n''} \leq t^S_{ulpn}) \wedge (t^S_{ulp''n''} \leq t^S_{ulp'n'})\} \tag{10}$$

$$\{(t^S_{ulp''n''} \geq t^S_{ulpn}) \wedge (t^S_{ulp''n''} \geq t^S_{ulp'n'})\} \tag{11}$$

is true. In an MILP model we could use an auxiliary binary variable $\alpha_{ulpnp'n'}$

$$\alpha_{ulpnp'n'} = \begin{cases} 1, \text{ if batch } B_{u'l'p'n'} \text{ is the successor to batch } B_{ulpn} \\ 0, \text{ otherwise} \end{cases}$$

to indicate a predecessor-successor relation of two batches B_{ulpn} and $B_{u'l'o'n'}$ and to describe this feature.

2 Cleaning processes

Sometimes a batch process requires a cleaning process C^P_{ulpn}. We define a slack variable $\eta_{ulpn'} \geq 0, n' \leq N_{ulp}$ measuring the machine time of production line R^L_{ul} that is not used for production or cleaning processes between a batch B_{ulpn} and its successor. The auxiliary binary variable β_{ulpn} is a decision variable with

$$\beta_{ulpn} := \begin{cases} 1, \text{ if a batch } B_{ulpn} \text{ is followed by a cleaning process} \\ 0, \text{ otherwise} \end{cases} \tag{12}$$

For all $\alpha_{ulpnp'n'} = 1$ we require that

$$e_{ulpn} + \beta_{ulpn}(F^C T_{ulp} + \eta_{ulpn}) = t^S_{ulp'n'} \tag{13}$$

where F^C is a factor used to describe the cleaning time: $F^C T_{ulp}$.

Two conditions enforce cleaning processes ($\beta_{ulpn} = 1$): (*i*) if $\alpha_{ulpnu'n'} = 1$ and $p \subset p'$ then a cleaning process has to be included. (*ii*) if a batch is the last batch process on a processing line a cleaning process has to follow.

Remark 1. A sequence of products with decreasing quality requirements can help to avoid cleaning processes. This viewpoint motivates condition (*i*) in our example. For this reason a schedule should prefer batch sequences $p \geq p'$ for predecessor-successor-pairs ($\alpha_{ulpnp'n'} = 1$).

Remark 2. Normally, cleaning time depends on the specific batch product and production line. For the sake of simplicity we assume a cleaning time that depends only on the duration of the predecessor batch and on a given factor F^C.

Remark 3. Production lines which are not in use should be cleaned to avoid ongoing reactions of residues. Equation (13) models this requirements because (13) can only be fulfilled if a positive slack time is combined with a cleaning process.

Remark 4. Different batch processes cannot run on the same production line simultaneously, i.e., $t^E_{ulpn} \leq t^S_{ulp'n'}$ for all $(u, l, p, n, p', n' | \alpha_{ulpnp'n'} = 1)$.

3 Stock conditions and plant input/output

The storable products P_{up} are given by the index set $(u, p) \in \mathcal{S}$ with

$$\mathcal{S} = \{(1,1), (2,2), (3,1), (4,2), (4,3), (4,4), (6,1), 6, 3)\} \tag{14}$$

The products P_{41}, P_{51}, P_{52}, P_{62} cannot be stored.

The quantity of product P_{up} stored at time t is described by the stock function $j_{up}(t)$. $j_{up}(t)$ is completely defined by the initial value $j_{up}(0)$ and the cumulative mass inflow $q^i_{up}(t)$ and outflow $q^o_{up}(t)$ following from the schedule of all batches $B_{u'l'p'n'}$ of all products:

$$q^i_{up}(t) = f(t, s_{ulpn}) \quad , \quad q^o_{up}(t) = f(t, e_{ulpn}) \tag{15}$$

with

$$f(t, t_r) := \sum_{l=1}^{L_u} \sum_{n=1}^{N_{ulp}} H(t, t_r) a_{ulpn}, \quad H(t, t_r) = \begin{cases} 1, \text{ if } t \geq T \\ 0, \text{ else} \end{cases} \tag{16}$$

$j_0(t)$ describes the time evolution of raw material requirement, *i.e.*, $j_0(t) = q^i_{11}(t)$. The product outflow j_{7p} of end products is given by $j_{7p}(t) = q^o_{7p}(t)$ for all $p = P_1, \ldots, P_7$.

4 Stock balance equations

The current quantity of a certain product in stock is given by its initial quantity and the cumulative mass inflow and outflow associated with the production of that product. If we explicitly resolve the production scheme we get

$$j_{11}(t) = j_{11}(0) + q^o_{11}(t) - q^i_{21}(t) + \sum_{n=1}^{N_{311}} (1 - F^{o1}_{311n}) a_{311n} H_{e_{311n}}(t) \tag{17}$$

$j_{11}(t)$ takes into account the re-usage of P_{32} as product P_{11}

$$j_{21}(t) = j_{21}(0) + \sum_{n=1}^{N_{211}} F^{o1}_{211n} a_{211n} H_{e_{211n}}(t) - q^i_{41}(t) - q^i_{42}(t) - q^i_{51}(t) \tag{18}$$

$$j_{22}(t) = j_{22}(0) + \sum_{n=1}^{N_{211}} (1 - F^{o1}_{211n}) a_{211n} H_{e_{211n}}(t) - q^i_{31}(t) \tag{19}$$

$j_{21}(t) = j_{22}(t)$ share the output of R_2

$$j_{31}(t) = j_{31}(0) + \sum_{n=1}^{N_{311}} F^{o1}_{311n} a_{311n} H_{e_{311}}(t) - q^i_{43}(t) - q^i_{44}(t) - q^i_{52}(t) \tag{20}$$

In j_{31} the stock input is reduced by the separated quantity of P_{32} (see j_{11}).

$$j_{4k}(t) = j_{4k}(0) + q^o_{4k}(t) - q^i_{6,k-1}(t) \quad , \quad k = 2, 3, 4$$

$$j_{61}(t) = j_{61}(0) + q^o_{61}(t) - \sum_{l=1}^{L_7} \sum_{n=1}^{N_{713}} (1 - F^{i1}_{713n}) a_{713n} H_{s_{713n}}(t)$$

$$j_{63}(t) = j_{63}(0) + q^o_{63}(t) - q^i_{75}(t)$$

The products which are not storable, lead to the equality constraints

$$q^o_{51}(t) = q^i_{71}(t), \quad q^o_{52}(t) = q^i_{72}(t) q^o_{62}(t) = q^i_{74}(t), \quad q^o_{41}(t) = F^{i1}_{713n} q^i_{73}(t)$$

The last equation uses the fact that F^{i1}_{713n} is a constant and does not depend on n.

5 Objective functions

Now, several tasks, i.e., optimization problems, can be formulated using the definitions provided in Section 5.

5.1 Task 1: Minimize makespan – compute a pre-schedule. The makespan to be minimized corresponds to the latest completion time of any of the processing tasks or cleaning operation. The rationale for this objective function is that the plant may be reconfigured at the earliest possible time. The demand D^A_{7p} to be satisfied in time period t^{T_1}, is given in Table 3. At time t^{T1} the stock function $j_{7p}(t^{T1})$ has to be equal or greater than the demand D^A_{7p}, $\upsilon = 1, \ldots, 5$. In Task 1, we set $F^{o1}_{311n} = 1$ for all n, i.e., the re-usage of P_{32} as P_{11} is not considered. Furthermore, cleaning processes ($F^C = 0$) are neglected.

Bounds, $J^-_{up} \leq j_{up}(t) \leq J^+_{up}$, on stock and initial stock conditions for all storable products $(u, p) \in \mathcal{S}$ (see Section 3 of this appendix) are given in Table 4.

Task. Compute the smallest value t^{T_1} for feasible batch plans consistent with all previous constraints!

Table 3. Product demand

j	1	2	3	4	5
D^A_{7j}/kg	0	0	90	50	40
D^B_{7j}/kg	30	30	40	20	40

Table 4. Stock conditions ($J^-_{up} = 0$ for all)

[kg]	j_{11}	j_{21}	j_{22}	j_{31}	j_{42}	j_{43}	j_{44}	j_{61}	j_{63}
$j_{up}(0)$	20	20	0	20	0	0	0	0	0
J^+_{up}	30	30	15	30	10	10	10	10	10

5.2 Task 2: Task 1 considering cleaning times. In addition to Task 1, the cleaning processes described in Section 2 of this appendix should be considered with F^C being set to $F^C = 0.5$. Determine t^{T_2} as small as possible to satisfy the demands A specified in Table 3.

5.3 Task 3: Task 1 consideration of a cyclic product flow. In addition to Task 1, take into account re-usage of product P_{32} as product P_{11} as described in Section 1 of this appendix and thereof set $F^{o1}_{311n} = 0.6923$ for all n. Neglect cleaning processes $(F^C = 0)$ and minimize $t^{T_{23}}$.

5.4 Task 4: Minimize makespan – compute a complete schedule. Take into account cleaning processes and a cycle in product flow and set $F^C = 0.5$ and $F^{o1}_{311n} = 0.6923$ for all n. The production requirement D^B_{7p} specified in Table 3 has to be fulfilled exactly $j_{7p}(t^{T_4}) = D^B_{7p}$ for all $p = 1, \ldots, 5$. Stock limitations are the same as in Task 1, but unlike Task 1, here, initial stock values $j_{up}(0)$ are not prescribed.

Task. Minimize t^{T_4} so that $j_{up}(0) = j_{up}(t^{T4})$ for all $(u,p) \in \mathcal{S}$ with $u \subset 7$! All batch processes and cleaning processes should be finished at cycle time t^{T_4}.

5.5 Task 5: Maximize profit. Take the parameter values as in Task 4 (cyclic production, requirement of Table 5 etc.). Here, a time interval of 2 days is defined $(t^{T_5} = 2d)$. A product requirement which has to be satisfied is not prescribed. Instead, market prices $C^p(P_{up})$ in units of \$/kg for raw material P_0 and end products P_{71}, P_{72}, P_{73}, P_{74}, P_{75} are given as $[5, 10, 10, 30, 20, 15]$.

Task. Find a product mix and a cyclic schedule which maximize the profit function w

$$w = -C^p(P_0)j_0(t^c) + \sum_{p=1}^{5} C^p(P_{70})j_{70}(t^c) \tag{21}$$

5.6 Task 6: Minimal cost capacity design. Take parameter values as in Task 4. Again, product requirement is given by Table 5. Find appropriate stock capacities

J^+_{up} and appropriate production rates r^P_{ul} for each production line l. Note that the stock capacities J^+_{up} are variable now.

Here, the production rate of a production line should be the same for each product manufactured on this line, i.e., the batch process time T_{ulp} does not depend on product index p. The batch process time should now be redefined by the maximum batch size A^+_u and the production rate r^P_{ul}, i.e., $T_{ulp} = A^+_u r^P_{ul}$. For the sake of simplicity we assume the same investment cost for all production rates (C^R =10 \$/kg/d) and the same investment cost of stock capacity for all products ($C^S = 3$ \$/kg).

Task. Determine optimal values for all r^P_{ul} and J^+_{up} so that demand specified in Table 5 under the constraints of Task 4 can be satisfied in time intervals of length $t^{T_6} = 0.5$ days and such that the investment cost c

$$c = \sum_{u=1}^{7} \sum_{l=1}^{L_u} C^R r^P_{ul} + \sum_{(u,p)\in\mathcal{S}} C^S J^+_{up} \tag{22}$$

are minimized, where $\mathcal{S}$ is the set of all indexes (u,p) of all storable products P_{up} defined in Section 3 of this appendix.

6 List of symbols

Symbol	*Description*
$l = 1, \ldots, L_u$	index referring to processing lines
$n = 1, \ldots, N_{ulp}$	index referring to batch processes
$p = 1, \ldots, P_u$	index referring to products
$u = 1, \ldots, U$	index referring to processing units
A^-_u, A^+_u	minimum and maximum batch size (fixed)
a_{ulpn}	batch size (variable)
B_{ulpn}	pointer to the nth batch of product P_{up} running on production line R^L_{ul}
$C^p(P_{up})$	product P_{up} (fixed)
C^r	investment cost for additional production capacity
C^S	investment cost for additional stock capacity
D_{Up}	demand for product p
F^{i1}_{ulpn} (variable)	fraction of first ingredient P_{vr} of batch B_{ulpn}
F^{o1}_{ulpn} (variable)	fraction of first output product P_{up} of batch B_{ulpn}
$F^C \geq 0$	factor describing the cleaning time: $F^C T_{ulp}$ (fixed)
$H_R(t)$	Heavyside function
J^-_{up}	minimum and maximum stock of product P_{up}
$j_{up}(t)$ (variable)	stock as a function of time t
$j_0(t)$	cumulative requirement of raw material
m^i_{ulpn1}, m^i_{ulpn2}	quantities of ingredients of batch B_{ulpn} (variable)

Symbol	*Description*
m^o_{ulpn1}, m^o_{ulpn2}	quantities of products of batch B_{ulpn} (variable)
N_{ulp} (variable)	number of batches of product P_{up} active on line R^L_{ul}
$P_{up}, p = 1, \ldots, P$	product p made in production unit R_u
$q^o_{up}(t), q^o_{up}(t)$	cumulative mass inflow and outflow (variable)
$R_u, u = 1, \ldots, U$	production unit related to a specific process step
$R^L_{ul}, l = 1, \ldots, L$	processing line (part of production unit R_u)
r^P_{ul}	production rates defined in Task 6 (variable)
$\mathcal{S}$	index set of storable products (fixed)
t^E_{ulpn}, t^S_{ulpn}	end and start time of batches (variable)
T_{ulp}	process time per batch process (variable in Task 6)
$\alpha_{ulpnp'n'}$	binary variable to indicate a predecessor-successor relation of two batches B_{ulpn} and $B_{ulp'n'}$
β_{ulpn}	auxiliary binary decision variable with $\beta_{ulpn} = 1$, if a batch B_{ulpn} is followed by a cleaning process
$\eta_{ulpn} \geq 0, n \leq N_{ulp}$	slack variable measuring the machine time unused for production between a batch B_{ulpn} and its successor.

Remark. The values of variables have to be positive or zero. Variables are usually defined by small letters; small Greek characters denote discrete variables. Exceptions: The maximum stock capacity, J^+_{up}, is a variable in Task 6. F^{o1}_{ulpn} is a variable for special values of (u, l, p, n) (see Table 2).

References

1. Aarts EHL, Korst JHM (1989) Simulated Annealing and Boltzmann Machines. Wiley, Chichester
2. Aarts EHL, Korst JHM, van Laarhoven PJM (1997) Simulated annealing. In: Aarts EHL, Lenstra JK (eds) Local search in combinatorial optimization, pp 91–120. Wiley, Chichester
3. Ahleff J (1995) Analyse von mathematischen Modellen zur Reihenfolgeplanung in der Prozessindustrie. Master thesis, RWTH Aachen, Aachen
4. Applequist G, Samikoglu O, Pekny J, Reklaitis GV (1997) Issues in the use, design and evolution of process scheduling and planning systems. ISA Trans 36: 81–121
5. Bartusch M, Möhring RH, Radermacher FJ (1988) Scheduling project networks with resource constraints and time windows. Annals of Opereration Research 16: 201–240
6. Ben-Tal A, Nemirovski A (2000) Robust solutions of linear programming problems contaminated with uncertain data. Mathematical Programming 88: 411–424
7. Berning G, Brandenburg M, Gürsoy K, Mehta V, Tölle FJ (2002) An integrated system solution for supply chain optimization in the chemical process industry. OR Spectrum 24, No. 4 (to appear)
8. Blömer F (1999) Produktionsplanung und -steuerung in der chemischen Industrie – Ressourceneinsatzplanung von Batchprozessen auf Mehrzweckanlagen. Gabler, Wiesbaden

9. Blömer F, Günther HO (1998) Scheduling of a multi-product batch process in the chemical industry. Computers in Industry 36: 245–259
10. Blömer F, Günther HO (2000) LP-based heuristics for scheduling chemical batch processes. International Journal of Production Research 35: 1029–1051
11. Borodin A, El-Yaniv R (1998) Online computation and competitive analyses. Cambridge University Press, Cambridge
12. Brockmüller B, Wolsey L (1995) BASF Problem 1 modelling – Production scheduling with batch-sizing and storage restrictions. Report of the ESPRIT project PAMIPS, Task 1.4 8/1995, BASF-AG, ZX/ZC-C13, D-67056 Ludwigshafen
13. Brucker P, Drexl A, Möhring M, Neumann K, Pesch E (1999) Resource-constrained project scheduling: Notation, classification, models and methods. European Journal of Operational Research 112: 3–41
14. Burkard RE, Hujter M, Klinz B, Rudolf R, Wennink M (2000) A process scheduling problem arising from chemical production planning. Optimization Methods & Software 10: 175–196
15. Burkard RE, Kocher M, Rudolf R (2000) Rounding strategies for mixed integer programs arising from chemical production planning. Yugoslav Journal of Operations Research 8: 9–23
16. Carøe CC, Schultz R (1999) Dual decomposition in stochastic integer programming. Operations Research Letters 24: 37–45
17. Cerda J, Gutierrez MVJM, Espulgas S, Mata J (1989) A new methodology for the optimal design and production schedule of multipurpose batch plants. Industrial Engeneering Chemical Research 28: 988–998
18. Christofides N, Alvarez-Valdes R, Tamarit JM (1987) Project scheduling with resource constraints: A branch and bound approach. European Journal of Operational Research 29: 262–273
19. DeReyck R, Herroelen W (1998) A branch and bound procedure for the resource-constrained project scheduling problem with generalized precedence relations. European Journal of Operational Research 111(1): 152–174
20. Domschke W, Scholl A, Voß S (1997) Produktionsplanung, 2nd ed. Springer, Berlin Heidelberg New York
21. Dorndorf U, Pesch E (1992) Genetic algorithms for job scheduling. In: Hansmann K-W (ed) Parallel problem solving from nature, 2. Operations Research Proceedings 1992, pp 243–250. Springer, Berlin Heidelberg New York
22. Dorndorf U, Pesch E, Phan Huy T (2000) A time-oriented branch-and-bound algorithm for resource-constrained project scheduling with generalized precedence constraints. Management Science 46: 1365–1384
23. Dorndorf U, Pesch E, Phan Huy T (2000) A time-oriented branch-and-bound algorithm for resource-constrained project scheduling. Mathematical Methods of Operations Research 52: 413–439
24. Drexl A, Haase K (1995) Proportional lotsizing and scheduling. International Journal of Production Economics 40: 73–87
25. Eglese RW (1990) Simulated annealing: A tool for operational research. European Journal of Operational Research 46: 271–281
26. Engell S, Märkert A, Sand G, Schultz R, Schulz C (2001) Online scheduling of multiproduct batch plants under uncertainty. In: Grötschel M, Krumke SO, Rambau J (eds) Online optimization of large scale systems, pp 649–676. Springer, Berlin Heidelberg New York
27. Fest A, Möhring RH, Stork F, Uetz M (1998) Resource constrained project scheduling with time windows: A branching scheme based on dynamic release dates. Technical Report 596/1998 (revised 1999), Fachbereich Mathematik, Technische Universität Berlin, Berlin

28. Fieldhouse M (1993) The pooling problem. In: Ciriani T, Leachman RC (eds) Optimization in industry: mathematical programming and modeling techniques in practice, pp 223–230. Wiley, Chichester
29. Glover F, Laguna M (1997) Tabu search. Kluwer Academic Publisher, Dordrecht
30. Grossmann IE, Caballero JA, Yeomans H (1999) Mathematical programming approaches for the synthesis of chemical process systems. Korean Journal Chemical Engeneering 68: 407–426
31. Grunow M, Günther H-O, Lehmann M (2002) Campaign planning for multi-stage batch processes in the chemical industry. OR Spectrum 24: 281–314
32. Harjunkoski I, Jain V, Grossmann IE (2000) Hybrid mixed-integer/constrained logic programming strategies for solving scheduling and combinatorial optimization problems. Computers and Chemical Engineering 24: 337–343
33. Heipcke S (1995) Resource constrained job-shop-scheduling with constraint nets. Diploma thesis, Katholische Universität Eichstätt, Mathem.-Geographische Fakultät, Universität Eichstätt, Eichstätt
34. Heipcke S (1999) Combined modelling and problem solving in mathematical programming and constraint programming. PhD thesis, School of Business, University of Buckingham, Buckingham
35. Holland JH (1975) Adaption in natural and artificial systems. University of Michigan Press, Ann Arbor, MI
36. Holland JH (1992) Adaption in Natural and Artificial Systems. MIT Press, Cambridge, MA
37. Jain V, Grossmann IE (2001) Algorithms for hybrid MILP/CP models for a class of optimization problems. IFORMS Journal on Computing 13: 258–276
38. Kallrath J (1999) The concept of contiguity in models based on time-indexed formulations. In: Keil F, Mackens W, Voss H, Werther J (eds) Scientific computing in chemical engineering ii, pp 330–337. Springer, Berlin Heidelberg New York
39. Kallrath J (2000) Mixed integer optimization in the chemical process industry: Experience, potential and future perspectives. Chemical Engineering Research and Design 78(6): 809–822
40. Kallrath J (2002) Combined strategic and operational planning – a MILP success story in chemical industry. OR Spectrum 24: 315–341
41. Kallrath J, Schreieck A (1995) Discrete optimization and real world problems. In: Hertzberger B, Serazzi G (eds) High-performance computing and networking, pp 351–359. Springer, Berlin Heidelberg New York
42. Kallrath J, Wilson JM (1997) Business optimisation using mathematical programming. Macmillan, Houndmills, Basingstoke
43. Kocher M (2000) A greedy approach for batch processing problems from process industry. Diploma thesis, TU Graz, Graz
44. Kondili E, Pantelides CC, Sargent RWH (1988) A general algorithm for scheduling of batch operations. In: Proceedings 3rd International Symposium on Process Systems Engeneering, pp 62–75. CACHE Corp., Sydney
45. Kondili E, Pantelides CC, Sargent RWH (1993) A general algorithm for short-term scheduling of batch operations – I. MILP formulation. Computers Chemical Engineering 17: 211–227
46. Marriot K, Hallo PJ (1998) Programming with constraints – an introduction. MIT Press, Boston
47. Mendez CA, Cerda J (2002) An MILP-based approach to the short-term scheduling of make-and-pack continuous production plants. OR Spectrum 24, No. 4 (to appear)
48. Michaelewicz Z, Fogel DB (2000) How to solve it: modern heuristics. Springer, Berlin Heidelberg New York

49. Möhring RH, Schulz AS, Stork F, Uetz M (2000) Solving project scheduling problems by minimum cut computations. Technical Report 680, Fachbereich Mathematik, Technische Universität Berlin, Berlin
50. Möhring RH, Schulz AS, Stork F, Uetz M (1999) On project scheduling with irregular starting time costs. Operations Research Letters (in press); also appeared in: Nešetřil J (ed) Algorithms — ESA'99, 7th Annual European Symposium, pp 139–150. Lecture Notes in Computer Science, Vol 1643. Springer, Berlin Heidelberg New York
51. Neumann K (1975) Operations Research Verfahren, Band III. Carl Hanser, München
52. Neumann K, Schwindt C, Trautmann N (2002) Advanced production scheduling for batch plants in process industries. OR Spectrum 24: 251–279
53. Neumann K, Schwindt C, Zimmermann J (2002) Project scheduling with time windows and scarce resources. Springer, Berlin Heidelberg New York
54. Neumann K, Trautmann N, Schwindt C (2001) Short-term planning of batch plants in process industries. In: Kischka P, Leopold-Wildburger I, Möhring RH, Radermacher FJ (eds) Models, methods and decision support for management, pp 211–226. Physica, Heidelberg
55. Pantelides CC (1994) Unified frameworks for optimal process planning and scheduling. In: Proceedings 2nd Conference Foundations of Comp Aided Proc Op, pp 253–274. CACHE Corp, Sydney
56. Patterson JH, Slowinski R, Talbot FB, Weglarz J (1989) An algorithm for a general class of precedence and resource constrained scheduling problems. In: Slowinski R, Weglarz J (eds) Advances in project scheduling, pp 1–26. Elsevier, Amsterdam
57. Pritsker AAB, Watters LJ, Wolfe PM (1969) Multi project scheduling with limited resources: A zero-one programming approach. Management Science 16: 93–108
58. Rosenau B (1996) Bearbeitung eines Losgrössenproblems aus der chemischen Industrie mit Schnittebenenverfahren. Diploma thesis, University of Marburg, Marburg
59. Sand G, Engell S, Märkert A, Schultz R, Schulz C (2000) Approximation of an ideal online scheduler for a multiproduct batch plant. Computers and Chemical Engineering 24: 361–367
60. Sawik T (1987) Multilevel scheduling of multistage production with limited in-process inventory. Journal of Operational Research Society 38: 651–664
61. Schultz R (1995) On structure and stability in stochastic programs with random technology matrix and complete integer recourse. Mathematical Programming 70: 73–89
62. Schwindt C (1998) Verfahren zur Lösung des ressourcenbeschränkten Projektdauerminimierungsproblems mit planungsabhängigen Zeitfenstern. Shaker, Aachen
63. Schwindt C, Trautmann N (2000) Batch scheduling in process industries: an application of resource-constrained project scheduling. OR Spektrum 22(4): 501–524
64. Sgall J (1998) On-line scheduling. In: Fiat A, Woeginger GJ (eds) Online algorithms, pp 196–231. Springer, Berlin Heidelberg New York
65. Shah N (1999) Single- and multisite planning and scheduling: Current status and future challenges. AIChE Symposium Serics 320(94): 75–90
66. Shah N, Pantelides CC, Sargent RWH (1993) A general algorithm for short-term scheduling of batch operations – 2. Computational issues. Computers Chemical Engineering 17: 229–244
67. Shah N, Pantelides CC, Sargent RWH (1993) Optimal periodic scheduling of multipurpose batch plants. Annals of Operations Research 42: 193–228
68. Timpe C (2002) Solving planning and scheduling problems with combined integer and constraint programming. OR Spectrum 24, No. 4 (to appear)
69. Trautmann N (2001) Anlagenbelegungsplanung in der Prozessindustrie. Gabler, Wiesbaden
70. Wolsey LA (1998) Integer programming. Wiley, New York

Advanced production scheduling for batch plants in process industries*

Klaus Neumann, Christoph Schwindt, and Norbert Trautmann

Institut für Wirtschaftstheorie und Operations Research, Universität Karlsruhe, 76128 Karlsruhe, Germany
(e-mail: {neumann,schwindt,trautmann}@wior.uni-karlsruhe.de)

Abstract. An Advanced Planning System (APS) offers support at all planning levels along the supply chain while observing limited resources. We consider an APS for process industries (e.g. chemical and pharmaceutical industries) consisting of the modules network design (for long–term decisions), supply network planning (for medium–term decisions), and detailed production scheduling (for short–term decisions). For each module, we outline the decision problem, discuss the specifics of process industries, and review state–of–the–art solution approaches. For the module detailed production scheduling, a new solution approach is proposed in the case of batch production, which can solve much larger practical problems than the methods known thus far. The new approach decomposes detailed production scheduling for batch production into batching and batch scheduling. The batching problem converts the primary requirements for products into individual batches, where the workload is to be minimized. We formulate the batching problem as a nonlinear mixed–integer program and transform it into a linear mixed–binary program of moderate size, which can be solved by standard software. The batch scheduling problem allocates the batches to scarce resources such as processing units, workers, and intermediate storage facilities, where some regular objective function like the makespan is to be minimized. The batch scheduling problem is modelled as a resource–constrained project scheduling problem, which can be solved by an efficient truncated branch–and–bound algorithm developed recently. The performance of the new solution procedures for batching and batch scheduling is demonstrated by solving several instances of a case study from process industries.

Key words: Detailed production scheduling – Process industries – Batch production – Advanced Planning Systems (APS)

---

* This research has been supported in part by the Deutsche Forschungsgemeinschaft under Grant Ne 137/4 and in part by the SAP AG, Walldorf.
Correspondence to: K. Neumannn

1 Introduction

In process industries, e.g. chemical or pharmaceutical industries, final products are typically produced through several successive chemical or physical transformation processes (also called *tasks*) such as heating, filtration, or packaging. A task requires different types of production resources: a *processing unit* (e.g. a reactor or filter) and *storage facilities* for input and output products (e.g. tanks or silos). All resources are available in limited capacity only.

Different software vendors such as SAP, AspenTech, i2 Technologies, or J.D. Edwards offer Advanced Planning Systems (APS) that support all planning sections along the supply chain while observing limited resources. In this paper, we concentrate on the production section of the supply chain. Long–term decisions are made in the *network design* phase, where plant locations are determined and production systems are configured. Mid–term decisions are made in the *supply network planning* phase, which provides the primary requirements for the final products to be produced at individual plants on the basis of *demand planning* data. The short–term allocation of individual production resources over time to the production of the primary requirements is performed during the *detailed production scheduling* phase.

The purpose of this paper is to present a new solution approach to the detailed production scheduling phase. At first we put the detailed production scheduling phase into the framework of an APS. Based on a generic architecture of an APS, we state the decision problems arising in the network design, supply network planning, and detailed production scheduling phases, discuss the specifics of process industries, and briefly review state–of–the–art solution approaches.

The new approach to detailed production scheduling is discussed in the main part of the paper. The approach is tailored to the minimization of some regular objective function (i.e. a function nondecreasing in the start times of operations) like makespan or mean tardiness in the case of batch production, where all material flows are discontinuous. The basic idea is to decompose the detailed production scheduling problem into a batching and a batch scheduling problem. *Batching* converts the primary requirements for products into individual batches. A *batch* combines a task and a production quantity. The time needed for processing the batch is independent of the production quantity. The objective of batching is to minimize the workload, i.e. the total amount of work to be performed on the processing units. Constraints result from given bounds on batch sizes and proportions of input and output products, storage capacities, perishable products, and the given primary requirements. We formulate the batching problem as a nonlinear mixed–integer program, which is subsequently transformed into a linear mixed–binary program. The processing of a batch is called an *operation*. *Batch scheduling* allocates scarce resources like processing units, workers, and intermediate storage facilities to the operations arising from the batching step such that the regular objective function of detailed production scheduling is minimized. We model the batch scheduling problem as a resource–constrained project scheduling problem, where an operation corresponds to an activity of the project. Perishable products give rise to minimum and maximum time lags between activities. Each processing unit represents a renewable resource of unit capacity with sequence–dependent cleaning times. Alternative processing units are modelled by alternative execution modes for the activities. Storage facilities represent so–called cumulative resources. After a detailed description of the project scheduling model, we show how to solve the project scheduling problem using a truncated branch–and–bound algorithm. The basic idea of this algorithm is to relax the (hard) resource constraints and to replace them by a disjunction of additional precedence constraints between activities.

The detailed production scheduling problem and the specifics of process industries are explained by an example from chemical industry. This example is part of a case study presented by Westenberger and Kallrath (1995) and will be called *WK example* in what follows. In the WK example, we consider a batch production plant producing five different final products. The problem is to find a feasible production schedule with minimum makespan for given primary requirements. We compare the decomposition approach proposed in this paper with the best solution approach known from literature so far by solving several instances that result from varying the primary requirements.

The remainder of this paper is organized as follows: Section 2 is concerned with the architecture of APS for process industries, the individual modules, and the information flow between different modules. Section 3 illustrates the detailed production scheduling problem in chemical industry using the WK example. The decomposition of detailed production scheduling into batching and batch scheduling is explained in Section 4. The latter subproblems and respective solution methods are discussed in Sections 5 and 6. Section 7 provides the results of the experimental performance analysis. In Section 8 we show how to integrate further constraints such as renewable resources of capacity greater than one (several identical processing units or manpower), calendars, or campaigns into the batch scheduling procedure. Section 9 is devoted to concluding remarks and directions for further research.

2 Components of Advanced Planning Systems

Advanced Planning Systems offer support for long–term, mid–term, and short–term planning of the supply chain, which consists of the sections procurement, production, distribution, and sales. In contrast to enterprise resource planning (ERP) systems like SAP R/3, limited availability of resources is taken into account in all planning phases. A monolithic supply chain planning is impracticable due to the complexity and size of the individual decision problems. Furthermore, the different planning horizons motivate a division into long–term, mid–term, and short–term planning. This is the reason why Advanced Planning is generally decomposed into planning modules where the interdependencies between the modules have to be respected.

Figure 1 shows the structure of a generic APS (cf. Meyr et al., 2000) consisting of modules for the phases network design, demand planning, supply network planning, detailed production scheduling, distribution & transport planning, demand fulfillment & available to promise (ATP), and inventory management. This structure reflects the common hierarchy of planning tasks that is implemented in the

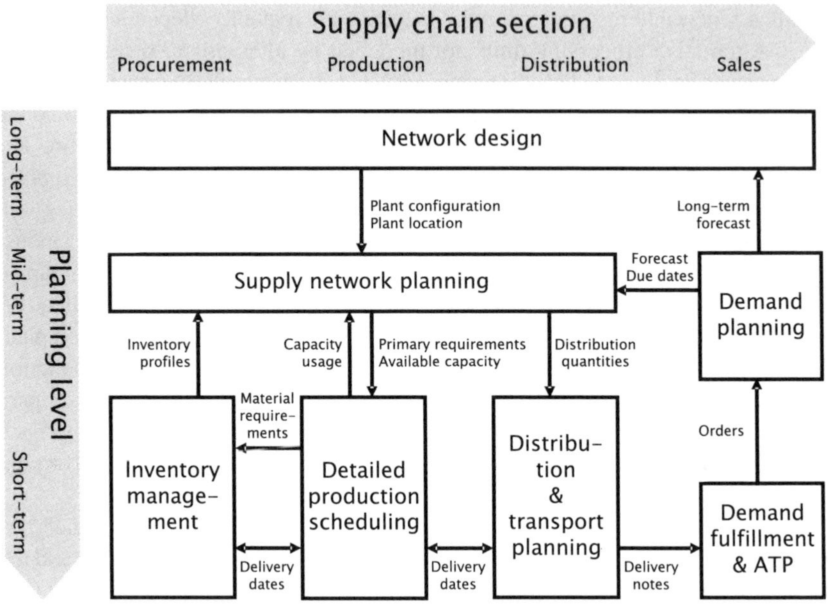

Fig. 1. Structure of an APS

APS offered by different software vendors such as SAP, AspenTech, i2 Technologies, or J.D. Edwards. Each box in Figure 1 corresponds to a module of the APS. The arcs between boxes represent information flows between the modules. In what follows, we concentrate on the production section in the supply chain consisting of the modules network design, supply network planning, and detailed production scheduling. We briefly discuss each module and state the corresponding decision problem, where we place special emphasis on specific requirements arising in process industries. The shorter the planning horizon, the greater is the impact of those specifics on the planning tasks.

Basic concepts of APS and the architecture of commercial APS are presented in Knolmayer et al. (2002) and Stadtler and Kilger (2000). For special requirements of process industries we refer to Applequist et al. (1997), Honkomp et al. (2000), and Pinto and Grossmann (1998).

2.1 Network design

Network design is concerned with the physical structure of the supply chain and thus deals with planning locations of plants and warehouses or other storage facilities and with designing the layout of the individual plants. Typically, the planning horizon ranges from three to ten years.

The *facility location problem* can be described as follows (cf. Tsiakis et al., 2001). We consider a network comprising possible locations for plants and warehouses, fixed locations of customer zones, and transportation facilities between individual locations (e.g. road networks, shipping lanes, sea routes, and air corridors). Long–term forecasts for potential sales of product groups in the customer zones are given. The objective is to determine the number, locations, and capacities of plants and warehouses such that the total annualized cost of the network is minimized. The cost comprises infrastructure cost related to establishing a plant or a warehouse at a certain location, production cost at plants, material handling cost at warehouses, and transportation cost. Binary decision variables indicate whether a warehouse or a plant is built at a certain location, whether a plant supplies a warehouse, or whether a warehouse is assigned to a customer zone. For each plant and each warehouse, a continuous decision variable is introduced representing the respective production or storage capacity. Different types of constraints have to be observed: Network structure constraints say that a link between a plant and a warehouse or a warehouse and a customer zone can exist only if the corresponding warehouses and plants are built. In addition, a flow of material can take place only if the corresponding link is established. Furthermore, so–called single sourcing constraints (cf. Holmberg et al., 1999) may be imposed, according to which a customer zone must be assigned to a single warehouse. Material balance constraints ensure that the production rates at the individual plants must coincide with the flow of the products from the plants to the warehouses, and that the amount of a product received by a customer zone matches the corresponding demand. Further constraints arise from lower and upper bounds on production, storage, and transportation capacities. Solution approaches to the facility location problem known from literature are based on formulations as linear mixed–integer programs combined with heuristic methods, standard solvers and problem decomposition, e.g. Benders' decomposition. For an overview of recent literature, we refer to Tsiakis et al. (2001) and Vidal and Goetschalckx (1997). Issues of network design and plant configurations are also treated in Kallrath (2002b).

The *layout design problem* is solved for each plant individually and involves the spatial arrangement of processing units and intermediate storage facilities. If the plant under consideration consists of a single floor, a typical problem is to divide the given set of equipment into groups of units that reflect the partitions created by aisles or corridors such that the intersection flow cost are minimized. Jayakumar and Reklaitis (1994) show how to model this problem as a graph partitioning problem and present a corresponding heuristic solution procedure. Constraints arise from process sequences, preferred directions of material flow, as well as safety and other restrictions. If the plant consists of multiple floors, the analogous problem is to divide the given set of equipment into subsets and additionally assign each of the subsets to a single level. In this case, the cost associated with the flow between two processing units or storage facilities depends on the distance in between and may be different for horizontal, downward, and upward flow. Jayakumar and Reklaitis (1996) formulate this problem as a nonlinear integer program and convert it to a linear mixed–integer program. For a detailed literature review on chemical plant layout problems, we review to Jayakumar and Reklaitis (1994, 1996).

In process industries, network design includes, for each processing unit, the decision whether it is operated in continuous, semi–continuous, or batch production mode. Plants operating in continuous production mode are dedicated to one product group, e.g. refineries in oil industry. In semi–continuous production mode, the input products are loaded continuously into the processing unit and the output products arise discontinuously or vice versa. Batch production mode is typical of chemical or pharmaceutical industries if small amounts of a large number of products are required, which corresponds to the increasing trend to low–volume production of speciality goods in customer–specific packaging. In what follows, we will concentrate on process industries operating in batch production mode. The different products are processed on multi–purpose equipment, where the production plant is configured according to the required final products. Before processing the next set of final products, the plant has to be reconfigured. This requires the termination of all operations.

Network design is based on long–term changes in the product program, sales figures, and production technologies. Thus, main input factors are long–term forecasts for potential sales of product groups in specific regions, which result from the **demand planning** module. Advanced Planning Systems perform demand planning in three steps (cf. Wagner, 2000): First, statistical forecasting captures the main characteristics of time series of the past and creates forecasts using time–series analysis and causal models (cf. Silver et al., 1998). Second, information is added to the time series that had not been taken into account within statistical forecasting, e.g. life cycles of product groups (cf. Wright and Goodwin, 1998). Third, collaboration between different functional areas (sales, production, procurement) is supported in order to get a consensus on the forecast.

2.2 Supply network planning

The goal of supply network planning is an efficient utilization of the production and storage capacities determined by the network design. The planning horizon has to cover at least one seasonal cycle to be able to balance all demand peaks. It is divided into weekly or monthly *time buckets*. Due to changes in input data, supply network planning is incrementally updated, e.g. in weeks or months, in order to take updated information like actual inventory profiles, capacity usages, and new demand data into consideration. Because of the problem size, it is often necessary to concentrate on bottleneck resources.

The decision variables of supply network planning are purchasing, production, and transportation quantities of final products, and the amount of additionally required capacities (e.g. overtime) in individual time buckets. The objective is to minimize the cost of production, transportation, and inventory holding, cost for delayed or cancelled fulfillment of customer orders, and cost for additional capacities. Constraints arise from given delivery times for customer orders, perishability of products, production, storage and transportation capacities, calendars, and lower and upper bounds on batch sizes. Supply network planning settles capacity bottlenecks by producing in earlier or later periods, at alternative sites or in alternative production modes, by working overtime, by buying products from external suppliers, or by late or cancelled delivery.

Solution approaches to supply network planning problems known from literature are based on formulations as mixed–integer programs, cf. e.g. Kallrath (1999,

2000), Oxe (1997), and Timpe and Kallrath (2000). Kallrath (1999) describes a model that includes the determination of so–called *campaigns*. A campaign consists of several batches of the same task that must be executed consecutively. Typically only campaigns of a minimum size can be produced. Between different campaigns, a sequence–dependent setup–time has to be considered. The problem of scheduling campaigns in distributed chemical batch plants is studied in Berning et al. (2002). For a survey on campaign planning in chemical industry we refer to Papageorgiou and Pantelides (1996). Grunow et al. (2002) present a three-stage decomposition approach to campaign planning and scheduling.

If only one plant is considered, supply network planning is similar to classical master production scheduling (MPS) with capacity constraints (cf. Nahmias, 1997). Note that certain features like cancelling delivery are not considered in MPS.

Supply network planning is based on weekly or monthly forecasts for at most one year including fixed customer orders, the effects of mid–term marketing events, and promotions on sales. This information has to be considered in the statistical forecasting step of demand planning. The customer orders are registered in the **demand fulfillment & ATP** module, which also determines a first due date for each order (cf. Kilger and Schneeweiss, 2000). Further input factors of supply network planning are the plant configuration, i.e. the plant layout, and the plant locations. Both are determined in the network design module.

The results of supply network planning are inputs to other modules. **Detailed production scheduling** is based on primary requirements for final products and on production capacities (remember that the production capacities for the time buckets result from supply network planning). **Distribution & transportation planning** refers to the distribution quantities computed.

2.3 Detailed production scheduling

Detailed production scheduling deals with the short–term allocation of resources over time to the production of the primary requirements determined by supply network planning. Thus, the problem is to explode the primary requirements into batches and to schedule those batches on scarce resources. The secondary requirements for raw materials arising from the batches determined are then reported to the **inventory management**, where appropriate lot sizes for purchasing are to be found.

Detailed production scheduling is performed individually for each plant. The planning horizon corresponds to the bucket size of supply network planning. The goal is to compute a feasible production plan that minimizes a regular objective function (i.e. a function nondecreasing in the start times of operations). The objective of makespan minimization is particularly important in the context of batch production mode in order to enable flexible responses to demand changes, especially if the whole plant has to be reconfigured before producing another set of final products. Another regular objective function that is often considered in practice is the mean tardiness with respect to given due dates for the delivery of customer–ordered final products.

Batches are processed on *multi–purpose processing units*, which can operate several kinds of tasks. At some production levels, there may be alternative processing units. In general, the processing time depends on the particular processing unit used. Recall, however, that in batch production mode the execution time of a task is independent of the respective batch size. For each task, a lower and an upper bound on the respective batch size are given. The minimum batch size generally arises from a minimum filling level, whereas the maximum batch size coincides with the capacity of the corresponding processing unit. A special feature of process industries is the need for cleaning processing units in order to guarantee purity of products. The cleaning times often depend on the sequence of batches and the quality of the products to be produced. In certain applications, a processing unit must also be cleaned when no other batch is started immediately after the completion of the preceding batch. At any point in time, at most one batch or cleaning operation can be executed on one and the same processing unit.

Each task consumes and produces different products. In batch production mode, the input of a task is consumed at its start and the output arises at its completion. Certain intermediate products can be buffered in *storage facilities* of given capacity. Other intermediate products are perishable and cannot be stored and thus must be consumed without any delay. In that case, producing tasks must be assigned to consuming tasks such that no perishable product is in stock at any time.

The storage facilities and processing units of a batch plant are linked by divergent, convergent, or cyclic material flows. Accordingly, recipes are analytic (e.g. in basic materials industry), synthetic (e.g. in pharmaceutical industry), or cyclic (e.g. if catalysts are present). For details we refer to Loos (1997). For each task, the number of executions and the production quantities have to be chosen such that the quantity produced of each product is sufficient to match its gross requirement. Note that in the case of cyclic material flows, some residual of the products belonging to a cycle cannot be used for production. Proportions of input and output goods of tasks are often fixed. However, the input and output proportions of a task may also vary within given bounds, that is, the output can be allotted to different products according to the specific demand situation.

Most solution procedures known from literature are based on formulations of the detailed production scheduling problem as a time–indexed mixed–integer program depending on the time grid chosen, see e.g. Blömer and Günther (1998, 2000), Burkard et al. (1998), Kondili et al. (1993), Pinto and Grossmann (1995, 1998), and Shah (1998). For a literature review we refer to Blömer and Günther (1998) and Shah (1998). Grunow et al. (2002) and Timpe (2002) successfully apply hybrid approaches involving mixed–integer and constraint programming.

3 A case study from chemical industry

Based on an existing plant, Westenberger and Kallrath (1995) have presented a case study dealing with detailed production scheduling in chemical industry. Here, we consider the third out of six "tasks" (see Kallrath, 2002a) where additionally the cleaning times are included. The data of the case study and a bibliogra-

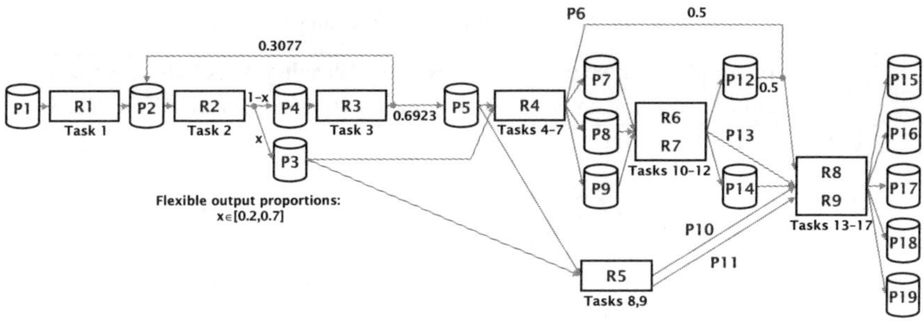

Fig. 2. Production flow of WK example

phy of papers referring to the WK example are available at http://www.wior.uni-karlsruhe.de/neumann/forschung/wk95/wk95.html.

Figure 2 illustrates the production flow of the WK example, which consists of 15 storage facilities for the storable products $P1, \ldots, P5, P7, P8, P9, P12, P14, \ldots,$ $P19$ and 9 multi–purpose processing units (reactors) $R1$ to $R9$. A representation of the production flow as a *state–task–network* can be found in Blömer and Günther (2000). Intermediate products $P6$, $P10$, $P11$, and $P13$ cannot be stocked. The storage facilities and processing units are linked by divergent, convergent, or cyclic material flows. There are primary requirements of 30, 30, 40, 20, and 40 units for the final products $P15$ to $P19$ that arise from supply network planning. The primary requirements have to be produced within 6 days with 24 hours of working time each. The problem is to explode the primary requirements into batches and to schedule these batches on the scarce production resources such that the makespan is minimized.

At some production levels, there are alternative processing units, e.g. $R6$ and $R7$. Table 1 summarizes lower and upper bounds on batch sizes, processing and cleaning times for processing units, and the materials consumed and produced by each task. The time unit is one hour. Recall that all processing times are independent of the respective batch sizes. Each task uses one processing unit. In Figure 2, the processing units are labelled with the tasks executed on them. The tasks are numbered according to increasing quality of output products, i.e., cleaning of a processing unit after the completion of a task becomes necessary if one passes to a task with a higher number or if there is no task on the processing unit which is started immediately after the completion of the preceding one.

Each task consumes and produces materials as given in Table 1. Tasks 2 and 3 produce two output products each and task 15 consumes two input products. The respective proportions of the output products are shown in Figure 2. The output proportions of task 2 are variable. $100x\%$ of the total output is allotted to product $P3$ and the remaining part to product $P4$, where x can vary between 0.2 and 0.7.

For storable products, the inventory is bounded from above (cf. Table 2). Sufficient capacity is available to store the required raw material $P1$ and final products $P15$ to $P19$. There is sufficient initial stock of $P1$, and there is no initial stock of $P15$ to $P19$. At each point in time, the inventories of products $P1$ to $P19$ must be nonnegative.

Table 1. Task settings of WK example

Task	Batch size [min, max]	Alternative processing units	Alternative processing times	Cleaning times	Materials consumed	Materials produced
1	[3,10]	$R1$	2	1	$P1$	$P2$
2	[5,20]	$R2$	4	2	$P2$	$P3$, $P4$
3	[4,10]	$R3$	2	1	$P4$	$P2$, $P5$
4	[4,10]	$R4$	4	2	$P3$	$P6$
5	[4,10]	$R4$	4	2	$P3$	$P7$
6	[4,10]	$R4$	4	2	$P5$	$P8$
7	[4,10]	$R4$	4	2	$P5$	$P9$
8	[4,10]	$R5$	6	3	$P3$	$P10$
9	[4,10]	$R5$	6	3	$P5$	$P11$
10	[3,7]	$R6$ / $R7$	4 / 5	2 / 2.5	$P7$	$P12$
11	[3,7]	$R6$ / $R7$	5 / 6	2.5 / 3	$P8$	$P13$
12	[3,7]	$R6$ / $R7$	6 / 6	3 / 3	$P9$	$P14$
13	[4,12]	$R8$ / $R9$	4 / 6	2 / 3	$P10$	$P15$
14	[4,12]	$R8$ / $R9$	4 / 6	2 / 3	$P11$	$P16$
15	[4,12]	$R8$	4	2	$P6$, $P12$	$P17$
16	[4,12]	$R8$ / $R9$	6 / 6	3 / 3	$P13$	$P18$
17	[4,12]	$R8$ / $R9$	6 / 6	3 / 3	$P14$	$P19$

Table 2. Bounds on inventories of WK example

	$P1$	$P2$	$P3$	$P4$	$P5$	$P7$	$P8$	$P9$	$P12$	$P14$	$P15$	$P16$	$P17$	$P18$	$P19$
Initial stock	∞	20	20	0	20	0	0	0	0	0	0	0	0	0	0
Max. stock	∞	30	30	15	30	10	10	10	10	10	∞	∞	∞	∞	∞

4 Decomposition of detailed production scheduling

In contrast to long–term and mid–term planning, short–term planning has to take into account a large number of constraints that are specific of process industries. That is why detailed production scheduling requires tailor–made models and solution procedures. The particular feature of our approach for detailed production scheduling is the decomposition into a *batching problem* (BP) and a *batch scheduling problem* (BSP). Batching provides the number and sizes of batches for all tasks, and the subsequent batch scheduling assigns a processing unit and an execution time interval to each batch. In Section 5, we show how to model (BP) as a nonlinear mixed–integer program and how to transform it into a linear mixed–integer program of moderate size that can be solved using standard software, e.g. CPLEX or XPRESS–MP. The essential decision variables of (BSP) represent the start times of the operations. Efficient solution procedures for resource–constrained project scheduling can be used for solving large instances of (BSP), see Section 6. A similar decomposition approach has been used by Brucker and Hurink (2000) for solving a special case of detailed production scheduling.

The number of decision variables of the time–indexed MIP models for detailed production scheduling from literature depends on the time grid chosen and is not polynomial in the number of operations. This is the reason why, in contrast to our approach, only relatively small instances can be solved by the latter models.

Obviously, the material flow establishes a causal relationship between the batching and batch scheduling problems (BP) and (BSP). This means that the optimal start times of operations depend on the optimal number and sizes of batches and vice versa. Consequently, when solving at first (BP) and then the corresponding (BSP) we are not assured to obtain an optimal solution to the original detailed production scheduling problem. Due to the limited capacity of intermediate storages, the decomposition method may even fail in finding any feasible solution. As will be shown in the performance analysis of Section 7, the decomposition nevertheless proves useful for heuristically solving problem instances of practical size.

5 Batching

Batching deals with converting the primary requirements for products into sets of batches for each task. The goal is to minimize the workload to be scheduled. Constraints result from given bounds on batch sizes and the number of task executions and from limited storage capacities. In Section 5.1, we present a nonlinear mixed–integer formulation of the batching problem, a preliminary version of which can be found in Neumann et al. (2001a). In Section 5.2, we transform that nonlinear program into a linear mixed–binary program of polynomially related size.

5.1 Formulation as a nonlinear mixed–integer program

Let $\mathcal{T}$ be the set of all tasks and let β_τ be the batch size and ε_τ be the number of batches for task $\tau \in \mathcal{T}$ executed within the given planning horizon $[0, \overline{d}]$ for detailed production scheduling. Thus, $\beta_\tau \varepsilon_\tau$ is the amount produced by task τ. Prescribed minimum and maximum batch sizes $\underline{\beta}_\tau$ and $\overline{\beta}_\tau$, respectively, for task τ provide the constraints

$$\underline{\beta}_\tau \leq \beta_\tau \leq \overline{\beta}_\tau \qquad (\tau \in \mathcal{T}) \tag{1}$$

Let $\mathcal{U}_\tau$ be the set of alternative processing units on which task τ can be carried out and $p_{\tau k}$ be the processing time of task τ on processing unit $k \in \mathcal{U}_\tau$. Then $\overline{\varepsilon}_\tau := \sum_{k \in \mathcal{U}_\tau} \overline{d}/p_{\tau k}$ is an upper bound on the number of executions of tasks τ, and we have the constraints

$$\left.\begin{array}{l} 0 \leq \varepsilon_\tau \leq \overline{\varepsilon}_\tau \\ \varepsilon_\tau \in \mathbb{Z} \end{array}\right\} \quad (\tau \in \mathcal{T}) \tag{2}$$

A task may have several input and output products, e.g., task 2 in the WK example (cf. Section 3). For task $\tau \in \mathcal{T}$, let $\alpha_{\tau\pi} > 0$ be the proportion of output product π and $-\alpha_{\tau\pi} > 0$ be the proportion of input product π. If π is neither input nor output product of task τ, we set $\alpha_{\tau\pi} := 0$. Obviously,

$$\sum_{\pi \in \mathcal{P}_\tau^+} \alpha_{\tau\pi} = - \sum_{\pi \in \mathcal{P}_\tau^-} \alpha_{\tau\pi} = 1 \qquad (\tau \in \mathcal{T}) \tag{3}$$

where $\mathcal{P}_\tau^+$ and $\mathcal{P}_\tau^-$ are the sets of output and input products, respectively, of task τ. Let $\underline{\alpha}_{\tau\pi}$ and $\overline{\alpha}_{\tau\pi}$ be given lower and upper bounds on $\alpha_{\tau\pi}$. Then

$$\underline{\alpha}_{\tau\pi} \leq \alpha_{\tau\pi} \leq \overline{\alpha}_{\tau\pi} \qquad (\tau \in \mathcal{T}, \pi \in \mathcal{P}_\tau) \tag{4}$$

where $\mathcal{P}_\tau := \mathcal{P}_\tau^+ \cup \mathcal{P}_\tau^-$ is the set of products which are either input or output of task τ. If $\alpha_{\tau\pi}$ is fixed, we have $\underline{\alpha}_{\tau\pi} = \overline{\alpha}_{\tau\pi}$.

Next, we consider inventory constraints. For a product $\pi \in \mathcal{P}_\tau$, $\alpha_{\tau\pi}\beta_\tau\varepsilon_\tau$ or $-\alpha_{\tau\pi}\beta_\tau\varepsilon_\tau$ is the total amount of π produced or consumed, respectively, by task τ. Let ρ_π be the given primary requirement minus the initial stock for product π. Consider a recycled product π which belongs to a cycle in the product structure and is the output product of a task τ inside the cycle and of a different task τ' outside the cycle, e.g., product $P2$ in the WK example (see Fig. 2). In that case the amount of product π recycled after the last execution of task τ necessarily remains on stock because it is not consumed. To guarantee that a sufficient quantity of product π is available as input of consuming tasks $\tau'' \in \mathcal{T}_\pi^-$, ρ_π has to be enlarged by the maximum residual stock of π. To meet all primary requirements, it then has to hold that

$$\sum_{\tau \in \mathcal{T}_\pi} \alpha_{\tau\pi}\beta_\tau\varepsilon_\tau \geq \rho_\pi \qquad (\pi \in \mathcal{P}) \tag{5}$$

where $\mathcal{T}_\pi$ is the set of tasks producing or consuming product π and $\mathcal{P} = \cup_{\tau\in\mathcal{T}}\mathcal{P}_\tau$ is the set of all products. $\sum_{\tau\in\mathcal{T}_\pi} \alpha_{\tau\pi}\beta_\tau\varepsilon_\tau - \rho_\pi$ is the final inventory of product π after satisfying the primary and secondary requirements. This final inventory must not exceed the given capacity σ_π of the storage facility for π. That is,

$$\sum_{\tau \in \mathcal{T}_\pi} \alpha_{\tau\pi}\beta_\tau\varepsilon_\tau \leq \rho_\pi + \sigma_\pi \qquad (\pi \in \mathcal{P}) \tag{6}$$

If product π cannot be stored, then $\sigma_\pi := 0$.

The last constraints refer to perishable products, which cannot be stored. Let $\mathcal{T}_\pi^+$ and $\mathcal{T}_\pi^-$ be the sets of tasks producing and consuming, respectively, product π. We assume that for a perishable product π, the amount produced by a batch of a task $\tau \in \mathcal{T}_\pi^+$ must equal the amount consumed by a batch of any task $\tau' \in \mathcal{T}_\pi^-$, that is,

$$\alpha_{\tau\pi}\beta_\tau = -\alpha_{\tau'\pi}\beta_{\tau'} \qquad (\pi \in \mathcal{P}^p, (\tau,\tau') \in \mathcal{T}_\pi^+ \times \mathcal{T}_\pi^-) \tag{7}$$

where $\mathcal{P}^p$ is the set of perishable products. This means that in contrast to the formulation by Neumann et al. (2001), we do not consider solutions to the batching problem where batches of several tasks $\tau \in \mathcal{T}_\pi^+$ and batches of several tasks $\tau' \in \mathcal{T}_\pi^-$ are matched in such a way that the total amount of product π produced equals the total amount consumed. The reason why we restrict ourselves to the one–to–one correspondence (7) between tasks $\tau \in \mathcal{T}_\pi^+$ and $\tau' \in \mathcal{T}_\pi^+$ is that otherwise, the completion of several batches producing π and the start of several batches consuming π would have to occur simultaneously. The latter requirement would considerably reduce the set of feasible solutions to the corresponding batch scheduling problem. Moreover, in practice the implementation of schedules where several operations must be *completed* simultaneously is generally impossible because already small differences between predicted and realized processing times lead to infeasibility.

Let $\overline{p}_\tau := \sum_{k \in \mathcal{U}_\tau} p_{\tau k}/|\mathcal{U}_\tau|$ be the mean processing time of task τ on any of the alternative processing units. To minimize the workload, the objective function to be minimized is chosen to be the total mean processing time $\sum_{\tau \in \mathcal{T}} \overline{p}_\tau \varepsilon_\tau$ (recall that the processing time of a batch is independent of the batch size). The batching problem then takes the form

$$\left.\begin{array}{ll} \text{Min.} & \sum\limits_{\tau \in \mathcal{T}} \overline{p}_\tau \varepsilon_\tau \\ \text{s.t.} & (1) \text{ to } (7) \end{array}\right\} \qquad \text{(BP)}$$

(BP) represents a nonlinear mixed–integer program with the integral decision variables ε_τ and the continuous decision variables β_τ and $\alpha_{\tau\pi}$ ($\tau \in \mathcal{T}$, $\pi \in \mathcal{P}_\tau$). Next we show the NP–hardness of batching problem (BP).

Proposition. The feasibility problem for (BP) is NP–hard even if $\mathcal{P}^p = \emptyset$, $\sigma_\pi = \infty$ for all $\pi \in \mathcal{P}$, and $\overline{\varepsilon}_\tau = \infty$, $\underline{\beta}_\tau = \overline{\beta}_\tau$ for all $\tau \in \mathcal{T}$.

Proof. We provide a polynomial transformation from 3–PARTITION (cf. Garey and Johnson, 1979). Let $B \in \mathbb{N}$ be a bound and let A be a set of 3ν elements λ with associated sizes $s(\lambda) \in \mathbb{N}$ such that $B/4 < s(\lambda) < B/2$ and $\sum_{\lambda \in A} s(\lambda) = \nu B$. The question is whether or not A can be partitioned into ν sets $A_1, \ldots, A_\nu$ such that $\sum_{\lambda \in A_\mu} s(\lambda) = B$ for all $\mu = 1, \ldots, \nu$.

For each $\lambda \in A$, we introduce a raw material π_λ with initial stock $s(\lambda)$, and each index $\mu = 1, \ldots, \nu$ is assigned to a final product π_μ with primary requirement B. For each raw material π_λ and each final product π_μ, we define one task $\tau_{\lambda\mu}$

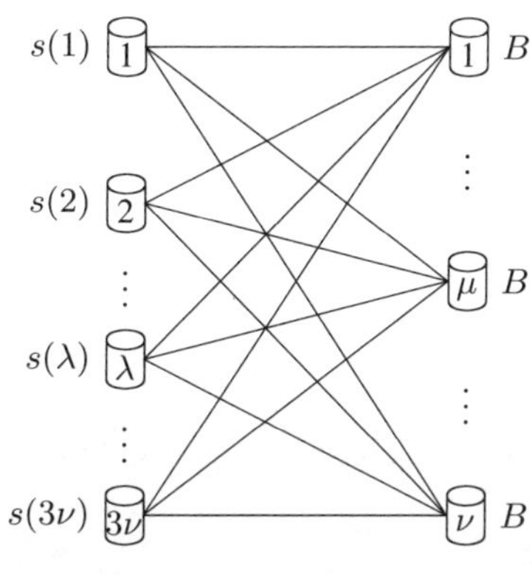

Fig. 3. Batching problem of the proof to the Proposition

transforming π_λ into π_μ with $\underline{\beta}_{\lambda\mu} = \overline{\beta}_{\lambda\mu} = s(\lambda)$ (see Fig. 3). Because of the scarcity of raw materials, at most one batch of tasks $\tau_{\lambda\mu}$ can be executed for each $\lambda \in A$. Since the cumulative requirements νB for all final products equal the total inventory $\sum_{\lambda \in A} s(\lambda)$ of all raw materials, exactly one batch consuming product π_λ is executed for each $\lambda \in A$, say, a batch of task $\tau_{\lambda\mu}$. There is a feasible batching precisely if $\sum_{\lambda \in A} s(\lambda)\varepsilon_{\tau_{\lambda\mu}} = B$ holds for all $\mu = 1, \ldots, \nu$. Assuming that $\lambda \in A_\mu$ if and only if $\varepsilon_{\tau_{\lambda\mu}} = 1$, this condition is met exactly if the 3–PARTITION instance is a yes–instance. □

It can easily be shown that the feasible region belonging to the continuous relaxation of problem (BP) is generally nonconvex. Thus, the computation of an optimal solution to (BP) is not only intractable from the complexity point of view but even worse, it poses a serious algorithmic problem. Thus in the following subsection, we develop a formulation of the batching problem as an equivalent linear mixed–binary program. The number of binary variables in the latter problem equals the upper bound $\sum_{\tau \in \mathcal{T}} \overline{\varepsilon}_\tau$ on the number of batches which can be carried out within the planning horizon.

5.2 *Formulation as a linear mixed–binary program*

To obtain an equivalent linear formulation of (BP), we first introduce the continuous variables

$$\xi_{\tau\pi} = \alpha_{\tau\pi}\beta_\tau \qquad (\tau \in \mathcal{T}, \pi \in \mathcal{P}_\tau)$$

representing the negative amount of product π consumed or the amount of product π produced, respectively, by a batch of task τ. Since the batch size of a task equals the sum of all input quantities, we have

$$\beta_\tau = \sum_{\pi \in \mathcal{P}_\tau^+} \xi_{\tau\pi} \qquad (\tau \in \mathcal{T})$$

Thus, the lower and upper bound constraints (4) on proportions $\alpha_{\tau\pi}$ can be written as

$$\underline{\alpha}_{\tau\pi} \sum_{\pi' \in \mathcal{P}_\tau^+} \xi_{\tau\pi'} \leq \xi_{\tau\pi} \leq \overline{\alpha}_{\tau\pi} \sum_{\pi' \in \mathcal{P}_\tau^+} \xi_{\tau\pi'} \qquad (\tau \in \mathcal{T}, \pi \in \mathcal{P}_\tau) \tag{8}$$

The equations

$$\sum_{\pi' \in \mathcal{P}_\tau^+} \xi_{\tau\pi'} = - \sum_{\pi' \in \mathcal{P}_\tau^-} \xi_{\tau\pi'} \qquad (\tau \in \mathcal{T}) \tag{9}$$

correspond to the mass balance constraints (3), and the batch size constraints (1) now read

$$\underline{\beta}_\tau \leq \sum_{\pi \in \mathcal{P}_\tau^+} \xi_{\tau\pi} \leq \overline{\beta}_\tau \qquad (\tau \in \mathcal{T}) \tag{10}$$

Similarly, the batch size coupling conditions (7) for perishable products can be formulated as

$$\xi_{\tau\pi} = -\xi_{\tau'\pi} \qquad (\pi \in \mathcal{P}^p, (\tau,\tau') \in \mathcal{T}_\pi^+ \times \mathcal{T}_\pi^-) \tag{11}$$

In order to eliminate the nonlinear term $\alpha_{\tau\pi}\beta_\tau\varepsilon_\tau = \xi_{\tau\pi}\varepsilon_\tau$ occurring in the inventory constraints (5) and (6), we replace $\xi_{\tau\pi}\varepsilon_\tau$ with the sum of $\overline{\varepsilon}_\tau$ continuous variables $\xi_{\tau\pi}^\mu$ ($\mu = 1,\dots,\overline{\varepsilon}_\tau$), where

$$\xi_{\tau\pi}^\mu = \begin{cases} \xi_{\tau\pi}, & \text{if } \mu \le \varepsilon_\tau \\ 0, & \text{otherwise} \end{cases}$$

The number ε_τ of batches for task τ is represented by the sum of $\overline{\varepsilon}_\tau$ binary variables θ_τ^μ ($\mu = 1,\dots,\overline{\varepsilon}_\tau$), which equal one exactly if $\mu \le \varepsilon_\tau$, that is,

$$\theta_\tau^\mu = \begin{cases} 1, & \text{if } \sum_{\pi \in \mathcal{P}_\tau^+} \xi_{\tau\pi}^\mu > 0 \\ 0, & \text{otherwise} \end{cases}$$

The linking between variables $\xi_{\tau\pi}^\mu$ and θ_τ^μ can be achieved as follows if π is output product of task τ (i.e., $\xi_{\tau\pi} > 0$):

$$\left.\begin{array}{rcl} 0 \le & \xi_{\tau\pi}^\mu & \le \xi_{\tau\pi} \\ \xi_{\tau\pi} - (1-\theta_\tau^\mu)\overline{\alpha}_{\tau\pi}\overline{\beta}_\tau \le & \xi_{\tau\pi}^\mu & \le \overline{\alpha}_{\tau\pi}\overline{\beta}_\tau\theta_\tau^\mu \end{array}\right\} \quad \begin{array}{l} (\tau \in \mathcal{T}, \pi \in \mathcal{P}_\tau^+, \\ \mu = 1,\dots,\overline{\varepsilon}_\tau) \end{array} \tag{12}$$

If $\xi_{\tau\pi}^\mu > 0$, then $\xi_{\tau\pi}^\mu \le \overline{\alpha}_{\tau\pi}\overline{\beta}_\tau\theta_\tau^\mu$ implies $\theta_\tau^\mu > 0$. For $\xi_{\tau\pi}^\mu = 0$, inequality $\xi_{\tau\pi} - (1-\theta_\tau^\mu)\overline{\alpha}_{\tau\pi}\overline{\beta}_\tau \le \xi_{\tau\pi}^\mu$ provides $\theta_\tau^\mu = 0$. $\theta_\tau^\mu = 1$ and inequality $\xi_{\tau\pi} - (1-\theta_\tau^\mu)\overline{\alpha}_{\tau\pi}\overline{\beta}_\tau \le \xi_{\tau\pi}^\mu$ imply $\xi_{\tau\pi}^\mu = \xi_{\tau\pi}$. For $\theta_\tau^\mu = 0$, it follows from $\xi_{\tau\pi}^\mu \le \overline{\alpha}_{\tau\pi}\overline{\beta}_\tau\theta_\tau^\mu$ that $\xi_{\tau\pi}^\mu = 0$. Hence, constraints (12) provide the equivalences

$$\theta_\tau^\mu = 1 \quad \Leftrightarrow \quad \xi_{\tau\pi}^\mu > 0 \quad \Leftrightarrow \quad \xi_{\tau\pi}^\mu = \xi_{\tau\pi}$$

Constraints (13) show the analogous conditions for tasks τ and input products π (i.e., $\xi_{\tau\pi} < 0$):

$$\left.\begin{array}{rcl} \xi_{\tau\pi} \le & \xi_{\tau\pi}^\mu & \le 0 \\ \underline{\alpha}_{\tau\pi}\overline{\beta}_\tau\theta_\tau^\mu \le & \xi_{\tau\pi}^\mu & \le \xi_{\tau\pi} - (1-\theta_\tau^\mu)\underline{\alpha}_{\tau\pi}\overline{\beta}_\tau \end{array}\right\} \quad \begin{array}{l} (\tau \in \mathcal{T}, \pi \in \mathcal{P}_\tau^-, \\ \mu = 1,\dots,\overline{\varepsilon}_\tau) \end{array} \tag{13}$$

The linear ordering

$$\theta_\tau^1 \ge \theta_\tau^2 \ge \dots \ge \theta_\tau^{\overline{\varepsilon}_\tau} \qquad (\tau \in \mathcal{T}) \tag{14}$$

of the binary variables θ_τ^μ associated with one and the same task τ ensures that precisely the first ε_τ binary variables $\theta_\tau^1,\dots,\theta_\tau^{\varepsilon_\tau}$ equal one. The latter condition is not necessary for reducing the nonlinear mixed–integer program (BP) to a linear mixed–binary program. However, it reduces the size of the feasible region considerably without loss of generality.

Now we are ready to formulate the inventory constraints (5) and (6) as linear inequalities:

$$\rho_\pi \le \sum_{\tau \in \mathcal{T}_\pi} \sum_{\mu=1}^{\overline{\varepsilon}_\tau} \xi_{\tau\pi}^\mu \le \rho_\pi + \sigma_\pi \qquad (\pi \in \mathcal{P}) \tag{15}$$

The mixed–binary programming formulation of the batching problem is then as follows:

$$\left.\begin{array}{ll} \text{Min.} & \sum\limits_{\tau\in\mathcal{T}} \overline{p}_\tau \sum\limits_{\mu=1}^{\overline{\varepsilon}_\tau} \theta_\tau^\mu \\ \text{s.t.} & (8) \text{ to } (15) \\ & \theta_\tau^\mu \in \{0,1\} \qquad (\tau\in\mathcal{T}, \mu=1,\dots,\overline{\varepsilon}_\tau) \end{array}\right\} \quad (\widetilde{\text{BP}})$$

$(\widetilde{\text{BP}})$ represents a linear mixed–binary program with continuous decision variables $\xi_{\tau\pi}$ $(\tau\in\mathcal{T}, \pi\in\mathcal{P}_\tau)$, $\xi_{\tau\pi}^\mu$ $(\tau\in\mathcal{T}, \pi\in\mathcal{P}_\tau, \mu=1,\dots,\overline{\varepsilon}_\tau)$, and binary decision variables θ_τ^μ $(\tau\in\mathcal{T}, \mu=1,\dots,\overline{\varepsilon}_\tau)$.

The batching problem $(\widetilde{\text{BP}})$ belonging to the WK example (cf. Section 3) with 876 continuous and 768 binary variables has been solved to optimality within 4 seconds using CPLEX 6.0 on an 800 MHz Pentium personal computer. The corresponding optimal solution gives rise to 78 operations to be scheduled on the processing units in the batch scheduling step.

6 Batch scheduling

In this section, we deal with scheduling the processing of the batches, i.e. the operations. Recall that each batch is given by the corresponding task τ, the batch size β_τ, and the proportions $\alpha_{\tau\pi}$. Scheduling means that the start time of each operation and the processing unit on which it is to be carried out have to be determined. The objective is to minimize a regular objective function, where temporal and resource constraints have to be taken into account. In Sections 6.1 and 6.2, we show how the batch scheduling problem can be modelled as a resource–constrained project scheduling problem. In Section 6.3 we are concerned with a solution procedure which exploits concepts of resource–constrained project scheduling. Preliminary versions of the latter procedure have been devised by Schwindt and Trautmann (2000) and Neumann et al. (2001a).

6.1 Batch scheduling and project scheduling

We briefly discuss some basic concepts from project scheduling related to our batch scheduling problem. For more details we refer to Brucker et al. (1999) or Neumann et al. (2001b). Suppose that n operations $1,\dots,n$ have to be scheduled. We additionally introduce two dummy operations 0 and $n+1$ representing the beginning and termination, respectively, of the production process, which is regarded as a *project*. $\widetilde{V} := \{1,\dots,n\}$ is the set of *real operations*, and $V := \{0,1,\dots,n+1\}$ is the set of all operations. In project scheduling, the term *activity* is often used instead of operation. In our problem of detailed production scheduling, we have $n = \sum_{\tau\in\mathcal{T}} \varepsilon_\tau$ and $\widetilde{V} = \cup_{\tau\in\mathcal{T}} V_\tau$, where V_τ with $|V_\tau| = \varepsilon_\tau$ is the set of all operations or batches of task τ.

Let $S_i \geq 0$ be the *start time* of operation i. In particular, we set $S_0 := 0$ and we have $S_{n+1} \leq \overline{d}$, where S_{n+1} coincides with the makespan. $S = (S_i)_{i\in V}$ is called a *schedule*. The *processing time* of operation i is denoted by p_i and the *cleaning time* after operation i by c_i. In particular, $p_0 = p_{n+1} = 0$ and $c_0 = c_{n+1} = 0$.

In project scheduling, *temporal constraints* for operations of the type $S_j \geq S_i + d_{ij}$ $((i,j) \in E)$ with $E \subseteq V \times V$ are generally prescribed. If $d_{ij} \geq 0$, then $d_{ij}^{min} := d_{ij}$ represents a *minimum time lag* between the start of operations i and j. If $d_{ij} < 0$, then $d_{ji}^{max} := -d_{ij}$ is a *maximum time lag* between the start of operations j and i. In particular, the maximum time lag $d_{0,n+1}^{max} = \overline{d} = -d_{n+1,0}$ guarantees that inequality $S_{n+1} \leq \overline{d}$ is satisfied. Moreover, the following time lags between real operations $i, j \in \widetilde{V}$ may occur in detailed production scheduling, where $\tau(i)$ denotes the task corresponding to operation $i \in \widetilde{V}$.

(a) Operation j has to wait for the completion of operation i because both operations are carried out on the same processing unit or because an output product of $\tau(i)$ represents an input product of $\tau(j)$. Then we set $d_{ij}^{min} := p_i$. If the processing unit has to be cleaned before processing operation j, we have $d_{ij}^{min} := p_i + c_i$.
(b) Operation j cannot be completed until the start of some operation i frees some storage space required for an output product of $\tau(j)$. We set $d_{ji}^{max} := p_j$.
(c) Operation i provides a perishable product that has to be consumed without delay by some operation j. Then we set $d_{ij}^{min} := d_{ij}^{max} := p_i$.

The time lags discussed in (a) and (b) will be introduced in the course of the solution procedure presented in Section 6.3.

It is well–known that an *activity–on–node project network* N with node set V, arc set E, and arc weights d_{ij} can uniquely be assigned to the project in question (see Neumann and Schwindt, 1997). If there is no path in N from node 0 to node $i \in V, i \neq 0$, with nonnegative length, we add an arc $(0, i)$ with weight $d_{0i} = 0$. If there is no path in N from node $i \in V, i \neq n+1$, to node $n+1$ whose length is at least $p_i + c_i$, we introduce an arc $(i, n+1)$ with weight $d_{i,n+1} = p_i + c_i$. The latter arcs ensure that all operations and cleanings are completed by time S_{n+1}. For simplicity, for the arc set and network enlarged in this way, we again use the symbols E and N, respectively.

A schedule S that satisfies

$$\left.\begin{array}{ll} S_0 = 0 & \\ S_j \geq S_i + d_{ij} & ((i,j) \in E) \end{array}\right\} \qquad (16)$$

is called *time–feasible*. It is well–known that a time–feasible schedule exists exactly if N does not contain any cycle of positive length.

6.2 Resource–constrained project scheduling

We now turn to the different resources. First, we deal with **processing units**. Each processing unit represents a *renewable resource* of capacity 1. Let $\mathcal{R}^\rho$ be the set of all renewable resources and $\mathcal{R}_i^\rho$ be the set of those alternative renewable resources on which operation i can be carried out. Note that $\mathcal{R}_i^\rho = \mathcal{U}_\tau$ for all $i \in V_\tau$. In project scheduling, a pair (i, k) with $k \in \mathcal{R}_i^\rho$ corresponds to an alternative *execution mode* of operation i (cf. e.g. De Reyck and Herroelen, 1999; Drexl et al., 1997; Heilmann, 2001). For $i \in \widetilde{V}$ and $k \in \mathcal{R}_i^\rho$, let

$$x_{ik} := \begin{cases} 1, & \text{if operation } i \text{ is carried out on resource } k \\ 0, & \text{otherwise} \end{cases}$$

Then we have

$$\left.\begin{array}{ll} x_{ik} \in \{0,1\} & (i \in \widetilde{V}, k \in \mathcal{R}_i^\rho) \\ \sum\limits_{k \in \mathcal{R}_i^\rho} x_{ik} = 1 & (i \in \widetilde{V}) \end{array}\right\} \quad (17)$$

because each real operation i must be carried out on exactly one of the alternative processing units for task $\tau(i)$. Note that a processing unit can operate several different tasks (but only one at a time). Thus, we may have $\mathcal{R}_i^\rho \cap \mathcal{R}_j^\rho \neq \emptyset$ and $x_{ik} = x_{jk} = 1$ for $k \in \mathcal{R}_i^\rho \cap \mathcal{R}_j^\rho$, even if $\tau(i) \neq \tau(j)$. We speak of an *assignment* x of processing units to operations if for all $i \in \widetilde{V}$ and $k \in \mathcal{R}_i^\rho$, the binary variables x_{ik} are specified. An assignment x satisfying (17) is called *complete*.

Since a task and thus all of its operations can be carried out on more than one alternative processing unit, the processing and cleaning times generally depend on the processing unit selected. Let $p_{\tau k}$ be again the processing time of task τ and $c_{\tau k}$ be the cleaning time after performing task τ on processing unit k. Then

$$p_i(x) := \sum_{k \in \mathcal{R}_i^\rho} p_{\tau(i),k} x_{ik} \quad \text{and}$$
$$c_i(x) := \sum_{k \in \mathcal{R}_i^\rho} c_{\tau(i),k} x_{ik}$$

represent the processing time and cleaning time, respectively, for operation $i \in \widetilde{V}$ on the processing unit selected. Recall that $p_i = c_i = 0$ for $i = 0, n+1$. We assume that the inequality $c_{\tau k} \leq p_{\tau' k} + c_{\tau' k}$ is satisfied for all τ, τ' with $k \in \mathcal{U}_\tau \cap \mathcal{U}_{\tau'}$, which says that the time needed for cleaning processing unit k after an execution of some task τ does not exceed the processing time of any task τ' on k plus the cleaning time of k after τ'. Since the weight d_{ij} of arc $(i,j) \in E$ in project network N may depend on p_i, p_j, or c_i (see Section 6.1) and the latter quantities generally depend on assignment x, we rewrite constraints (16) as

$$\left.\begin{array}{ll} S_0 = 0 & \\ S_j \geq S_i + d_{ij}(x) & ((i,j) \in E) \end{array}\right\} \quad (18)$$

The cleaning of processing units generally depends on the sequence of operations to be performed. Let $P_k \subseteq \widetilde{V} \times \widetilde{V}$ denote the set of operation pairs (i,j) for which passing from i to j requires a cleaning of processing unit k. For the WK example from Section 3, we have $P_k = \{(i,j) \mid k \in \mathcal{R}_i^\rho \cap \mathcal{R}_j^\rho,\ \tau(j) > \tau(i)\}$. We suppose that relation P_k is *negatively transitive*, i.e., $(h,j) \in P_k$ implies $(h,i) \in P_k$ or $(i,j) \in P_k$ for all $i \in \widetilde{V}$. The latter condition means that when the transition from operation h to operation j requires a cleaning, then the cleaning of k between

h and j cannot be avoided by processing an intermediate operation i. Furthermore, we assume that in order to guarantee purity of products a processing unit must also be cleaned between operations i and j with $(i,j) \notin P_k$ if there is an idle time between the completion of i and the start of j, i.e., $S_j > S_i + p_i(x)$. To formulate the conditions necessitating a cleaning, we introduce the following relation in the set $\widetilde{V}$ of real operations given a processing unit k, a schedule S, and an assignment x:

$$O_k(S,x) := \{(i,j) \in \widetilde{V} \times \widetilde{V} \mid i \neq j, k \in \mathcal{R}_i^\rho \cap \mathcal{R}_j^\rho, S_i \leq S_j, x_{ik} = x_{jk} = 1\}$$

It holds that $(i,j) \in O_k(S,x)$ if $S_j \geq S_i$ and both operations i and j are carried out on processing unit k. Moreover, we define

$$\begin{aligned} C_k(S,x) &:= \{(i,j) \in O_k(S,x) \mid (i,j) \in P_k \text{ or } S_j > S_i + p_i(x)\} \\ \overline{C}_k(S,x) &:= O_k(S,x) \setminus C_k(S,x) \end{aligned}$$

$C_k(S,t)$ is the set of all pairs (i,j) for which resource k has to be cleaned if operation j is carried out after operation i. $\overline{C}_k(S,t)$ is the set of all pairs (i,j) where operation j has to be started immediately after the completion of operation i. A schedule S is then called *process–feasible* with respect to assignment x if

$$\left.\begin{aligned} S_j &\geq S_i + p_i(x) + c_i(x), \text{ if } (i,j) \in C_k(S,x) \\ S_j &= S_i + p_i(x), \text{ if } (i,j) \in \overline{C}_k(S,x) \end{aligned}\right\} \quad (k \in \mathcal{R}^\rho) \tag{19}$$

Second, we deal with **storage facilities**, which represent so–called *cumulative resources* (cf. Neumann and Schwindt, 1999). For each nonperishable product $\pi \in \mathcal{P} \setminus \mathcal{P}^p$, there is one cumulative resource keeping its inventory. Let $\mathcal{R}^\gamma$ be the set of all cumulative resources. For each $k \in \mathcal{R}^\gamma$, there are a prescribed minimum inventory $\underline{R}_k$ (safety stock) and a given maximum inventory $\overline{R}_k$ (storage capacity). Each operation $i \in V$ has a demand r_{ik} for resource $k \in \mathcal{R}^\gamma$. If $r_{ik} \geq 0$, the inventory of resource k is replenished by r_{ik} units at time $S_i + p_i(x)$. If $r_{ik} < 0$, the inventory is depleted by $-r_{ik}$ units at time S_i. Let π be the product stocked in storage facility k. Then $\underline{R}_k = 0$, $\overline{R}_k = \sigma_\pi$, and $r_{ik} = \alpha_{\tau(i),\pi}\beta_{\tau(i)}$ for $i \in \widetilde{V}$. Moreover, r_{0k} represents the initial stock of product π and $r_{n+1,k} = 0$.

Let $V_k^+ := \{i \in V \mid r_{ik} > 0\}$ and $V_k^- := \{i \in V \mid r_{ik} < 0\}$ be the sets of operations replenishing and depleting, respectively, the inventory of $k \in \mathcal{R}^\gamma$. Given schedule S and assignment x,

$$\mathcal{A}_k(S,x,t) := \{i \in V_k^+ \mid S_i + p_i(x) \leq t\} \cup \{i \in V_k^- \mid S_i \leq t\}$$

is called the *active set* of activities replenishing or depleting resource $k \in \mathcal{R}^\gamma$ by time $t \in [0, \overline{d}]$. Schedule S is said to be *storage–feasible* with respect to assignment x if

$$\underline{R}_k \leq \sum_{i \in \mathcal{A}_k(S,x,t)} r_{ik} \leq \overline{R}_k \qquad (k \in \mathcal{R}^\gamma, 0 \leq t \leq \overline{d}) \tag{20}$$

Clearly, there exists a storage–feasible schedule only if $\underline{R}_k \leq \sum_{i \in V} r_{ik} \leq \overline{R}_k$ for all $k \in \mathcal{R}^\gamma$ (cf. Neumann and Schwindt, 1999).

A schedule which is time–, process–, and storage–feasible with respect to some assignment x is called *feasible* with respect to x. A pair (S, x) is called a *feasible solution* if x is a complete assignment and if S is feasible with respect to x. The batch scheduling problem consists of finding a feasible solution that minimizes a regular objective function f and can be formulated as follows:

$$\left.\begin{array}{ll} \text{Min.} & f(S) \\ \text{s.t.} & (17) \text{ to } (20) \end{array}\right\} \qquad \text{(BSP)}$$

The decision variables of batch scheduling problem (BSP) are the continuous variables S_i $(i \in V)$ and the binary variables x_{ik} $(i \in \widetilde{V}, k \in \mathcal{R}^{\rho})$. By transformation from the flow shop problem $F||C_{\max}$ it can easily be shown that the feasibility problem for (BSP) is NP–hard. Next, we briefly sketch the generation scheme of a branch–and–bound algorithm for solving the resource–constrained project scheduling problem (BSP).

6.3 Solving the batch scheduling problem

The constraints that make the batch scheduling problem (BSP) intractable arise from the scarcity of renewable and cumulative resources. By relaxing the corresponding constraints (17), (19), and (20) we obtain a *temporal scheduling problem*, where $d_{ij}(x)$ in (18) is replaced with $d_{ij} = \min\{d_{ij}(x) \mid x \text{ satisfies } (17)\}$ (i.e., for each arc $(i, j) \in E$, we replace the mode-dependent arc weight $d_{ij}(x)$ by the minimum weight d_{ij} of (i, j) with respect to all complete mode assignments). The temporal scheduling problem represents a longest path problem in the (single-mode) project network N with arc weights d_{ij} and can be solved by standard network algorithms (cf. Ahuja et al., 1993). Since the temporal scheduling problem is a relaxation of problem (BSP), the longest path length from node 0 to node $n + 1$ in N provides a lower bound on the minimum makespan of feasible schedules.

An optimal solution to this relaxation may be infeasible for (BSP) due to two reasons: (i) there may be operations $i \in \widetilde{V}$ for which no renewable resource $k \in \mathcal{R}_i^{\rho}$ has been selected so far and thus equation (17) is violated, or (ii) one of the resource constraints (19) or (20) may not be met. When checking inequalities (20), we replace $p_i(x)$ in the definition of active set $\mathcal{A}_k(S, x, t)$ with $\min\{p_i(x) \mid x \text{ satisfies } (17)\}$. In case (i), we select some $k \in \mathcal{R}_i^{\rho}$ for processing i and replace $\mathcal{R}_i^{\rho}$ with $\{k\}$. This means that from now on for any assignment x satisfying (17) we have $x_{ik} = 1$ and $x_{ik'} = 0$ for all $k' \in \mathcal{R}_i^{\rho} \setminus \{k\}$. In case (ii), violations of the resource constraints can be avoided by introducing appropriate time lags between some operations and adding the corresponding arcs to project network N.

By assigning renewable resources to the execution of operations and adding time lags, we obtain a new problem of type (BSP) with a tighter relaxation belonging to a reduced set of complete assignments and an expanded project network again denoted by N. The selection of renewable resources and the addition of time lags is continued until assignment x is complete and the solution to the temporal scheduling problem provides a feasible schedule S or until the temporal scheduling problem is unsolvable because N contains a cycle of positive length. Figure 4

FOR $i = 1, \ldots, n$ **DO**
  IF $|\mathcal{R}_i^\rho| > 1$ **THEN** set $x_{ik} := 0$ for all $k \in \mathcal{R}_i^\rho$;
  ELSE set $x_{ik} := 1$ for all $k \in \mathcal{R}_i^\rho$;
END (∗ FOR ∗)
REPEAT
  Compute longest path lengths ℓ_{0i} from 0 to all nodes $i \in V$ in the (expanded) project network N;
  IF N contains cycle of positive length **THEN** terminate;
  ELSE set schedule $S := (\ell_{0i})_{i=0}^{n+1}$;
  Scan $\{S_1, S_1 + p_1, \ldots, S_n, S_n + p_n\}$ for the earliest point in time t for which constraints (19) or (20) are not satisfied;
  IF there is such a time t **THEN** select appropriate arc (i,j) with weight $d_{ij}(x)$ and add (i,j) to (expanded) project network N;
  ELSE (∗ S is feasible with respect to x ∗)
    Scan $\{S_1, \ldots, S_n\}$ for the earliest point in time t at which some operation i with $|\mathcal{R}_i^\rho| > 1$ is started;
    IF there is such a time t **THEN** select some renewable resource $k \in \mathcal{R}_i^\rho$ and set $x_{ik} := 1$, $\mathcal{R}_i^\rho := \{k\}$;
    ELSE RETURN (S, x); (∗ (S, x) is feasible solution ∗)
  END (∗ IF ∗)
END (∗ REPEAT ∗)

Fig. 4. Generation scheme for batch scheduling

shows a generation scheme of a branch–and–bound algorithm, where violations of the resource constraints (19) and (20) are resolved in chronological order and a renewable resource is selected each time schedule S satisfies the resource constraints (19) or (20) with respect to current assignment x. The complete branch–and–bound algorithm is obtained from the generation scheme by first, branching over the arcs (i,j) to be added to N and the renewable resources $k \in \mathcal{R}_i^\rho$ to be selected for processing the real activities $i \in \widetilde{V}$ and second, fathoming enumeration nodes for which the longest path length in N is not less than the makespan of the best schedule found.

We now consider in more detail how to determine appropriate time lags for resolving resource conflicts. Let us assume that schedule S is not process–feasible with respect to x. Then there exist pairs $(i,j) \in O_k(S,x)$ such that

(a) $(i,j) \in P_k$ and $S_j < S_i + p_i(x) + c_i(x)$, or
(b) $(i,j) \notin P_k$ and $S_j < S_i + p_i(x)$, or
(c) $(i,j) \notin P_k$ and $S_i + p_i(x) < S_j < S_i + p_i(x) + c_i(x)$.

We first assume that $(j,i) \in P_k$. Case (a), where $(i,j) \in P_k$, can be dealt with by considering the two alternatives depicted in Figure 5a, where the shaded boxes correspond to cleanings between operations i and j. First, we may introduce the minimum time lag $d_{ij}^{min}(x) := p_i(x) + c_i(x)$ delaying operation j up to the point in time where the cleaning of resource k is terminated. Second, the conflict can be settled by adding the minimum time lag $d_{ji}^{min}(x) := p_j(x) + c_j(x)$ saying that operation i cannot be started before operation j has been completed (note that in this

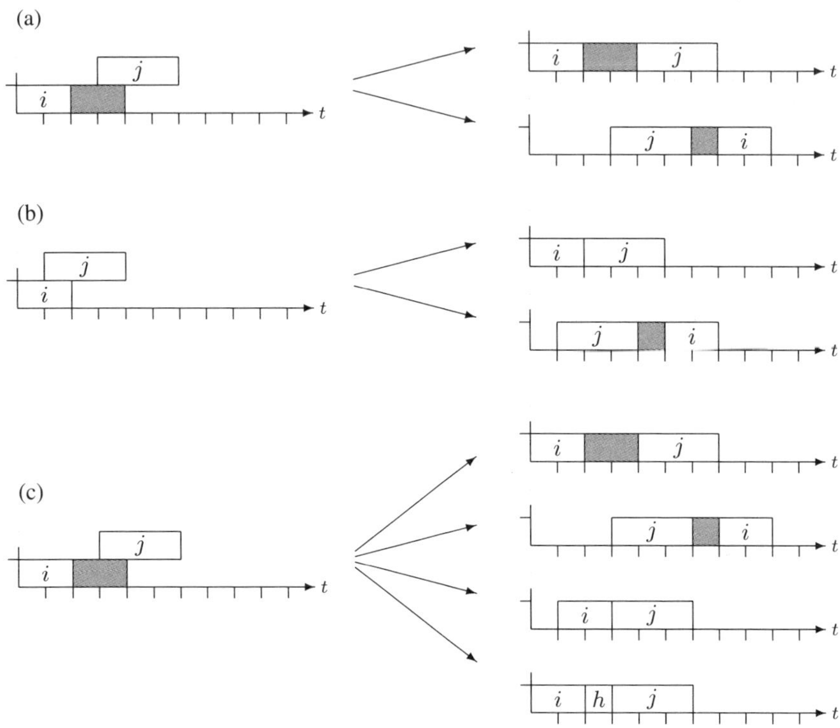

Fig. 5. How to resolve process–infeasibility

case, cleaning of k is also necessary). We now consider the case where $(i, j) \notin P_k$. Case (b) can be handled by interchanging the roles of operations i and j: The two alternatives consist of introducing the minimum time lags $d_{ij}^{min}(x) := p_i(x)$ or $d_{ji}^{min}(x) := p_j(x)+c_j(x)$ (see Fig. 5b). The conditions of case (c) say that there is an idle time between the completion of operation i and the start of operation j which is not sufficiently large for cleaning k (see Fig. 5c). Then the first alternative is to delay j up to the end of the cleaning after the execution of i, i.e., $d_{ij}^{min}(x) := p_i(x)+c_i(x)$. Analogously, i may be delayed up to the end of the cleaning after the execution of j, i.e., $d_{ji}^{min}(x) := p_j(x)+c_j(x)$. A third alternative consists of avoiding the cleaning of k before j starts by introducing the maximum time lag $d_{ij}^{max}(x) := p_i(x)$. Finally, the cleaning of k between i and j can also be avoided by delaying some operation h with $k \in \mathcal{R}_h^\rho$, $S_h < S_i + p_i(x)$, $(i, h) \notin P_k$, $(h, j) \notin P_k$, and $p_h(x) < c_i(x)$ up to the completion of operation i, i.e., $d_{ih}^{min} := p_i(x)$.

If $(j, i) \notin P_k$, no cleaning is required when passing from j to i, i.e., minimum time lag d_{ji}^{min} in cases (a), (b), and (c) is equal to $p_j(x)$.

The cumulative-resource constraints (20) are violated if the inventory of some product falls below zero or exceeds the storage capacity. At first, we consider the case where at some time $t \geq 0$ the inventory of a product π kept in cumulative resource k is negative. We choose an operation i with $S_i + p_i(x) > t$ and $r_{ik} > 0$ and an operation j with $S_j \leq t$ and $r_{jk} < 0$ (cf. Fig. 6a). We then delay the start

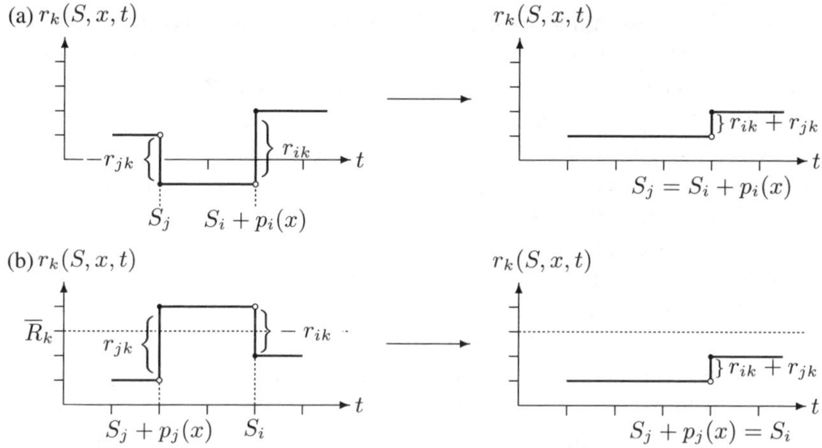

Fig. 6. How to resolve storage–infeasibility

of operation j (i.e. the depletion of k) up to the completion of operation i (i.e. the replenishment of k) by introducing the minimum time lag $d_{ij}^{min}(x) := p_i(x)$. The case where at time $t \geq 0$ the inventory of π exceeds the capacity of k is illustrated in Figure 6b. We select an operation i with $S_i > t$ and $r_{ik} < 0$ and an operation j with $S_j + p_j(x) \leq t$ and $r_{jk} > 0$. The completion of operation j (i.e. the replenishment of k) is delayed up to the start of operation i (i.e. the depletion of k) by introducing the maximum time lag $d_{ji}^{max}(x) := p_j(x)$.

A branch–and–bound algorithm based on the above generation scheme has been implemented in C under MS–Visual C++ 6.0. The batch scheduling problem with 78 operations resulting from an optimal solution to batching problem $(\widetilde{\text{BP}})$ for the WK example has been approximately solved using a *beam search* procedure, that is, a truncated version of the branch–and–bound algorithm. The basic idea of beam search is to select, at each enumeration node, a given number of most promising child nodes for further branching and to fathom the remaining child nodes (cf. Pinedo, 1995). On a Pentium personal computer with 800 MHz clock pulse operating under MS Windows 2000, we have obtained a feasible solution with a makespan of 88 units of time within an imposed running time limit of 56 seconds. Thus, including the 4 seconds for solving $(\widetilde{\text{BP}})$, approximately solving the entire detailed production scheduling problem has taken one minute of computing time. The solution found is the best known thus far for the WK example.

7 Experimental performance analysis

In this section, we are going to test the decomposition approach using 22 instances which have been generated by varying the primary requirements for the final products of the WK example. This test set has been used by Blömer and Günther (1998) as well as Blömer (1999) for evaluating different solution procedures which are based on a monolithic mixed–integer linear programming formulation of the detailed production scheduling problem. We compare our decomposition approach (denoted by B+BS) with the most effective of those methods, the so–called time grid heuristic (TGH). In the performance analysis by Blömer (1999), the cleaning times of tasks 1 to 12 have been assumed to be equal to the respective processing times. Furthermore, the initial stock of products $P2$, $P3$, and $P5$ has been chosen to be equal to 10. For comparison purposes, we have used the modified cleaning times and initial stocks as well.

For each instance, we have imposed a running time limit of one minute to B+BS, have solved the corresponding batching problem to optimality using CPLEX 6.0, and have allotted the remaining computation time to the beam search procedure for the batch scheduling problem. For each instance, Table 3 shows the makespans and CPU times belonging to methods TGH and B+BS. Column "Best makespan" refers to the best feasible solution which could be obtained by some of the heuristics based on mixed–integer linear programming (the data for those procedures have been communicated by Günther, 1999). An asterisk after the makespan for the method B+BS indicates that either a solution proven to be optimal has been found or the best solution known thus far could be improved. The last two columns compare the makespans belonging to methods TGH and B+BS in the case where cleaning of processing units is ignored.

The results displayed in Table 3 indicate that in comparison with the time grid heuristic, the decomposition method generally (for 21 out of 22 instances) finds a better solution within a markedly shorter amount of time. In particular, the results for the four largest instances 19 to 22 show that, compared to the monolithic approaches, procedure B+BS scales quite well. It is worth noting that all batching problems could be solved to optimality, where the computation times vary from 2 seconds (instance 1) to 31 seconds (instance 11). The heuristics based on mixed–integer linear programming perform considerably better when cleanings are neglected. However, the decomposition method still finds solutions of comparable quality within a shorter amount of time.

The good performance of the decomposition method is mainly due to the strong correlation between the workload to be scheduled (i.e., the objective function of the batching problem) and the makespan that can be achieved in the course of batch scheduling. The makespan can generally be decreased only by reducing the workload to be processed in an *active chain* of the schedule found consisting of critical operations whose delay would lead to an increase in the makespan. As a rule, the percentage of critical operations grows as the utilization of the processing units increases. That is why we guess that the scarcer the renewable resources and thus the harder the instances are, the smaller is the loss of accuracy when using the decomposition method.

8 Supplements

The following additional features, which frequently occur when coping with real–world problems, have been integrated into the beam search procedure for batch scheduling:

Table 3. Computational results

Instance	Primary requirements	TGH ms[a]	TGH time[b]	Best ms	B+BS ms	B+BS time[c]	TGH ms[d]	B+BS ms[d]
1	(20,20,20,0,0)	36	1110	36	36 *	3	30	30
2	(20,20,0,20,0)	42	2247	38	38 *	13	34	35
3	(20,20,0,0,20)	42	2487	38	38 *	17	34	34
4	(20,0,20,20,0)	48	1550	38	39	60	35	33
5	(20,0,20,0,20)	44	1778	36	41	60	34	32
6	(20,0,0,20,20)	48	3605	48	43 *	60	41	42
7	(0,20,20,20,0)	52	2587	42	38 *	60	35	36
8	(0,20,20,0,20)	48	3123	42	39 *	60	36	36
9	(0,20,0,20,20)	54	3607	54	53 *	60	44	45
10	(0,0,20,20,20)	60	3607	56	50 *	60	41	42
11	(10,10,20,20,30)	68	3605	66	66	60	48	53
12	(30,20,20,10,10)	60	3605	52	52	60	42	40
13	(10,20,30,20,10)	64	3604	61	50 *	60	47	46
14	(18,18,18,18,18)	66	3606	66	57 *	60	48	48
15	(15,15,30,30,45)	148	3622	112	114	60	78	72
16	(45,30,30,15,15)	124	3628	76	80	60	58	62
17	(15,30,45,30,15)	112	3621	88	91	60	71	69
18	(27,27,27,27,27)	124	3631	88	91	60	72	73
19	(20,20,40,40,60)	208	5152	208	135 *	60	100	92
20	(60,40,40,20,20)	184	3638	184	100 *	60	70	74
21	(20,40,60,40,20)	184	3643	124	112 *	60	82	85
22	(36,36,36,36,36)	214	3635	172	134 *	60	88	88

[a] Makespan.
[b] CPU time in seconds on a Pentium–266 PC including memory freeing, enumeration stopped after one hour if feasible solution could be found.
[c] CPU time in seconds on a Pentium–800 PC, stopped after one minute.
[d] Without cleanings.

(a) Besides processing units also *workers* are needed for processing tasks, e.g. for operating a processing unit or checking the quality of an intermediate product. Workers with the same skills are grouped to a pool that is modelled as a renewable resource whose capacity equals the number of workers in the corresponding pool. For the case where the number of available workers is not constant over time, we refer to Schwindt and Trautmann (2000). Möhring and Uetz (2001) have applied resource–constrained project scheduling models and methods to the problem of scheduling tasks with time–varying manpower demand.

(b) If several *identical processing units* are available, those processing units can also be grouped to a pool and modelled as a renewable resource. The process–feasibility of a schedule can then be checked by solving an assignment problem. The basic idea is to examine whether the operations $i \in \widetilde{V}$ can be assigned to the individual processing units $k \in \mathcal{R}_k^\rho$ such that constraints (19) are fulfilled. Grouping identical processing units considerably reduces the number of alternative modes of the activities and thus offers advantages in computation performance, especially for large instances. For details we refer to Schwindt and Trautmann (2000), who deal with the case of identical processing units that do not require cleaning after idle times.

(c) *Break calendars* provide time intervals (breaks) where some resources are not available. Some activities may be interrupted during a break, others must not be interrupted. The corresponding temporal scheduling problem can be solved efficiently by a polynomial longest path algorithm of type label correcting, where start and completion times of activities are appropriately delayed until all calendar constraints are met (cf. Neumann et al., 2001b; Trautmann, 2001b).

(d) A *quarantine time* $q \geq 0$ says that a product can be consumed q units of time after its production at the earliest. A *shelf life time* $s \geq 0$ implies that a product must be consumed s units of time after its production at the latest. Since one and the same product may be produced by several operations and may be consumed by several operations, it is generally not possible to model quarantine and shelf life times by minimum and maximum time lags between producing and consuming operations. Schwindt and Trautmann (2002) show how to integrate quarantine and shelf life times into the batch scheduling model discussed in Section 6 using fictitious cumulative resources and fictitious activities.

(e) *Campaigns* are used for grouping operations belonging to similar products in order to reduce cleaning times. Once the first operation of a campaign has been started, it is not allowed to start any operation not belonging to the campaign on any processing unit used until the last operation of the campaign on this processing unit has been completed. This leads to an additional type of resource conflicts. For details we refer to Trautmann (2001a).

9 Conclusions

An Advanced Planning System has been considered comprising the modules network design, supply network planning, and detailed production scheduling. The decision problems related to the first two modules have been sketched briefly. The module of detailed production scheduling has been discussed in detail for batch production in process industries. A new approach to solving the corresponding optimization problem has been proposed, which consists of decomposing the problem into batching and batch scheduling. The batching problem can be reduced to a linear mixed–binary program of moderate size and solved by standard software. The batch scheduling problem can be modelled as a resource–constrained project scheduling problem and solved by an efficient beam search procedure. The new approach is markedly superior to the monolithic solution methods known thus far. This is also demonstrated by an experimental performance analysis based on a case study from chemical industry.

Important areas of future research are, for example, the development of an iterative solution procedure for detailed production scheduling, where the batching and the batch scheduling problems are solved alternately, and of efficient decomposition methods for (approximately) solving very large instances with thousands of operations. Moreover, new solution methods for the module of supply network planning should be developed whose performance is comparable to that of the new approach to detailed production scheduling presented.

References

Ahuja RK, Magnanti TL, Orlin JB (1993) Network flows. Prentice Hall, Englewood Cliffs

Applequist G, Samikoglu O, Pekny J, Reklaitis GV (1997) Issues in the use, design and evolution of process scheduling and planning systems. ISA Transactions 36:81–121

Berning G, Brandenburg M, Gürsoy K, Mehta V, Tölle FJ (2002) An integrated system solution for supply chain optimization in the chemical process industry. OR Spectrum 24, No. 4 (to appear)

Blömer F (1999) Produktionsplanung und -steuerung in der chemischen Industrie – Ressourceneinsatzplanung von Batchprozessen auf Mehrzweckanlagen. Gabler, Wiesbaden

Blömer F, Günther HO (1998) Scheduling of a multi–product batch process in the chemical industry. Computers in Industry 36:245–259

Blömer F, Günther HO (2000) LP–based heuristics for scheduling chemical batch processes. International Journal of Production Research 35:1029–1052

Brucker P, Drexl A, Möhring RH, Neumann K, Pesch E (1999) Resource–constrained project scheduling: notation, classification, models, and methods. European Journal of Operational Research 112:3–41

Brucker P, Hurink J (2000) Solving a chemical batch scheduling problem by local search. Annals of Operations Research 96:17–36

Burkard RE, Kocher M, Rudolf R (1998) Rounding strategies for mixed integer programs arising from chemical production planning. Yugoslav Journal of Operations Research 8:9–23

De Reyck B, Herroelen WS (1999) The multi–mode resource–constrained project scheduling problem with generalized precedence relations. European Journal of Operational Research 119:538–556

Drexl A, Hartmann S, Sprecher A (1997) An exact algorithm for project scheduling with multiple modes. OR Spektrum 19:195–203

Garey MR, Johnson DS (1979) Computers and intractability. Freeman, New York

Grunow M, Günther HO, Lehmann M (2002) Campaign planning for multi–stage batch processes in the chemical industry. OR Spectrum 24:281–314

Günther HO (1999) Personal communication

Heilmann R (2001) Resource–constrained project scheduling: A heuristic for the multi–mode case. OR Spektrum 23:335–357

Holmberg K, Roennqvist M, Yuan D (1999) An exact algorithm for the capacitated facility location problems with single sourcing. European Journal of Operational Research 113:544–559

Honkomp SJ, Lombardo S, Rosen O, Pekny J (2000) The curse of reality—why process scheduling optimization problems are difficult in practice. Computers & Chemical Engineering 24:323–328

Jayakumar S, Reklaitis GV (1994) Chemical plant layout via graph partitioning — I. single level. Computers & Chemical Engineering 18:441–458

Jayakumar S, Reklaitis G (1996) Chemical plant layout via graph partitioning — II. multiple levels. Computers & Chemical Engineering 20:563–578

Kallrath J (1999) Mixed–integer nonlinear programming applications. In: Ciriani T, Gliozzi S, Johnson E, Tadei R (eds) Operational research in industry, pp 42–76. Ichor Business Books, West Lafayette

Kallrath J (2000) Mixed integer optimization in the chemical process industry — experience, potential and future perspectives. Transactions of the Institution of Chemical Engineers 78:809–822

Kallrath J (2002a) Planning and scheduling in the process industry. OR Spectrum 24:219–250

Kallrath J (2002b) Combined strategic and operational planning — A MILP success story in chemical industry. OR Spectrum 24:315–341

Kilger C, Schneeweiss L (2000) Demand fulfillment and ATP. In: Stadtler H, Kilger C (eds) Supply chain management and advanced planning, pp 135–148. Springer, Berlin Heidelberg New York

Knolmayer G, Mertens P, Zeier A (2002) Supply chain management based on SAP systems. Springer, Berlin Heidelberg New York

Kondili E, Pantelides CC, Sargent RWH (1993) A general algorithm for short–term scheduling of batch operations — I. MILP Formulation. Computers & Chemical Engineering 17:211–227

Loos, P (1997) Produktionslogistik in der chemischen Industrie. Gabler, Wiesbaden

Meyr H, Wagner M, Rohde J (2000) Structure of advanced planning systems. In: Stadtler H, Kilger C (eds) Supply chain management and advanced planning, pp 75–78. Springer, Berlin Heidelberg New York

Möhring RH, Uetz M (2001a) Scheduling scarce resources in chemical engineering. In: Kischka P, Leopold–Wildburger U, Möhring RH, Radermacher FJ (eds) Models, methods and decision support in management, pp 195–210. Physica, Heidelberg

Nahmias S (1997) Production and operations analysis. Irvin, Homewood

Neumann K, Schwindt C (1997) Activity–on–node networks with minimal and maximal time lags and their application to make–to–order production. OR Spektrum 19:205–217

Neumann K, Schwindt C (1999) Project scheduling with inventory constraints. Report WIOR–572, University of Karlsruhe

Neumann K, Schwindt C, Trautmann N (2001a) Short–term planning of batch plants in process industries. In: Kischka P, Leopold–Wildburger U, Möhring RH, Radermacher FJ (eds) Models, methods and decision support in management, pp 211–226. Physica, Heidelberg

Neumann K, Schwindt C, Zimmermann J (2001b) Project scheduling with time windows and scarce resources. Lecture Notes in Economics and Mathematical Systems, Vol 508. Springer, Berlin Heidelberg New York

Oxe G (1997) Reducing overcapacity in chemical plants by linear programming. European Journal of Operational Research 97:337–347

Papageorgiou LG, Pantelides CC (1996) Optimal campaign planning/scheduling of multipurpose batch/semicontinuous plants. 1. Mathematical formulation. Industrial & Engineering Chemical Research 35:488–509

Pinedo M (1995) Scheduling: theory, algorithms and systems. Prentice Hall, Englewood Cliffs

Pinto M, Grossmann IE (1995) A continuous mixed integer linear programming model for short term scheduling of multistage batch plants. Industrial & Engineering Chemical Research 34:3037–3051

Pinto M, Grossmann IE (1998) Assignment and sequencing models for the scheduling of process systems. Annals of Operations Research 81:443–466

Schwindt C, Trautmann N (2000) Batch scheduling in process industries: An application of resource–constrained project scheduling. OR Spektrum 22:501–524

Schwindt C, Trautmann N (2002) Storage problems in batch scheduling. In: Chamoni P, Leisten R, Martin A, Minnemann J, Stadtler H (eds) Operations Research Proceedings 2001, pp 213–217. Springer, Berlin Heidelberg New York

Shah N (1998) Single- and multisite planning and scheduling: current status and future challenges. In: Pekny J, Blau GE (eds) Foundations of computer–aided process operations, pp 75–90. Amer. Inst. Chem. Eng., New York

Silver EA, Pyke DF, Peterson R (1998) Inventory management and production planning and scheduling. Wiley, New York

Stadtler H, Kilger C (2000) Supply chain management and advanced planning. Springer, Berlin Heidelberg New York

Timpe C (2002) Solving planning and scheduling problems with combined integer and constraint programming. OR Spectrum 24, No. 4 (to appear)

Timpe C, Kallrath J (2000) Optimal planning in large–site production networks. European Journal of Operational Research 126:422-435

Trautmann N (2001a) Anlagenbelegungsplanung in der Prozessindustrie. Gabler, Wiesbaden

Trautmann N (2001b) Calendars in project scheduling. In: Fleischmann B, Lasch R, Derigs U, Domschke W, Rieder U (eds) Operations Research Proceedings 2000. Springer, Berlin Heidelberg New York

Tsiakis P, Shah N, Pantelides CC (2001) Design of multi–echelon supply chain networks under demand uncertainty. Industrial & Engineering Chemical Research 40:3585–3604

Vidal CJ, Goetschalckx M (1997) Strategic production–distribution models: A critical review with emphasis on global supply chain models. European Journal of Operational Research 98:1–18

Wagner M (2000) Demand planning. In: Stadtler H, Kilger C (eds) Supply chain management and advanced planning, pp 97–115. Springer, Berlin Heidelberg New York

Westenberger H, Kallrath J (1995) Formulation of a job shop problem in process industry. Unpublished working paper, Bayer AG, Leverkusen, and BASF AG, Ludwigshafen

Wright G, Goodwin P (1998) Forecasting with judgment. Wiley, New York

Campaign planning for multi-stage batch processes in the chemical industry

Martin Grunow, Hans-Otto Günther, and Matthias Lehmann

Deptartment of Production Management, Technical University Berlin,
Wilmersdorfer Straße 148, 10585 Berlin, Germany
(e-mail: {m.grunow, ho.guenther, m.lehmann}@pm-berlin.net)

Abstract. Inspired by a case study from industry, the production of special chemical products is considered. In this industrial environment, multi-purpose equipment is operated in batch mode to carry out the diverse processing tasks. Often, extensive set-up and cleaning of the equipment are required when production switches between different types of products. Hence, processes are scheduled in campaign mode, i.e. a number of batches of the same type are processed in sequence. The production of chemical products usually involves various stages with significant cumulative lead times. Typically, these production stages are assigned to different plants. A hierarchical modelling approach is presented which co-ordinates the various plant operations within the entire supply network. In the first stage, the length of the campaigns, their timing, the corresponding material flows, and equipment requirements have to be determined. At this stage, an aggregation scheme based on feasibility constraints is employed in order to reflect the limited availability of the various types of production equipment. The second stage consists of an assignment model, which allocates the available equipment units between the production campaigns determined in the first stage of the solution procedure. In the third stage, resource conflicts are resolved, which may occur if clean-out operations and minimal campaign lengths have to be considered. The proposed hierarchical approach allows a more compact model formulation compared to other approaches known from the literature. As a result, a very efficient and flexible solution approach is obtained. In particular, commercially available standard solvers can be used to solve a wide range of campaign planning problems arising in the chemical industry.

Key words: Campaign planning – Supply networks – Chemical industry – Linear optimization model – Hierarchical production planning

1 Introduction

A typical manufacturer of low-volume, high-value-added specialty chemicals runs various plants in order to serve the international markets. In most cases, production requires a number of consecutive processing stages and involves various intermediate products which are supplied to downstream facilities for further processing. This often results in a complex supply network which comprises a number of individual facilities. In some cases, these facilities are located in the vicinity. In other cases, e.g. in the pharmaceutical industry, processing stages are geographically distributed across production sites at different locations, even in different countries. As a consequence, the strategic design of such networks, for instance the opening and closure of plants and capacity considerations, are issues of major practical importance (e.g. Kallrath, 2002a). Moreover, production operations within such networks have to be well co-ordinated in order to avoid excessive inventories, inefficient capacity utilization and poor customer service (see, e.g., Berning et al., 2002, for an integrated system solution to a complex industrial supply chain optimization problem).

In this paper, we consider the case of a network of chemical plants which produce a variety of final products through a number of processing stages. Figure 1 displays a merely small section of the entire supply network. According to the number of production sites involved, so-called *local* and *chain products* can be distinguished. Local products are completely produced at a single plant. In contrast, the various stages required to produce a chain product are assigned to different plants. For instance, intermediate $A1$ is produced from raw materials $a1$ and $a11$ and further processed in plant B where the remaining processing stages are performed. Finally, end-products $B7$ and $B10$ are obtained through intermediates $B5$, $B6$, $B8$, and $B9$ as well as raw materials $b6$ to $b10$. Another example of a chain product is represented by end-product $D1$, which involves processing at all four plants within the supply network. From a management point of view, it is interesting to note that there is a permanent resource competition between local and chain products at the individual plants. Since plant management might favour local products and pay less attention to chain products, an overall co-ordination of activities within the entire supply network is a critical factor.

In the chemical industries, a major distinction can be made between *continuous* and *batch production* systems. Plants which produce a limited number of products each in a relatively high volume typically use special purpose equipment, allowing an almost continuous flow of material. An example is the production of petrol. On the other hand, should small amounts of a large number of specialized products be required, multi-purpose equipment is used and plants are operated in batch mode. Batch processing is suitable for producing a wide range of products whose individual demand is not large enough to justify the construction of a dedicated plant. In these cases, it is economical to use multi-purpose equipment to carry out the diverse processing tasks. Moreover, batch processing offers the advantage of an increased flexibility with respect to product variety, production volume, and the range of recipes that can be processed by an individual piece of equipment. In contrast to continuous production systems, however, production scheduling is

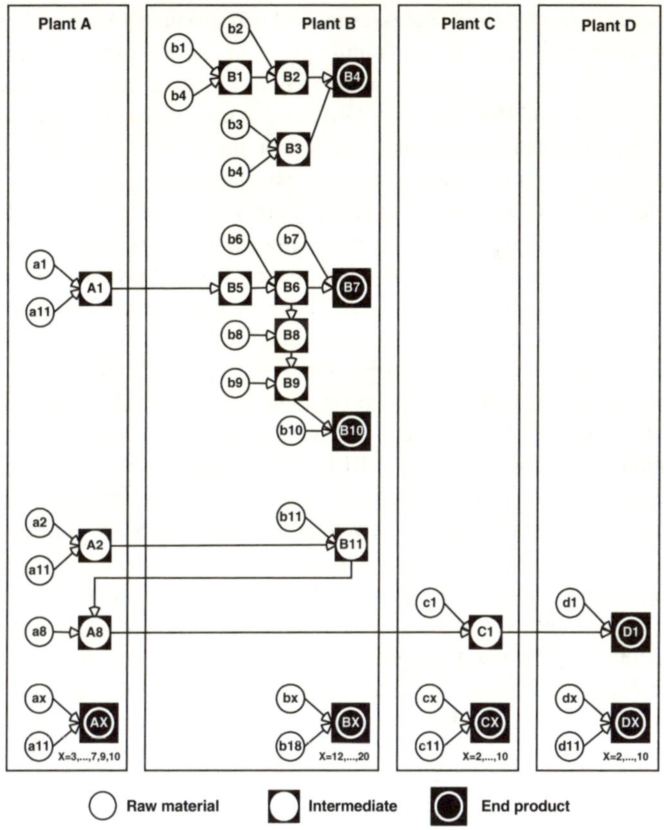

Fig. 1. Section of the entire production network

significantly complicated by the large number of batches involved, the dissimilarity of the production paths and the short-term variations in product demand.

In the industrial application considered, a fairly small range of high-value-added specialty chemicals is produced. Production is carried out in batch mode using highly flexible equipment. The complexity of production scheduling is determined by a large number of feasible task/equipment assignments and the possibility of using *temporary configurations* of equipment units for performing a specific group of processing tasks. An illustrative example will be given in Section 3. Due to the high set-up and cleaning times involved, processes are operated in campaign mode, i.e. similar processing tasks are repeated before production switches to another type of product.

Planning and scheduling problems arising in chemical batch production have been studied extensively in the last decade and numerous models and solution approaches have been developed. For comprehensive literature reviews on this subject, see for instance Rippin (1993), Reklaitis (1996), Pinto and Grossmann (1998), Blömer (1999), and Kallrath (2002b). The key problem in chemical batch production is campaign planning, i.e. determining the duration and timing of production runs for the various chemical processes and related products. A general mathematical formulation of the campaign planning problem and a comprehensive literature review are given by Papageorgiou and Pantelides (1996a). In their companion paper (Papageorgiou and Pantelides, 1996b) computational issues are discussed. Another early and still pursued research focus is on the development of planning and scheduling approaches for single-stage processors (e.g. Dessouky et al., 1999; Oh and Karimi, 2001a,b). An important aspect in this research is the consideration of sequence-dependent changeovers due to cleaning and set-up requirements. Another area of research is the so-called flowshop plant where all products are processed sequentially through a number of stages. These plants process a family of products with a high degree of similarity (e.g. Pinto and Grossmann, 1994; Voudouris and Grossmann, 1996; Mendez and Cerda, 2002). More recently, the batch sizing and scheduling problem arising in multi-product/multi-purpose chemical batch plants has gained considerable attention (cf. Grossmann et al., 1996; Pinto and Grossmann, 1998; Timpe, 2002) for a review of MILP models, related solution techniques and applications). Also solution procedures based on project scheduling methods seem to be promising (e.g. Schwindt and Trautmann, 2000; Neumann et al., 2002).

The major intention of this paper is to develop a novel mixed-integer linear programming (MILP) model formulation for campaign planning within multi-plant supply networks and to demonstrate its applicability in a chemical batch production environment. The main features of the MILP model formulation are: (1) A hierarchical modelling framework that subdivides the entire decision problem into three consecutive levels. (2) A new approach for modelling the assignment of equipment units to processing tasks and the derivation of feasibility constraints which allow a much more compact model formulation. (3) A disaggregation procedure for assigning individual equipment units to processing tasks, based on the regime of batches and campaigns obtained at the aggregate level.

The remainder of this paper is organized as follows. In the next section, a representation scheme of chemical batch processing is introduced. This is followed by a detailed problem statement in Section 3. In the subsequent section, a hierarchical modelling approach is presented. The mixed-integer linear optimization model of the campaign planning problem under consideration is developed in Section 5. Also, possible extensions are introduced to demonstrate the applicability of the suggested hierarchical approach to real industrial applications. Section 6 covers the problem of assigning equipment units to the campaigns. In Section 7, repair algorithms are outlined which have to be applied in the case of non-standard problem assumptions. Finally, results of the numerical investigation are summarized in Section 8.

2 Representation of chemical batch processes

To support operational planning of production and logistics, computer-based production planning and control (PPC) systems are used in many industries. [An overview of the current state of PPC systems, their architecture and future perspectives is given by Zijm (2000)]. PPC systems as well as recent developments of so-called *advanced planning and scheduling (APS) systems* (cf. Stadtler and

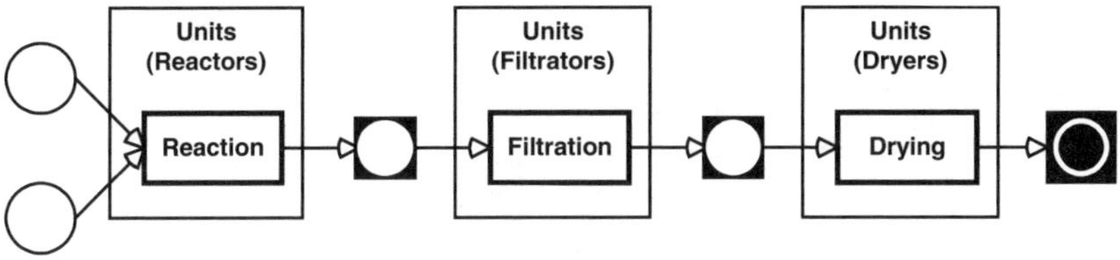

Fig. 2. State-task-network representation

Kilger, 2000) are based on a hierarchical structure of the various planning activities. Especially, in the chemical industry, APS systems have been adopted by many enterprises. These systems subdivide the entire planning phase into several levels, e.g. demand planning, supply network planning (SNP), and production planning and detailed scheduling (PPDS) (Meyr et al., 2000). SNP and PPDS differ by their degree of aggregation. Accordingly, different representation schemes of chemical processes are required in order to reflect production activities at a level of detail which is appropriate for the specific planning function.

Basically, chemical plants are characterized by two key elements: the *physical configuration* and the *network of processing tasks*. The physical configuration consists of various equipment units (e.g. reactors and tanks) and the network of devices (e.g. pipelines) available to transfer material between the various production or storage units. The entire chemical process consists of a network of processing tasks linked by individual material flows. Typically, various final products are produced out of a number of raw materials and intermediate products through a sequence of processing tasks.

In order to represent chemical batch processes, we use a modification of the *state-task-network* (STN) concept introduced by Kondili et al. (1993). A STN is a directed graph which consists of three elements:

- *state nodes*, representing the feeds, intermediates and final products,
- *task nodes*, representing the process operations which transform material from one or more input states into one or more output states,
- and *arcs*, indicating the flow of material.

From a logistics point of view, a state of the chemical process corresponds to an individual product, which is either fed into or received from a task. There are several attributes which characterize the various elements of a STN. A set of feeds and output products, the batch processing time, and a set of feasible equipment units is associated with each task. States as elements of a state-task-network representation are characterized by their storage condition (e.g. no intermediate storage allowed, finite, or unlimited intermediate storage), the feasible storage tanks, the quantities received from external suppliers at various times and the order quantities to be delivered at certain due dates. Figure 2 exhibits an illustrative example. In this figure, state and task nodes are denoted by circles and rectangles, respectively. In the process considered, a final product is produced from two raw materials through three processing stages. In each stage, there are several equipment units available which are capable of performing either the reaction, filtration or drying task.

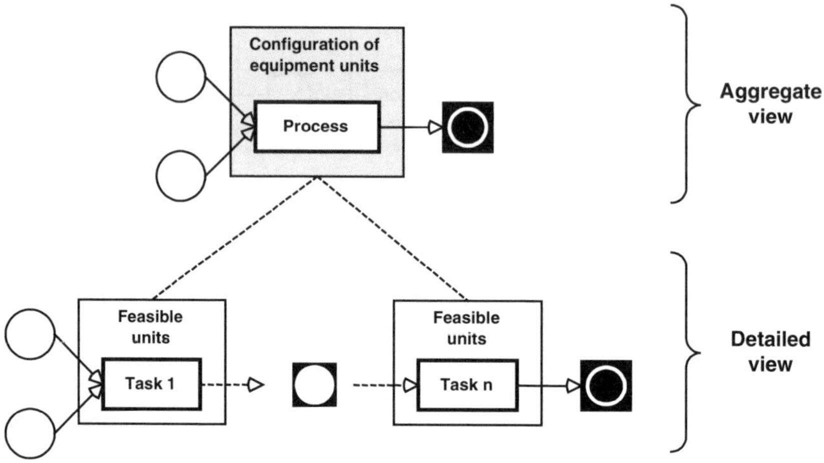

Fig. 3. Aggregate and detailed view of physical units and processing stages

Depending on the specific planning function, representation schemes of different levels of aggregation are used. At the production planning and detailed scheduling level a detailed representation of the underlying chemical processes and the production equipment is required. Supply network planning, however, is based on a superstructure that aggregates individual equipment units into corresponding configurations and networks of processing tasks into distinct process types. Figure 3 illustrates the aggregate and detailed view of the physical units and processing stages.

In this paper, the focus is on chemical batch processing. Thus, we define a *batch* as the quantity of material that undergoes processing by a single chemical process using an individual equipment unit. In an aggregate view, a batch is linked to a distinct process type and is produced through a feasible configuration of equipment units. As a single batch usually is not sufficient to satisfy a complete customer order or a given period demand and since there are considerable changeover costs between different types of chemical processes, production is generally carried out in campaign mode. Accordingly, a *campaign* is defined as a sequence of batches of the same process type to be processed without interruption. To process a campaign, specific configurations of individual equipment units are used depending on the type of the chemical process. In chemical batch production, very often a multi-purpose plant strategy is employed and the production units are assigned to *temporary resource configurations* which are reconfigured for another type of process after a campaign has been completed.

3 Problem statement

This paper focuses on scheduling chemical batch processing within a network of plants using multi-purpose equipment. Based on weekly demand figures and given

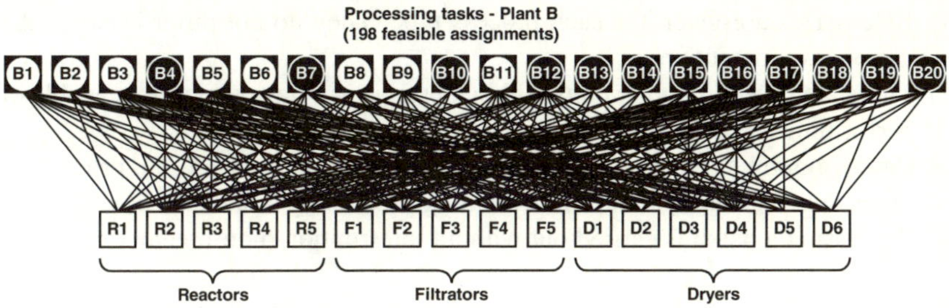

Fig. 4. Feasible task/equipment assignments

customer orders over a fixed horizon, a production schedule for the entire network of plants has to be determined. The major scheduling task is the determination of the number, sequence, and length of campaigns for each plant and the corresponding products and processing stages. The output of a campaign is usually distributed between several other campaigns at downstream plants or several customer orders. Co-ordinating production runs and material flows between plants is a major issue in this planning step. Minimizing total production and logistics costs within the supply network is considered as the objective function.

Since set-up costs are very high, customer orders which should be delivered at different due dates are combined into one production campaign. A cyclical production policy is usually inadequate to follow varying market demands. Therefore, a non-regular pattern is adopted. Scheduling of campaigns also involves the determination of detailed start and finish times for the individual processing tasks. Essentially, the length of a campaign is defined by the number of successive batches to be produced in each of the corresponding processing tasks.

Moreover, feasible equipment units have to be assigned to the individual processing tasks. Figure 4 illustrates the feasible assignments for plant B from the example of Figure 1 (where the task nodes were omitted for the sake of clarity). The chemical facilities in plant B produce 20 products by use of three groups of equipment units, i.e. reactors, filtrators, and dryers. According to the state-task-network representation of the production process introduced in Figures 2 and 3, each of the processes generating products $B1$ through $B20$ may be split up into a reaction task, a filtration task, and a drying task and thus requires one of the reactors $R1$ through $R5$, one of the filtrators $F1$ through $F5$, and one of the dryers $D1$ through $D6$. These three equipment units are simultaneously blocked for the duration of the process. In order to respond to changes in demand, the plant equipment is normally re-organized after a certain period of time according to the individual end-product requirements. Therefore, task/equipment assignments have to be made for each period within the planning horizon. Hence, production scheduling is significantly complicated by the large number of feasible assignments over the planning horizon.

Additional assumptions of the chemical batch manufacturing environment considered are the following.

(1) Batch sizes are given for each process type. They do not differ between the various equipment units.
(2) There are several parallel units within a production stage which are capable of performing a certain subset of the processing tasks.
(3) Processing times are assumed to be constant per batch. As a result, processing times depend solely on the corresponding processing task.
(4) Processing a batch is carried out without interruption (non-preemptive operation mode).
(5) Several batches of the same process type are allowed to start at the beginning of a period depending on the number of available equipment units.
(6) There are sufficient capacities to store all the required raw materials, intermediates and final products. All products are assumed to be storable.
(7) The fractions of all inputs and outputs are specified as fixed proportions of the batch size.
(8) All feeds and products are transferred at the beginning and the end of a processing task, respectively.
(9) Processing tasks are only allowed to start and to finish at the period boundaries so that the time required to process a batch always covers one or more periods.

In the basic model formulation presented in Section 5.1, we further assume that there are no set-up or clean-out times required when changing to another process type. Also campaigns may be of arbitrary length. In real industrial applications, however, both assumptions are fairly unrealistic. Therefore, extensions of the basic model formulation are presented in Section 5.2, which consider clean-out times and given minimal campaign lengths. However, first we introduce our basic modelling approach.

4 Modelling approach

4.1 Model hierarchy

From a practical point of view, the major difficulty with mathematical programming models is that the number of constraints and binary variables in the model may become quite large, especially if a dense time grid is used. One option to reduce the computational burden associated with optimally solving large scale optimization models for scheduling chemical batch processes is to use efficient simplification strategies (e.g. Blömer and Günther, 1998, 2000). Another option is to employ adequate *aggregation strategies*. For the scheduling problem considered, they allow a much more compact model formulation. In this paper the latter strategy is adopted. We develop a hierarchical modelling approach for planning campaigns and scheduling operations within supply networks in the chemical industry.

In the past, hierarchical planning concepts have played an important role in the academic literature. Based on the framework of Anthony (1965) and the seminal work of Hax and Meal (1975) many different procedures for hierarchical production planning have been developed (see e.g. Schneeweiss, 1999, for an overview of related concepts and Günther, 1986, for a practical application). The framework of hierarchical decision making relies on the observation that production planning and scheduling encompasses a large number of decisions and involves managers at various echelons within the organization. The decisions differ by complexity, scope, and time horizon. Moreover, relevant data differ by their degree of certainty and aggregation. For instance, decisions on co-ordinating production campaigns within a supply network are made at an upper planning level while decisions on scheduling the different tasks of a chemical process and the assignment of the required equipment units are typically left to local plant management. Theoretically, a monolithic model comprising all detailed decision problems over the entire planning horizon could be formulated. This approach, however, is neither consistent with the nature of the managerial decision process nor is it practical from a computational point of view.

In hierarchical production planning, decisions are made in sequence. On the higher level, aggregate decisions are made. At this level, it is usually not possible to consider every detail of the production system. Hence, relevant items (e.g. equipment units or processing tasks) have to be aggregated (see Section 2). The aggregate plan is then disaggregated into a detailed schedule for individual items and production units. In the ideal case, the disaggregation procedure is *consistent* with the aggregate plan (cf. Axsäter, 1986), i.e. the detailed schedule satisfies all constraints of the aggregate plan.

Figure 5 displays the architecture of the proposed hierarchical planning and scheduling system. In the first stage, an initial campaign schedule is derived. The underlying aggregate model formulation (see Sect. 5.1) does not explicitly consider requirements for the various types of equipment units. Hence, feasibility constraints are introduced which implicitly ensure that the corresponding capacity constraints are observed (see Sect. 4.2). The second stage consists of a simple assignment model (see Sect. 6). Based upon the number and timing of batches and aggregate equipment requirements obtained in the first stage, a feasible disaggregation is achieved by assigning individual equipment units to batches. However, if certain extensions of the basic campaign planning model have to be considered, e.g. clean-out operations after completion of a campaign or minimal campaign lengths, conflicts in the assignment of equipment units to chemical tasks may occur. Therefore, in the third stage of the solution procedure, resource conflicts are resolved by time-shifting individual operations (see Sect. 7). Our numerical investigation presented in Section 8 reveals that conflicting assignments rarely arise and, thus, a feasible task/equipment assignment is already achieved after step 2 of the solution procedure. It should be noted that in these cases an overall optimal solution is obtained. In the case of resource conflicts the deviation from the overall optimum is fairly small depending on the cost penalties associated with shifting individual operations for a limited number of periods.

4.2 Derivation of feasibility constraints

In the practical production planning problem under consideration, aggregation focuses on the representation of the physical units of the production system. To carry out the various campaigns, temporary configurations of equipment units have to

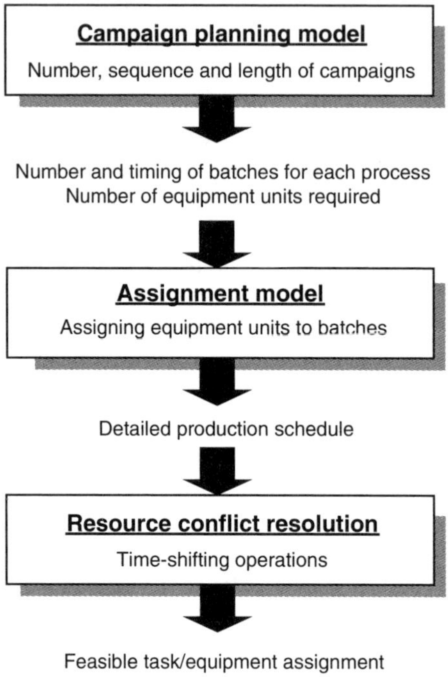

Fig. 5. Hierarchical modelling approach

be determined for each period $t \in T$. Essentially, the composition of these configurations depends on the particular type of chemical process. Thus, assignments between the individual equipment units i and process types j have to be made for each period t. This, naturally, leads to the definition of triple-indexed binary decision variables x'_{jit}, where $x'_{jit} = 1$, if process j is assigned to the individual unit i in period t, and $x'_{jit} = 0$, else. Model formulations based on triple-indexed decision variables may be found in the papers by Kondili et al. (1993) and Blömer and Günther (1998, 2000), for instance. In the example of Figure 4, almost 200 variables of this type are required to model the equipment assignments for a single period.

Since the computational effort required to optimally solve the resulting large-scale MILP model is prohibitive, an efficient concept of feasibility constraints is proposed. Simply speaking, we try to replace the triple-indexed decision variables x'_{jit} by double-indexed variables x_{jt}. In the example of Figure 4, the number of assignment variables can thus be reduced from 200 to less than 20 per period. However, additional feasibility constraints in the aggregate campaign planning model are needed.

To illustrate the proposed concept of feasibility constraints, we use the illustrative example depicted in Figure 6. There are three types of chemical processes $(j = 1, 2, 3)$ and three individual equipment units $(i = a, b, c)$. Since we only consider a single period in this example, the period index t is omitted. Arcs in

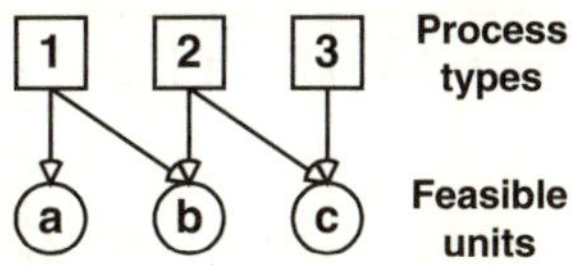

Fig. 6. Illustrative example

Figure 6 indicate feasible assignments. For instance, batches of process 1 can be processed using equipment units a or b, and process 2 is feasible on units b and c, while process 3 requires unit c. In the case of multiple feasible units, also multiple batches may be processed at the same time. Hence, decision variables x_j indicate the number of batches of process type j to be started at the beginning of a period.

At first glance, three simple formulations of the feasibility constraints seem appealing.

- Restricting the total number of batches to three (the number of equipment units available) through $x_1 + x_2 + x_3 \leq 3$. Obviously, $x_3 = 3$ satisfies this constraint, but is not feasible according to Figure 6.
- Restricting the number of batches of each process type to the number of feasible equipment units through $x_1 \leq 2$ (process 1), $x_2 \leq 2$ (process 2), and $x_3 \leq 1$ (process 3). Again, infeasible solutions satisfying the above constraints occur (e.g. $x_1 = 2$ and $x_2 = 2$).
- Restricting the number of batches for each equipment unit to the number of feasible process types through $x_1 \leq 1$ (unit 1), $x_1 + x_2 \leq 1$ (unit 2), and $x_2 + x_3 \leq 1$ (unit 3). Unfortunately, these constraints are too restrictive (e.g. $x_1 = 1$ imposes $x_2 = 0$, allowing only two out of three possible batches).

Also, combinations of the above constraints do not model the feasible equipment assignments correctly. Hence, in order to prevent possible conflicts between resources, feasibility constraints have to be defined for all distinct combinations of equipment units.

In a more general case, a process j is made up of several tasks (see Figs. 2 and 3) and there are different types of equipment units $u \in U$ corresponding to these tasks. Then, each chemical process $j \in J$ may require more than one unit type u (but at most one individual unit i per unit type u). In the example presented in Figure 4, the production processes for the products $B1$ through $B20$ could be disaggregated into three different tasks and thus require the selection of three individual processing units of three different types: a reactor, a filtrator, and a dryer.

Accordingly, all sub-sets $U_{iu}^{\subseteq}$ of the total set of equipment units have to be determined not only for one unit type such as in the example of Figure 6 but for all unit types $u \in U$. The resulting non-empty sub-sets are identified by the index $i \in I_u$. Feasibility constraints in their general form are given by

$$\sum_{j \in J : U_{ju} \subset U_{iu}^{\subseteq}} x_j \leq |U_{iu}^{\subseteq}| \qquad u \in U,\ i \in I_u \qquad (1)$$

where U_{ju} indicates the set of equipment units of type u feasible for process j.

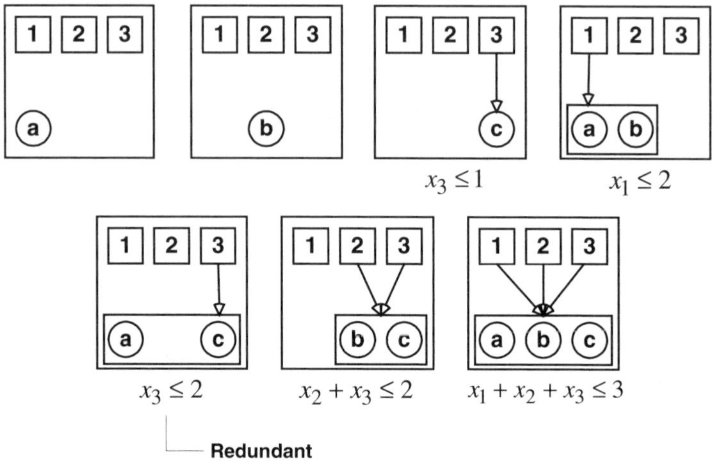

Fig. 7. Example of feasibility constraints

For the illustrative example stated above, the resulting feasibility constraints according to (1) are summarized in Figure 7. We are again omitting the indices u and t, as we are only regarding a single unit type and a single time period. In this example, the sub-sets of equipment units $\{a\}, \{b\}, \{c\}, \{a, b\}, \{a, c\}, \{b, c\}$, and $\{a, b, c\}$ are obtained. Each of these sub-sets is treated as a virtual equipment unit. According to the feasible assignments defined in Figure 6, the following feasibility constraints are obtained.

- Considering virtual unit $\{a\}$, there is no process type which is bounded by the capacity of this unit. Clearly, process type 1 can also use equipment unit b. Thus, no resource constraint is required for virtual unit $\{a\}$. The same holds for virtual unit $\{b\}$, since for both process types 1 and 2, alternative equipment units exist. According to (1), no constraints are generated.
- Virtual unit $\{c\}$ is the only one feasible for process 3. Thus, at most one batch of process type 3 is allowed to start at the beginning of period t. This is expressed by $x_3 \leq 1$.
- Capacity demand on the virtual unit $\{a, b\}$ is only imposed by process type 1. Thus, at most two batches of this process type, one on unit a and b each, can be processed. This is expressed by $x_1 \leq 2$. In a similar manner, the constraint $x_3 \leq 2$ is derived for the virtual unit $\{a, c\}$.
- As process types 2 and 3 utilize capacity of the virtual unit $\{b, c\}$, the total number of corresponding batches is restricted to two per period. Thus, the feasibility constraint $x_2 + x_3 \leq 2$ is obtained.
- Finally, all process types utilize capacity of the virtual unit $\{a, b, c\}$. Accordingly, the constraint $x_1 + x_2 + x_3 \leq 3$ restricts the total number of batches to the total number of equipment units available.

Apparently, redundant constraints (e.g. $x_3 \leq 2$) occur. These have to be eliminated from the model formulation. Pre-solve algorithms incorporated in modern standard optimization packages complete this job very efficiently.

For the example of Figures 6 and 7, the above constraints guarantee feasible task/equipment assignments even though the variables are not indexed over the individual units. In fact, Appendix A shows that, for the single period case, constraints (1) always permit feasible assignments. If, however, batch processing times comprise two or more periods, constraints (1) merely guarantee that there are enough units available of the required types. In other words, constraints (1) do not guarantee that the configuration of equipment units assigned to each batch remains the same over all corresponding periods. In that case, additional procedures are required in order to re-establish a feasible disaggregation (see Sect. 7).

5 Campaign planning model

5.1 Basic model formulation

The standard method to determine optimal production schedules in the process industry is mathematical programming. This approach has the advantage of being able to incorporate very easily the specific operating modes of the production system and the network of material flows between the various facilities. Therefore, we adopt this approach for developing an application-oriented solution methodology. The model formulation is based on the state-task-network representation introduced in Section 2. Decision variables refer to the number of batches of each process type to be started at the beginning of a period and the development of inventory over time.

Before introducing the proposed mathematical model formulation, the following notation has to be defined.

Indices, index sets

$j \in J$	types of chemical processes
$j \in J_p^{in}$	processes which use product p as input
$j \in J_p^{out}$	processes which produce product p
$p \in P_1$	intermediate products
$p \in P_2$	end-products
$u \in U$	types of equipment units
U_{ju}	equipment units of type u required for process j
$U_{iu}^{\subset}$	i^{th} sub-set of equipment units of type u
$i \in I_u$	index identifying the sub-sets $U_{iu}^{\subset}$
$t \in T$	periods ($t = 1..H$)

Parameters

α_{pj}^{in}	quantity of input of product p required per batch of process j
α_{pj}^{out}	quantity of output of product p generated per batch of process j
τ_j	processing time per batch of process j
c_j^b	set-up costs associated with processing a batch of process j
d_{pt}	external demand of end-product $p \in P_2$ at the end of period t
h_p	costs for holding one unit of product p in inventory for one period

Decision variables

x_{jt} number of batches of process j started at the beginning of period t
I_{pt} inventory of product p at the end of period t ($I_{p0} = given$)

The optimal campaign schedule at the aggregate level can be determined by solving the following mixed-integer linear programming (MILP) model. The objective function is to minimize total set-up costs associated with processing the various batches and costs incurred for holding inventory over time. As a result, the minimum number of batches required to satisfy the given external demand figures and the timing of batches are obtained. Note that fixed costs associated with individual campaigns are considered in the extended model formulation presented in Section 5.2.1.

$$\min \sum_{t \in T} \left[\sum_{j \in J} c_j^b \cdot x_{jt} + \sum_{p \in P_1 \cup P_2} h_p \cdot I_{pt} \right] \tag{2}$$

subject to the following constraints.

Availability of equipment units: The following inequalities represent the application-specific form of the feasibility constraints, previously defined in (1). In order to avoid resource conflicts in period $t \in T$, the number of batches of each process type is limited by the number of equipment units of type $u \in U$ within a virtual resource group $i \in I_u$. Once a unit is assigned, it is then blocked for a number of consecutive periods, depending on the duration τ_j of the particular batch process. We use the auxiliary time index s to reference the corresponding periods $s \in t - \tau_j + 1..t$.

$$\sum_{\substack{j \in J : U_{ju} \subset U_{iu}^{\subset} \\ s \in t - \tau_j + 1..t : t - \tau_j \geq 0}} x_{js} \leq \left| U_{iu}^{\subset} \right| \qquad u \in U, i \in I_u, t \in T \tag{3}$$

Inventory balance for intermediates: Inventory balances are expressed by the following constraints where the initial inventory I_{p0} is assumed to be known. The net change in inventory is determined by the amount of material produced and the quantities that are used as input in the various processes.

$$I_{pt} = I_{p,t-1} + \sum_{j \in J_p^{out}} \alpha_{pj}^{out} \cdot x_{j,t-\tau_j+1} - \sum_{j \in J_p^{in}} \alpha_{pj}^{in} \cdot x_{jt} \qquad p \in P_1, t \in T \tag{4}$$

$$\text{with} \quad x_{j,t-\tau_j} = 0 \quad \text{for} \quad t \leq \tau_j$$

Inventory balance for end-products: In the case of end-products, external demands are considered. Due to the non-negativity constraints on decision variables I_{pt} (see (7)), it is guaranteed that customer demand is satisfied without backorders.

$$I_{pt} = I_{p,t-1} + \sum_{j \in J_p^{out}} \alpha_{pj}^{out} \cdot x_{j,t-\tau_j+1} - d_{pt} \qquad p \in P_2, t \in T \tag{5}$$

$$\text{with} \quad x_{j,t-\tau_j} = 0 \quad \text{for} \quad t \leq \tau_j$$

Variable domains: Finally, the domains of the decision variables are defined as follows.

$$x_{jt} \in \{0, 1, 2, ...\} \qquad j \in J, t \in T \quad (6)$$

$$I_{pt} \geq 0 \qquad p \in P_1 \cup P_2, t \in T \quad (7)$$

The model formulation presented above is similar to a standard capacity-constrained lot-sizing and scheduling model (cf. Fleischmann and Meyr, 1997) with the following exceptions. First, the model formulation includes the assignment of distinct equipment units to tasks of the chemical processes. Task/equipment assignments are considered implicitly through the introduction of feasibility constraints. Second, a supply network consisting of multiple production stages is considered. Third, production quantities correspond to an integer number of batches. Finally, changeover costs do not depend on the sequence of batches.

For the determination of the specific task/equipment assignments, the assignment model of Section 6 is used. The number of equipment units of each type required in each period can easily be determined. These figures are expressed by a corresponding model parameter a_{jut} which indicates the number of equipment units of type u required for process j in period t (see Sect. 6). In the same way, equipment units, which are blocked in the various periods because of the necessary clean-out operations (see Sect. 5.2.1) are considered.

5.2 Extensions

The model formulation outlined above represents a generic model applicable to a wide class of chemical batch production processes. However, several features which may be relevant in specific industrial applications are not included in the model formulation. Possible extensions, such as consideration of clean-out operations after completion of a campaign and minimal campaign lengths are presented in the following. In these cases, the consistent disaggregation property laid out in Section 4.2 does no longer hold. As a result, the assignment model, which will be presented in Section 6, does not guarantee feasible task/equipment assignments for solutions obtained from the extended campaign planning model. Our numerical investigation, however, reveals that for a large number of problem instances a feasible disaggregation is still achieved (see Sect. 8). Even if this is not the case, a feasible assignment can always be generated through the repair step introduced in Section 7.

5.2.1 Clean-out operations after completion of campaigns

In batch plants, cleaning of equipment may take place for a number of reasons. For instance, cleaning is required after a given number of uses of a specific reactor in order to avoid the build-up of residues. In other cases, items of equipment need to be cleaned depending on the sequence of batches and the quality requirements of the products to be produced. In many situations a minor clean-out of equipment units is required after each batch. In this case, the corresponding clean-out time can be included in the batch processing time. When switching between different types of chemical processes or different products, however, equipment units generally require a major clean-out. If changeovers only occur between products of the same family, clean-out times do not depend on the sequence of products. This latter situation is considered in the following. It is shown how the campaign planning model has to be modified in order to consider clean-out operations after completion of a campaign.

The extended model formulation requires the following additional decision variables.

v_{jt} number of campaigns of process j finished at the end of period t

It should be noted that variables v_{jt} automatically take integer values in the optimal solution of the campaign planning model. Let γ_j denote the clean-out time after completing a campaign (a consecutive number of batches) of process j, expressed as an integer number of periods. The following additional constraints have to be included in the campaign planning model presented in Section 5.1.

Campaigns finished: The number of campaigns of a particular process finished at the end of period t is derived from the number of batches started at the beginning of period $t+1$ and the ones which are completed at the end of period t.

$$v_{jt} \geq x_{j,t-\tau_j+1} - x_{j,t+1} \qquad j \in J, t \in \tau_j..H-1 \quad (8)$$

Availability of equipment units: In order to consider clean-out operations, constraints (3) of the original model formulation have to be modified. The number of clean-out operations performed in period t depends on the number of campaigns completed. Clean-out operations may last several periods. Thus, also current clean-out operations which have been started in previous periods have to be taken into account. At any given time, a production unit may start at most one single operation, either processing or cleaning. The unit is then blocked for a number of consecutive periods, depending on the duration of the specific batch processing or cleaning operation.

$$\sum_{\substack{j \in J: U_{ju} \subset U_{iu}^C \\ s \in t-\tau_j+1...t: t-\tau_j > 0}} x_{js} + \sum_{\substack{j \in J: U_{ju} \subset U_{iu}^C \\ s \in t-\gamma_j..t-1: s \geq 1}} v_{js} \leq |U_{iu}^C| \qquad u \in U, i \in I_u, t \in T \quad (3a)$$

The additional costs incurred by clean-out operations are included in the objective function of the original model formulation.

$$\min \sum_{t \in T} \left[\sum_{j \in J} \left(c_j^b \cdot x_{jt} + c_j^c \cdot \sum_{s \in t-\gamma_j..t-1: s \geq 1} v_{js} \right) + \sum_{p \in P_1 \cup P_2} h_p \cdot I_{pt} \right] \quad (2a)$$

where c_j^c indicates costs associated with set-up and clean-out operations of a campaign of process j.

Variable domains complete the extended campaign planning model.

$$v_{jt} \geq 0 \qquad j \in J, t \in T \qquad (9)$$

5.2.2 Minimal campaign lengths

Another issue of considerable practical relevance is the consideration of a minimal campaign length. In many industrial applications, it is required that a campaign, once started, should continue for at least a pre-defined number of periods or include a minimum number of batches. This is regularly due to the high set-up and clean-out effort involved and the need to reconfigure the equipment after each campaign. A minimal campaign length can also be viewed as a so-called management parameter which expresses desired features of the final schedule obtained. Although fixed costs associated with a campaign are already included in the extended objective function (2a), maintaining a given minimal campaign length can also be considered as a "hard" constraint. Accordingly, the extended campaign planning model requires the following additional integer decision variables. Again, these variables automatically take integer values in the optimal solution of the campaign planning model.

u_{jt} number of campaigns of process j started at the beginning of period t

Let L_j denote the minimum number of batches per campaign of process j. Additional constraints to be included into the campaign planning model are the following.

Campaigns started: The number of campaigns of a particular process type started at the beginning of period t is derived from the number of batches started at the beginning of this period and the ones which are completed at the end of the previous period.

$$u_{jt} \geq x_{jt} - x_{j,t-\tau_j} \qquad j \in J, t \in T \qquad (10)$$

with $x_{j,t-\tau_j} = 0$ for $t \leq \tau_j$

Minimal campaign length: This constraint ensures that, for each campaign started in period t, an equivalent number of batches is started in the consecutive periods before the minimal length of the campaign is reached. It should be noted that the minimal campaign length is expressed in terms of the number of subsequent batches of a process type.

$$x_{j,t+(L_j-1)\cdot\tau_j} \geq \sum_{\substack{s \in t..t+(L_j-1)\cdot\tau_j: \\ s \equiv t \bmod \tau_j}} u_{js} \qquad j \in J, t \in 1..H-(L_j-1)\cdot\tau_j \quad (11)$$

Variable domains are defined as follows.

$$u_{jt} \geq 0 \qquad j \in J, t \in T \qquad (12)$$

5.2.3 Maximum number of parallel campaigns

In many practical applications, only a limited number of parallel campaigns is allowed. Especially, for end-products with low-volume requirements, only a single campaign at a time may be sufficient. Such issues are reflected by the following additional constraints of the campaign planning model.

$$\sum_{s \in t-\tau_j+1..t:t-\tau_j \geq 0} x_{js} \leq M_j \qquad j \in J, t \in T \qquad (13)$$

where M_j indicates the maximum number of parallel campaigns of process type j.

6 Assignment model

The underlying idea, leading to the development of the proposed hierarchical modelling concept, is to decompose the original planning and scheduling problem into smaller sub-problems which are solved successively. The motivation for our hierarchical solution procedure is to determine the number and timing of batches for each process type in the initial stage and then to assign the required equipment units in the second stage. From the optimal solution to the campaign planning model, i.e. the values of the variables x_{jt} (and in case of cleaning processes also v_{jt}), and the corresponding recipes of the various processes, the number of equipment units a_{jut} of type u required for process j in period t (either for processing or cleaning) can easily be determined. Based on these input data, the assignment model seeks to assign individual equipment units to the various campaigns and thus to define temporary configurations according to the specific types of chemical processes produced in each period. An essential feature of the assignment model is that the sequence and timing of batches of each process type remains unchanged.

Given the requirements for all types of equipment units over all periods from the campaign planning model, simple disaggregation rules could be applied in order to determine the specific task/equipment assignments. Following this approach, however, solutions could be obtained which frequently switch individual equipment units between different process types from period to period. If, for instance, a batch of some process type is completed at the end of period t and another batch (or clean-out operation) of the same process type is started at the beginning of period $t+1$, the assignment of equipment should not be changed. For obvious reasons, frequent reassignments of equipment units are not desired in most practical applications. Hence, the objective function of the proposed assignment model is to minimize the number of events where such reassignments occur.

In the formulation of the assignment model the following additional notation is used.

Indices, index sets

$k \in K_u$ equipment units of type u
$k \in U_{ju}$ equipment units of type u feasible for process j

Parameter

a_{jut} number of batches of process j being processed on unit type u in period t

Decision variables

z_{kujt} = 1, if equipment unit k of type u is assigned to process j in period t (0, otherwise)

r_{kujt} = 1, if process j is started or reassigned to another equipment unit at the beginning of period t (0, otherwise)

The complete formulation of the assignment problem as a pure binary linear optimization model is given as follows.

Objective function: The total number of reassignments is minimized.

$$\min \sum_{u \in U} \sum_{k \in K_u} \sum_{j \in J} \sum_{t \in T} r_{kujt} \tag{14}$$

subject to the following constraints.

Availability of equipment units: Each individual equipment unit is assigned at most once.

$$\sum_{j \in J: k \in U_{ju}} z_{kujt} \leq 1 \qquad u \in U, k \in K_u, t \in T \tag{15}$$

Requirement of equipment units: The individual equipment units required for each process type are assigned to the corresponding temporary configurations.

$$\sum_{k \in U_{ju}} z_{kujt} = a_{jut} \qquad u \in U, j \in J, t \in T \tag{16}$$

Reassignment of equipment units: The logical variable r_{kujt} indicates a change in the assignment of an individual equipment unit.

$$r_{kujt} \geq z_{kujt} - z_{kuj,t-1} \qquad u \in U, k \in K_u, j \in J, t = 2..H \tag{17}$$

Variable domains: All decision variables are restricted to binary values.

$$z_{kujt} \in \{0,1\} \qquad u \in U, k \in K_u, j \in J, t \in T \tag{18}$$

$$r_{kujt} \in \{0,1\} \qquad j \in J, t \in T \tag{19}$$

Due to its pure binary nature, the assignment model stated above requires only moderate computational effort. Optimal solutions to this model are obtained within fractions of a second of CPU time on a common personal computer using a standard software package for linear optimization.

As mentioned in Section 4.2, a disaggregation procedure being perfectly consistent with the constraints of the original decision model assumes batch processing times of a single period. Otherwise, it is not guaranteed that the same configuration of equipment units can be assigned to a batch throughout its entire processing time. Repair algorithms presented in Section 7, however, may be applied in order to determine feasible task/equipment assignments.

7 Solving resource conflicts

As mentioned before, the various extensions of the basic campaign planning model do not guarantee feasible task/equipment assignments. In particular, clean-out features and minimal campaign lengths may lead to resource conflicts in the detailed production schedule. An illustrative example considering two equipment units a and b and three campaigns 1, 2, and 3 is shown in Figure 8. As can be seen, campaign 2 starts processing on unit a and then switches to unit b. Thus, two distinct clean-out operations would be required which are not explicitly considered in the assignment model of Section 6. Moreover, if a campaign is split, the pre-defined minimal campaign length could be violated. Thus, with the aid of repair algorithms a schedule has to be determined which eliminates such infeasibilities. Principally, there are four options available to achieve a feasible assignment of equipment units.

- left-shifting the predecessor campaign 1 on unit b,
- right-shifting the successor campaign 3 on unit a,
- left-shifting campaign 2 such that the entire campaign can be carried out on unit a,
- right-shifting campaign 2 such that the entire campaign can be carried out on unit b.

The use of these options to solve a specific resource conflict depends on the available buffer times of the various campaigns. Because of the complex links of material flows between campaigns at different production stages within the supply network, it is not always easy to prove the feasibility of an individual shift operation. Nonetheless, repair algorithms based on shifting operations as well as splitting and merging campaigns are used in many commercially available software packages for planning and scheduling in the chemical industry. Apparently, simple rule-based heuristic algorithms are not always effective, especially in an environment with high capacity utilization. Hence, in order to solve resource conflicts resulting from the assignment model, we develop a compact model formulation using logical constraints. A major advantage of this type of model compared to heuristic repair algorithms is that the entire supply of material within the network is retained, even if individual campaigns are rescheduled. The model formulation is based on the following features.

- A *resource conflict* is defined as an instance in which a campaign is started and completed at different equipment units of the same type (see Fig. 8 as an exam-

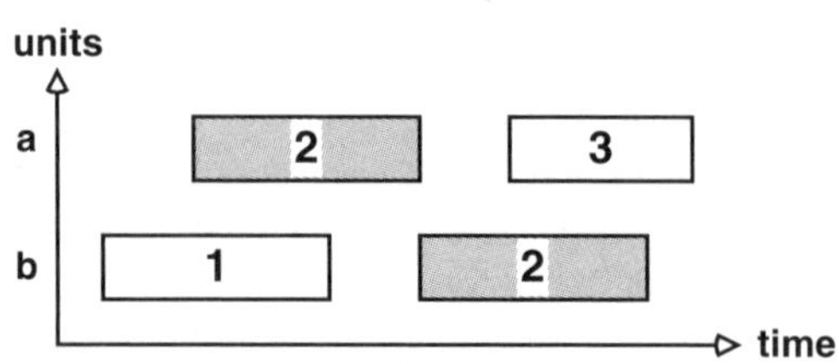

Fig. 8. Example of infeasible equipment assignments

ple). In the sequel, we only consider the case of a campaign switching between two different equipment units. It can be shown, however, that the suggested procedure is also applicable when campaigns switch between three or more equipment units.

- Prior to generating the problem-specific scheduling model, the set of *conflicting campaigns* $(i_1, i_2, i_3) \in C$ is determined. Let i_2 denote the *critical campaign*, which starts processing on some unit k_1 and finishes processing on some other unit k_2 ($i_2 = 2$, $k_1 = a$ and $k_2 = b$ in the example of Figure 8). Both units k_1 and k_2 are assumed to be of the same type (for instance both representing a reactor). Accordingly, i_1 indicates the immediate predecessor campaign of i_2 on unit k_1 and i_3 indicates the immediate successor campaign of i_2 on unit k_2 ($i_1 = 1$ and $i_3 = 3$ in the example of Fig. 8).
- Evidently, each campaign $i \in I$ uses specific equipment units k of unit type u. Corresponding *assignments* are defined as tuples $(u, k) \in K_i$. In the case of no resource conflicts, assignments (u, k) are unique for each type of equipment unit u. Otherwise, if resource conflicts occur, the index set K_i comprises all the conflicting assignments. For campaign 2 from Figure 8: $K_2 = \{(u, a), (u, b)\}$.
- For each campaign $i \in I$, *initial start-up periods* ts_i^o are obtained from the assignment model. Based on these data, the *float* f_i of a campaign can be determined. The float expresses the time span that could be used for right-shifting the campaign without causing shortages in the supply of other campaigns in the network or violating given demand figures and corresponding due-dates. In this definition it is assumed that all other campaigns are scheduled at their initial start-up period.
- *Holding costs changes* of h_i^l per period are caused by *left-shifting* a campaign. The output products have to be stocked for an extra period while the input products are stored for a period less. Similar holding costs changes of h_i^r apply when a campaign is *right-shifted* by one period.
- Additional *shortage costs* induced by shifting a campaign by more than f_i periods to the right are assumed to be given as e_i per period. In this case, inevitably, the supply between various campaigns is delayed according to the specific material flows. Hence, e_i has to be interpreted as a *penalty* for not meeting the planned due-dates of individual campaigns rather than out-of-pocket costs. In general, $e_i >> h_i^l, h_i^r$ will be assumed.

The notation used in the model formulation is summarized as follows.

Indices, index sets

$i \in I$	campaigns
$i \in I_2$	campaigns which produce end-products
$i_2 \in I_{i_1}^{in}$	campaigns (i_2) which are supplied by campaign i_1
$(i_1, i_2, i_3) \in C$	conflicting campaigns
$(k, u) \in K_i$	equipment units k of type u used for campaign i
$j(i)$	process assigned to campaign i

Parameters

f_i float of campaign i (i.e. available buffer time for right-shifting the campaign)
λ_i length of campaign i including clean-out time
h^l_i change of holding costs when left-shifting campaign i by one period
h^r_i change of holding costs when right-shifting campaign i by one period
e_i additional costs incurred for exceeding the due-date of campaign i by one period
ts^o_i initial start-up period of campaign i (obtained from the assignment model)

Decision variables

ts_i actual start-up period of campaign i
rs_i number of periods campaign i is shifted to the right
rs^+_i number of periods campaign i is shifted to the right exceeding the float f_i
ls_i number of periods campaign i is shifted to the left

The final campaign schedule can be determined by solving the following MILP model.

Objective function: Total costs incurred by left- and right-shifting campaigns are minimized.

$$\min \sum_{i \in I_2} e_i \cdot rs^+_i + \sum_{i \in I} h^r_i \cdot rs_i + \sum_{i \in I} h^l_i \cdot ls_i \tag{20}$$

subject to the following constraints.

Sequence of campaigns: Non-conflicting campaigns are scheduled according to their initial sequence obtained from the assignment model. Resource conflicts between campaigns, however, are solved through left- or right-shifting campaigns such that the critical campaign i_2 is either started after its immediate predecessor campaign i_1 or is completed before its immediate successor campaign i_3.

$$ts_{i_1} + \lambda_{i_1} \leq ts_{i_2} \qquad i_1, i_2 \in I : (K_{i_1} \cap K_{i_2} \neq \emptyset) \wedge (ts^o_{i_1} + \lambda_{i_1} \leq ts^o_{i_2}) \tag{21}$$

$$(ts_{i_1} + \lambda_{i_1} \leq ts_{i_2}) \vee (ts_{i_2} + \lambda_{i_2} \leq ts_{i_3}) \qquad (i_1, i_2, i_3) \subset C \tag{22}$$

Float of campaigns: Actual start times for a pair of consecutive campaigns may be readjusted according to the float of the head campaign.

$$ts_{i_2} - ts_{i_1} \geq ts^o_{i_2} - ts^o_{i_1} - f_{i_1} \qquad i_1, i_2 \in I : i_2 \in I^{in}_{i_1} \tag{23}$$

Left- and right-shifting campaigns: The number of periods a campaign is shifted to the left or the right, respectively, is calculated from the deviation between its initial and actual start time. For right-shifts of type rs_i^+, only periods in excess of the available float are considered.

$$ls_i \geq ts_i^o - ts_i \qquad i \in I \quad (24)$$
$$rs_i \geq ts_i - ts_i^o \qquad i \in I \quad (25)$$
$$rs_i^+ \geq ts_i - ts_i^o - f_i \qquad i \in I_2 \quad (26)$$

Variable domains: Finally, the domains of the decision variables are defined as follows.

$$ls_i, rs_i, rs_i^+ \in \{0, 1, 2, ...\} \qquad i \in I \quad (27)$$
$$ts_i \in \{1, 2, .., H - \lambda_i + 1\} \qquad i \in I \quad (28)$$

The model formulation stated above comprises logical constraints (22), which can easily be transformed into a pure mixed-integer model formulation so that standard MILP solvers could be applied. On the other hand, constraint-programming solvers are readily available today, which exploit the special structure of the scheduling model under consideration very efficiently and allow optimal solutions to be obtained with only very little computational effort.

8 Numerical investigation

The campaign planning model presented in Section 5.1 (including the extensions from Sect. 5.2), the assignment model introduced in Section 6, and the model for resource conflict resolution developed in Section 7 were implemented using ILOG OPL Studio 3.1. The individual models were interfaced through OPL's script language. The mixed-integer linear campaign planning model (step 1) and the integer linear assignment model (step 2) were solved through the standard optimisation software CPLEX 7.0. For eliminating the remaining resource conflicts (step 3), ILOG Solver 5.0 was used to solve the corresponding constraint logic model. All calculations were carried out on a PC with a 900 MHz AMD Athlon processor. In the following section, the performance of the presented approach is first analysed for a number of problem instances based on a hypothetical chemical batch process. Next, the case study presented in Figure 1 is considered. This specific problem instance is equivalent to real industrial applications in both, size and complexity.

8.1 Sample problems

The chemical production process considered in the sample problems consists of three plants a, b, and c equipped with two units each. Units are numbered $u1$ through $u6$. Some chemical processes can be performed at two units, e.g. process 2 at units $u1$ or $u2$, process 5 at units $u3$ or $u4$, and process 8 at units $u5$ or $u6$. No such options exist for process types 1, 3, 4, 6, 7, and 9. Several campaigns of the same process type may overlap if there are enough equipment units available. The structure of the chemical process under consideration is shown in Figure 9.

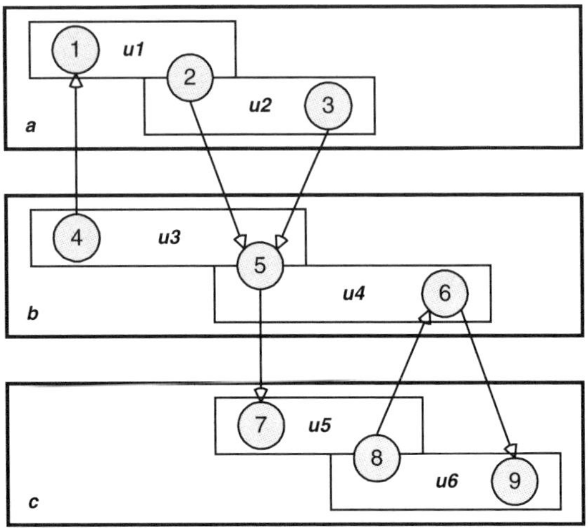

Fig. 9. Structure of the chemical process for the sample problems

Table 1. Demand of final products at the end of planning horizon

Problem	Demand of product [# of batches]		
instance	1	7	9
1	10	10	10
2	15	10	5
3	15	5	10
4	10	15	5
5	5	15	10
6	10	5	15
7	5	10	15

The arcs in Figure 9 indicate material flows of single batches. For instance, one batch of each of the processes 2 and 3 is required as input for processing a single batch of process 5. In turn, one batch of process 5 is required to feed each batch of process 7. Throughout the chemical batch processing network, an unlimited intermediate storage policy is applied. For the three end-products 1, 7, and 9 demand figures are given for the last period of the planning horizon.

Each of the sample problems is characterized by a given total requirement equivalent to 30 batches. For the generation of problem instances, the demand mix of the various end-products is varied so that the process bottleneck is shifted between different production units. In total, seven demand distributions are defined. It is assumed that sufficient capacity is available to store all of the required raw material and end-product quantities and that there is no initial stock of end-products available at the beginning of the planning horizon. In Table 1 an overview of the demand distributions used in the numerical experiments is given.

Table 2. Characteristics of the test sets

Parameters	Test set			
	A	B	C	D
Planning horizon [# of periods]	25	30	25	30
Processing time per patch [# of periods]	1	1	1	1
Parallel campaigns	✓	✓	✓	✓
Minimal camgaign length	–	–	4	4
Clean-out-time [# of periods]	–	1	–	1

In order to analyse the impact of the several extensions presented in Section 5.2, four different test sets are used. Test set A refers to the basic problem. In test set B, clean-out operations are considered. Test set C extends the basic problem by including minimal campaign lengths. Finally, in test set D, both minimal campaign lengths as well as clean-out operations are considered. The larger workload of the production units for test sets B and D due to additional clean-out operations is taken into account by an extended time horizon. A summary of the different parameters used in the test sets is given in Table 2. The full data and test sets may be obtained from the authors. In order to solve the various sample problems, the corresponding extensions of the basic model formulation (see Sect. 5.2) had to be applied. As a result, the number of variables and constraints required in the optimization models varies across thc test sets (see Tables 3 and 4).

Table 3. Number of variables for the sample problems

	Test set							
	A		B		C		D	
No. of variables	integer	total	integer	total	integer	total	integer	total
Campaign planning	225	450	270	810	225	675	270	1,080
Assignment	600	600	720	720	600	600	720	720
Ressource conflict resolution	–	–	33–39	33–39	33	33	30–39	30–39
Conventional model	300	525	720	990	300	525	720	990

As can be seen from Table 3, the number of variables, namely integer variables, in the campaign planning step is small compared to the assignment step. However, due to the pure binary nature of the assignment model, the complexity of the corresponding optimization problem if fairly low. For the third step, the number of both, the required variables as well as the constraints, depends on the number of resource conflicts that occur in the task/equipment assignment of the individual problem instances.

Conventional model formulations such as in Blömer and Günther (2000) and Kondili et al. (1993) do not aim at determining campaigns. Thus, clean-out op-

Table 4. Number of constraints for the sample problems

	Test set			
No. of constraints	A	B	C	D
Campaign planning	675	1,071	1,098	1,584
Assignment	663	798	663	798
Resource conflict resolution[a]	–	34–45	32–35	29–44
Conventional model	600	1,068	900	1,428

[a] Including no. of logical constraints

erations at the end of a campaign, minimal campaign lengths, and the maximal number of parallel campaigns are not accounted for. Thus, these basic models have to be complemented by additional variables and constraints similar to those presented in Section 5.2 of this paper, i.e. constraints (3a), (8), (11), and (13). However, since no aggregation is applied in the conventional formulation, the equivalents for the clean-out constraints (8) and the minimal campaign length constraints (11) are valid for all process/unit combinations rather than just for all processes as in the campaign planning step of our decomposition approach.

Compared to the conventional model formulation the number of binary variables in the critical campaign planning step is only slightly smaller for tests A and C. The reason for this limited reduction lies in the restricted number of feasible units for each process in the sample problems. In fact, only processes 2, 5, and 8 may be carried out on more than one unit (cf. Fig. 9). However, the difference is larger for test sets B and D as for these more realistic cases the conventional model formulation requires the introduction of additional variables modelling the clean-out operations. The resulting number of constraints in the campaign planning step is approximately the same as in the conventional model formulation. The number of feasibility constraints which are included in our decomposition approach (cf. Sect. 4.2) is roughly outweighed by the larger number of constraints needed to model the cleaning and minimum campaign length constraints in the conventional model formulation for each unit/process combination rather than just for each process. Clearly, the aggregation scheme proposed in this paper is less effective when there is only a small number of feasible task/equipment assignments.

For test set A, which requires constraints neither for clean-outs nor for minimal campaign lengths, only the first two steps of the approach apply. No resource conflicts need to be solved. Since the processing time for each batch is one period, the solution to the assignment model is always feasible (see Sect. 4.2). However, as resource conflicts due to longer processing times are treated exactly the same way as those arising from minimal campaign lengths and clean-out operations, our numerical results for these cases also hold for processing times longer than one time period. Table 5 indicates the number of instances in the various test sets that require the resolution of resource conflicts, i.e. step 3 of the solution procedure. The numerical experiments revealed that at most one resource conflict per problem instance occurred. Only in these cases, the proposed hierarchical solution approach

Table 5. Number of instances (out of seven) that require resolution ofd resource conflicts

	Test set			
	A	B	C	D
No. of instances	–	5	3	3

Table 6. CPU time for sample problems

CPU time	Test set											
	A			B			C			D		
[s]	min	⊘	max	min	⊘	max	min	⊘	max	min	⊘	max
Campaign planning	4.46	9.01	17.44	8.60	14.06	22.72	3.93	9.63	21.23	10.42	18.85	33.04
Assignment	0.05	0.06	0.07	0.06	0.07	0.08	0.06	0.06	0.07	0.06	0.07	0.08
Ressource conflict resolution[a]	–	–	–	0.05	0.06	0.06	0.03	0.05	0.07	0.04	0.04	0.05
Total	4.51	9.07	17.51	8.71	14.19	22.86	4.02	9.74	21.37	10.52	18.96	33.17

[a] Only instances that require ressource conflict resolution

does not guarantee optimal solutions to the overall decision problem. For all other problem instances, globally optimal solutions are obtained.

Table 6 summarizes the CPU times required for each of the three steps of the solution procedure. With total CPU times between 4 and 33 seconds, the approach suggested appears to be computationally very efficient. The CPU time is almost entirely used for step 1 (campaign planning). Despite step 2 requiring a far larger number of integer variables, it is solved within very short time. Obviously, the assignment problem can be treated very efficiently by standard optimisation software such as CPLEX. The constraint logic programming solver from ILOG which was employed in step 3 also worked very efficiently. All resource conflicts remaining from the previous step could be resolved with only little computational effort. CPU times for steps 2 and 3 never exceeded one tenth of a second. The comparison between the test sets shows that the computational time rises with an increase in problem complexity. The average CPU time for solving test set D instances is twice as high as the time required for test set A. Most of the increase is caused by the inclusion of additional constraints for the consideration of clean-out operations, as can be seen from the figures for test set B. On the contrary, the results for test set C suggest that including constraints for minimal campaign lengths has only a minor impact on the average CPU time.

8.2 Case study

Many sophisticated solution approaches, which have been successfully tested against hypothetical sample problems, perform rather poorly when they are confronted with real life applications. In order to show the practicality of the presented approach, we applied it to a case study with significant relevance to industrial practice. The case study was developed by a multinational manufacturer of chemical and pharmaceutical products with headquarters in Germany. Based on the planning problems in one of its business units, the case study is designed to encompass the major features which characterize the company's supply network and campaign planning environment.

The structure of the chemical production process upon which the case study is based has already been presented in Figure 1. There are four plants which produce fifty different types of products. Twelve of these are intermediates which are further processed. Chain products require processing at more than one of the four plants and may even cause cyclical material flows, as is the case in the production of end-product $D1$ for which intermediate $A2$ is transferred to plant B, where it is fed into the production of intermediate $B11$ which is then shipped back to plant A. Apart from producing intermediates for $D1$, plant B also performs the bulk of the processing of the other two chain products, $B7$ and $B10$. Customer demands for the various end-products, including all chain products and the complex local product $B4$, are given. They are distributed over the entire planning horizon of 60 periods. In contrast to the sample problems investigated in the previous section, parallel campaigns of the same chemical process are not allowed. This can be very simply incorporated into our approach by including constraint (13) presented in Section 5.2 into the campaign planning model. A detailed description of the entire case study is given in Berning et al. (2002).

In real life applications, it is frequently impossible to meet all customer demands on time. Therefore, backorders are allowed in the case study investigation. Also, minimal campaign lengths have to be considered. They differ between local products and chain products and amount to 6 and 11 batches, respectively. As in the sample problems, clean-out operations have to be performed after each campaign. Thus, the extensions introduced in Sections 5.2.1 and 5.2.2 are included. Moreover, transportation times for the transfer of material from one plant to another as well as delivery times to customer locations are taken into account. Table 7 summarizes the characteristics of the case study.

Determining the exact optimum or proving the optimality of a solution to a mixed-integer linear optimization model can be very time consuming. For many practical applications it is not necessary to find the exact optimum as long as a quality guarantee can be given. In this case study, a sufficient solution quality is assumed to be given, once a solution is found which deviates no more than 5% from the lower bound obtained during the branch-and-bound solution procedure.

According to the modelling approaches known from the literature so far, a model formulation with triple-indexed decision variables had to be used. Due to its computational complexity, such an approach condemns any solution attempt to failure regarding a total of 39,120 variables (36,120 integer) and 47,339 constraints which would have been required for the case study instance. Our experience with such model formulations revealed that no feasible solution could be obtained within reasonable computation time. In contrast, solving the case study problem with the approach proposed in this paper requires little more than 2 minutes of CPU-time.

Table 7. Characteristics of the case study

Parameters	Case study
Planning horizon [# of periods]	60
Processing time per batch [# of periods]	1
Parallel campaigns	–
Backorders	✓
Minimal campaign length [# of batches]	6, 11 resp.
Clean-out-time [# of periods]	3
Transportation time [# of periods]	1
Delivery time [# of periods]	2

Table 8. Number of variables, constraints, and CPU time for the case study problem

	Variables		Constraints	CPU
	Integer	Total		time [s]
Campaign planning	3,000	12,000	17,210	133.08
Assignment	36,120	36,120	26,279	4.12
Ressource conflict resolution	96	96	130	0.76
Total				137.96
Conventional model	36,120	39,120	47, 339	

Table 8 exhibits the details. The decomposition approach reduces the number of variables significantly to less than one tenth of the conventional approach. The number of constraints is also reduced substantially. The number of additional feasibility constraints (+780) according to (3a) is by far overcompensated by the larger number of constraints needed to model the cleaning (−14.809) and minimum campaign length constraints (−12.500) in the conventional model formulation for each unit/process combination rather than just for each process. The conventional model also requires some additional constraints ensuring that all units required for the different tasks of a process are assigned to a unique temporary configuration (−3.600). Contrary to the small sample problems in the previous section, the number of feasible combinations of individual units is large for all processes. For this realistic setting, the benefits of the decomposition approach become even more apparent.

The small computation time is still remarkable regarding the number of variables and constraints which remain for the campaign planning and assignment models. Apparently, the decomposition approach reduces the complexity of each sub-problem to such an extent that even large problem instances can be solved efficiently. After the assignment step, the solution obtained was not completely feasible since one of the clean-out operations could not be properly assigned to the equipment unit at which the preceding campaign was performed. This resource conflict, however, was rapidly repaired in step 3 of the procedure by re-assigning some of the processing tasks.

9 Summary and conclusions

A novel hierarchical framework for campaign planning and resource scheduling based on mixed-integer linear and constraint programming has been introduced. The approach suggested is especially designed to co-ordinate the various plant operations within multi-plant supply networks in the chemical industry. Basically, it relies on sub-dividing the entire planning problem into three decision levels. At the upper planning level, an aggregation scheme based on feasibility constraints is employed in order to reflect the limited number of the various types of production equipment. This concept permits a feasible disaggregation at the lower assignment level. Hence, a simple top-down solution approach can be applied and solutions are obtained with considerably reduced computational effort. Under certain modelling assumptions, theoretically optimal solutions to the overall planning problem are achieved. These assumptions do not always hold in real industrial applications. The disaggregation step can therefore result in partially infeasible schedules. In these cases, a third step follows that resolves the remaining resource conflicts.

Our numerical experiments show that the decomposition approach reduces the problem complexity in the campaign planning step significantly. This is primarily due to a substantial reduction in the number of integer variables compared to a standard model formulation which uses triple-indexed decision variables. In the subsequent step individual equipment units are assigned to the various campaigns. The resulting linear assignment model can be solved with only very little computational effort. In both steps, standard optimization tools were used. Finally, constraint logic programming was applied in order to resolve resource conflicts, which may occur if clean-out operations and minimal campaign lengths have to be considered. This is achieved with practically negligible CPU time. The proposed approach appears thus to be suitable for solving real industrial problems. This is tested in a case study from industry, in which a complex production network with multiple plants is considered. This case study was solved by our decomposition approach in little more than two minutes of CPU time.

In principle, we could refine the decomposition approach even further. In the second step, separate models for the task/equipment assignment in each plant could be used which would certainly be favourable from an organizational as well as computational point of view. However, our numerical experiments show that computational times in the assignment step are fairly low. Yet, setting up plant-specific assignment models could be attractive in the case of extremely large supply networks. Similarly, the conflict resolution scheme of the third step could also be refined. For instance, local shifts at a single plant could be considered before the scope is extended to the entire supply chain of plants which are affected by a specific campaign. Again, regarding the small computational times we observed, the potential benefits of such an approach appear to be very limited, but may be of interest in some special cases.

In practice, additional constraints occur. For instance, in order to meet the required specifications of end products, certain configurations of equipment units are preferred by plant management. Likewise, minimal campaign lengths have to be observed in order to avoid too frequent reconfigurations of resource groups.

Moreover, the limited availability of pool resources (e.g. personnel, utilities etc.) and storage devices imposes additional constraints. The framework of linear programming, however, is quite flexible. For some extensions we have already shown how they can be integrated into our approach. For others, the proposed model formulations may easily be adapted to accommodate specific features required by the individual application.

Appendix A

In Section 4.2 feasibility constraints for modelling task/equipment assignments in the campaign planning model have been introduced. This concept is based on the definition of a complete set of virtual units, i.e. all sub-sets of equipment units of a particular type. Finally, it is shown in this appendix that, for a set of processes scheduled in a certain period, there is a feasible assignment for all these processes to the available (real) equipment units, in which no more than one process is assigned to any unit.

For the ease of presentation we assume in this proof that at most one campaign of each process can be performed at the same time. The proof for the general case of n parallel campaigns of a process can easily be obtained by defining each of the parallel campaigns as a different process which has the same set of feasible units as the original process. The proof then follows the same scheme as below, i.e. the case in which parallel campaigns are not allowed.

Theorem. Let J be a set of processes scheduled for the considered period and meeting the resource constraints for all virtual units. Then there is in this period a feasible assignment for all processes of J to production units without assigning more than one process to any unit.

Proof (indirect). Since each process has at least one feasible unit, it follows that each process could be assigned to such a unit. It remains to be shown that no multiple assignments to the same unit are required when all processes are assigned.

For the sake of a contradiction let us assume that there is a set J of scheduled processes for which there is no assignment to feasible units such that multiple assignments could be avoided. Under this assumption we will now construct a sequence $U_0, U_1, U_2, \ldots$ of sets of units and a sequence $J_1, J_2, J_3, \ldots$ of sets of processes such that one the following cases occurs:

case a) For the sequence of sets of units holds: $U_0 \subsetneq U_1 \subsetneq U_2 \subsetneq \ldots$, i.e. each set is a *true* subset of its successor

case b) For the two sequences holds: $U_0 \subset U_1 \subset U_2 \subset \ldots$ and $J_1 \subset J_2 \subset J_3 \subset \ldots$ and there is an index k, such that:

1. $|J_{k+1} \backslash J_k| < |U_k \backslash U_{k-1}|$,
2. $|J_{i+1} \backslash J_i| \geq |U_i \backslash U_{i-1}|$, for all $1 < i < k$, and
3. $|J_1| > |U_0|$.

Figure 10 (left) illustrates case b) for k =2.

In the following, both cases are led to a contradiction.

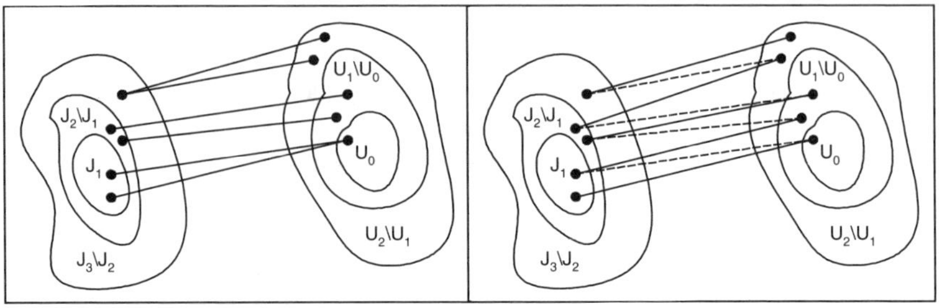

Fig. 10. An infeasible multiple assignment (left) is reduced by an exchange (right)

Among all assignments of the processes of J to feasible units let us consider the one with the smallest number of multiple assignments. Such an assignment always exists since the number of multiple assignments has a lower bound of zero.

Let u be a unit to which more than one process is assigned in the assignment under consideration. Let us define $U_0 := \{u\}$ and $J_1 := \{j | j$ is assigned to a unit from $U_0\}$ as the first members of the sequences to be constructed. Since more than one process is assigned to u, $|J_1| > |U_0|$ holds. Now let us define $U_1 := \bigcup_{j \in J_1} U_j$, with U_j being the set of feasible units for process j. Since all processes of J_1 meet the resource constraints for virtual units formulated in (3), $|J_1| \leq |U_1|$ must hold. This follows from the definition of U_1 and the design of the virtual units which requires one of the virtual units to be equal to U_1. Therefore, the cardinality of J_1 can be bounded from above. $|U_1| \geq |J_1| > |U_0|$ holds and, since $U_0 \subset U_1$ holds due to the definition of U_1, U_0 is a true subset of U_1.

Let us define now J_2 by $J_2 := \{j | j$ is assigned to a unit from $U_1\}$. Then, due to the definition of J_1 and J_2 and the relation between U_0 and U_1, the set J_2 includes J_1 as a true subset. Furthermore, $J_2 \backslash J_1$ describes the set of processes assigned to units of $U_1 \backslash U_0$. Comparing the cardinalities of these two sets, there are two possibilities. If $|J_2 \backslash J_1| < |U_1 \backslash U_0|$, case b) occurs which is led to a contradiction afterwards. Therefore, for the moment we only need to consider the case of $|J_2 \backslash J_1| \geq |U_1 \backslash U_0|$. Now, the following inequality holds:

$$|J_2| = |(J_2 \backslash J_1) \cup J_1| = |J_2 \backslash J_1| + |J_1| > |U_1 \backslash U_0| + |U_0| = |(U_1 \backslash U_0) \cup U_0| = |U_1|$$

Let U_2 be defined by $U_2 := \bigcup_{j \in J_2} U_j$. Again, from the resource constraints for virtual units in (3) it follows that $|J_2| \leq |U_2|$. Thus, $|U_2| \geq |J_2| > |U_1|$ holds, and because of the same argumentation as for the sets U_0 and U_1, U_1 must be a truc subset of U_2.

This procedure can be repeated iteratively. If case b) does not occur, the sequence of sets of units required for case a) is constructed. Since the cardinality of the sets increases by at least one for each member of the sequence, we will eventually get a member of the sequence with a cardinality higher than that of the set of available units. However, due to the definition of the sequence, all members must be subsets of the set of available units which leads to a contradiction. It remains to be shown that case b) also implies a contradiction.

Case b) is described by a situation as in Figure 10 (left). In the general case let k be the index for which case b) occurs for the first time. Then $|J_{k+1} \backslash J_k| <$

$|U_k \backslash U_{k+1}|$ means that no process is assigned to at least one unit from the set $U_k \backslash U_{k-1}$. For all indices i with $1 < i < k$, the number of processes assigned to the set $U_i \backslash U_{i-1}$ is at least as high as the number of units available in this set. There always exists a feasible process j from the set J_k which can be reassigned to a unit with no processes assigned from the set $U_k \backslash U_{k-1}$, since U_k is defined as an union of all sets of feasible units for the processes in J_k. If this is done, one would either reduce the number of multiple assignments (in case more than one process was assigned to the unit which j was originally assigned to) or leave one unit of a set $U_i \backslash U_{i-1}$ with $1 < i < k$ with no process assigned. The former case contradicts the choice of the original assignment as being the one with the minimum number of multiple assignments.

In the latter case one could continue the exchange process until either the former case occurs or finally a process of J_1 is assigned to a unit with no process assigned. In Figure 10 (right) this principle is illustrated for the given example. Due to the described exchange process the number of multiple assignments of units of the set U_0 would be reduced which results in a reduction of the overall number of multiple assignments. Again, this contradicts the choice of the original assignment as being the one with the minimum number of multiple assignments. □

References

Anthony RN (1965) Planning and control systems: a framework for analysis. Harvard University, Boston

Axsäter S (1986) On the feasibility of aggregate production plans. Operations Research 34: 796–800

Berning G, Brandenburg M, Gürsoy K, Mehta V, Tölle, FJ (2002) An integrated system solution for supply chain optimization in the chemical process industry. OR Spectrum 24, No. 4 (to appear)

Blömer F (1999) Produktionsplanung und –steuerung in der chemischen Industrie. Gabler – Deutscher Universitäts Verlag, Wiesbaden (in German)

Blömer F, Günther HO (1998) Scheduling of a multi-product batch process in the chemical industry. Computers in Industry 36: 245–259

Blömer F, Günther HO (2000) LP-based heuristics for scheduling chemical batch processes. International Journal of Production Research 38: 1029–1051

Dessouky M, Kijowski B, Verma S (1999) Simultaneous batching and scheduling for chemical processing with earliness and tardiness penalties. Production and Operations Management 8: 433–444

Fleischmann B, Meyr H (1997) The general lotsizing and scheduling problem. OR Spektrum 19: 11–22

Grossmann IE, Quesada I, Raman R, Voudouris VT (1996) Mixed-integer optimization techniques for the design and scheduling of batch processes. In: Reklaitis V, Sunol AK, Rippin DWT, Hortacsu O (eds) Batch processing systems engineering, pp 451–494. Springer, Berlin Heidelberg New York

Günther HO (1986) The design of an hierarchical model for production planning and scheduling. In: Axsäter S, Schneeweiss C, Silver E (eds) Multi-stage production planning and inventory control, pp 227–260. Springer, Berlin Heidelberg New York

Hax AC, Meal HC (1975) Hierarchical integration of production planning and scheduling. In: Geisler MA (ed) Studies in management science, Vol 1. Logistics, pp 53–69. Elsevier, Amsterdam

Kallrath J (2002a) Combined strategic and operational planning – a MILP success story in chemical industry. OR Spectrum 24: 315–341

Kallrath J (2002b) Planning and scheduling in the process industry. OR Spectrum 24: 219–250

Kondili E, Pantelides CC, Sargent RWH (1993) A general algorithm for short-term scheduling of batch operations – I. MILP formulation. Computers and Chemical Engineering 17: 211–227

Mendez CA, Cerda J (2002) A MILP-based approach to the short-term scheduling of make-and-pack continuous production plants. OR Spectrum 24, No. 4 (to appear)

Meyr H, Wagner M., Rohde J (2000) Structure of advanced planning systems. In: Stadtler H, Kilger C (eds) (2000) Supply chain management and advanced planning, pp 75–77. Springer, Berlin Heidelberg New York

Neumann K, Schwindt C, Trautmann N (2002) Advanced production scheduling for batch plants in process industries. OR Spectrum 24: 251–279

Oh HO, Karimi IA (2001a) Planning production on a single processor with sequence-dependent setups – part 1: determination of campaigns. Computers and Chemical Engineering 25: 1021–1030

Oh HO, Karimi IA (2001b) Planning production on a single processor with sequence-dependent setups – part 2: campaign sequencing and scheduling. Computers and Chemical Engineering 25: 1031–1043

Papageorgiou LG, Pantelides CC (1996a) Optimal campaign planning/scheduling of multipurpose batch/semicontinuous plants. 1. Mathematical formulation. Ind Eng Chem Res 35: 488–509

Papageorgiou LG, Pantelides CC (1996b) Optimal campaign planning/scheduling of multipurpose batch/semicontinuous plants. 2. A mathematical decomposition approach. Ind Eng Chem Res 35: 510–529

Pinto JM, Grossmann IE (1994) Optimal cyclic scheduling of multistage continuous multiproduct plants. Computers and Chemical Engineering 18: 797–816

Pinto JM, Grossmann IE (1998) Assignment and sequencing models for the scheduling of process systems. Annals of Operations Research 81: 433–466

Reklaitis GV (1996) Overview of scheduling and planning of batch process operations. In: Reklaitis V, Sunol AK, Rippin DWT, Hortacsu O (eds) Batch processing systems engineering, pp 660–705. Springer, Berlin Heidelberg New York

Rippin DWT (1993) Batch process systems engineering: a retrospective and prospective review. Computers and Chemical Engineering 17(Suppl.): S1–S13

Schneeweiss C (1999) Hierarchies in Distributed Decision Making. Springer, Berlin Heidelberg New York

Schwindt C, Trautmann N (2000) Batch scheduling in process industries: an application of resource-constrained project scheduling. OR Spektrum 22: 501–524

Stadtler H, Kilger C (eds) (2000) Supply chain management and advanced planning. Springer, Berlin Heidelberg New York

Timpe C (2002) Solving planning and scheduling problems with combined integer and constraint programming. OR Spectrum 24, No. 4 (to appear)

Voudouris VT, Grossmann IE (1996) MILP model for scheduling and design of a special class of multipurpose batch plants. Computers and Chemical Engineering 20: 1335–1360

Zijm WHM (2000) Towards intelligent manufacturing planning and control systems. OR Spektrum 22: 313–345

Solving planning and scheduling problems with combined integer and constraint programming*

Christian Timpe

BASF-AG, GVC/S-B9, 67056 Ludwigshafen, Germany
(e-mail: christian.timpe@basf-ag.de)

Abstract. There have been several reports on successful application of hybrid MIP/CP algorithms for solving otherwise difficult or intractable problems. However, the practical examples presented in many papers suffer from the fact that they are far away from being realistic, in terms of both their size and their complexity. This paper presents a realistic planning problem of the chemical industry, which has been successfully treated with a hybrid approach.

Realistic mid-term production planning in chemical industry very often contains decisions on lot-sizing, assignment and sequencing, which have to be made simultaneously. A widely-used approach for implementing decision-support systems for these problems is to decompose them into a lot-sizing and a sequencing subproblem, and to solve the two subproblems with different technologies.

While this approach is adequate in many cases, there are, however, some examples where the quality of the resulting schedule depends crucially on the simultaneity of the three different planning decisions. For a planning problem of this type we developed a very efficient hybrid MIP/CP solution strategy. The MIP acts as the master process, but either one of the solvers gets just a partial description of the problem. In this paper we present the planning problem, the solution technology and our practical experience with the implementation.

Key words: Hybrid methods – Decision support systems – Production planning – Scheduling

---

* This work was partially supported by the Fifth European Community Framework Programme (GROWTH project G1RD-1999-00034, LISCOS).
The author is thankful to S. Heipcke and A. Aggoun, whose fruitful comments facilitated a lot the development of the algorithm.

1 Introduction

Both Mixed-Integer Programming [14] and Constraint Logic Programming [13] have been applied successfully to industrial problems for quite some time. Although they both rely on the idea of branching and are principally able to solve a very similar class of problems, the efficiency of the two techniques on certain problems can be very different. This is often referred to as the "complementary strength" of the techniques. From the viewpoint of industrial applications, it is often more adequate to speak of complementary weaknesses, since many practical problems have "complementary aspects" which make them hard for either technique.

The application we describe here is the mathematical core of a production scheduling system. The system entails an interactive Gantt chart as a user interface and is linked to an SAP R/3 system. The planning process to be supported by the system entails lot-sizing, assignment and sequencing decisions. Because the quality of the resulting schedule depends crucially on the simultaneity of the three different planning decisions, but no single technology is able to solve the problem on it's own, a hybrid MIP/CP solution algorithm had to be developed. Neither of the two solvers has a full representation of the problem, instead the two solvers work in close co-operation. The MIP acts as the master process, which calls a CP solver every few nodes. If the CP model is infeasible, a ("no good", [10]) cut is added to the MIP model, otherwise some variables of the MIP model are fixed according to the result of the CP model.

The remainder of the paper is organized as follows. In Section 2 we show the relation of our work to current research. In Section 3 the planning problem is described in greater detail. In Section 4 we explain the different models involved. The aim here is to give an impression of the complexity of the problem, and to explain the ideas of the modeling approach. Practical experience is reported in Section 5. In the last section we outline future work.

2 Related work

Over the last five years, there has been a growing interest in the idea of combining the two techniques of Mixed-Integer Programming and Constraint Logic Programming, and a number of ideas have been developed how to do it. The first applications consisted more or less in the incorporation of one technique into the other, i.e. they used a linear relaxation as a *global* constraint, or applied constraint reasoning at MIP tree nodes ([4]). (The idea of a global constraint is, in a nutshell, to provide an easy way for modeling complex constraints on sets of variables, and to exploit semantic information to reduce the domains of the variables involved, see e.g.[1].) Recently, more refined ideas have been implemented, where two solvers run in parallel, with independent or synchronised search trees ([8, 11, 5]).

The latter approaches have the disadvantages that two models have to be built, maintained and executed, which does not only require knowledge of the two fields, or – in an industrial context even worse – two experts, but also leads to computational overhead. For some problems this can be justified, because there is no other way of solving them. We believe, however, that the key for successful application of hybrid solutions lies in efficient modeling. The hybrid model described here can be seen as an application of Mixed Logical/Linear Programming (MLLP) framework proposed by Hooker et. al. [9]. The algorithm extends the idea described in [11] in such a way, that already partial solutions to the MIP model are examined with a CP model and that cuts are derived not only from failed CP problems, but also from successful ones.

3 Description of the planning problem

Very often in the process industry, there are many more products than machines. (See e.g. [12] for an overview of planning and scheduling problems in the process industries.) Therefore products have to be produced alternatingly. The process of changing from one product to another is called a *changeover.* Changeovers are generally expensive both in time and money. The cost usually consists of two components. The first is for the physical work that has to be performed during changeover, like cleaning, changing tools, etc. The second component captures the risk, that machines, tools or products suffer from the changeover. The first component is proportional to the duration of the changeover, the second to the difficulty. If changeover duration is considered explicitly in the machine capacity constraint, the loss of productive time must not be included in the changeover cost, since this is already entailed in the penalty for late production.

The production of one particular product on a machine between two changeovers is called a *campaign*. While the total amount of a product to be produced in the scheduling horizon (2 to 3 months) is known in advance, the splitting of this total amount into campaigns of arbitrary lengths (lot-sizing), the machine, the campaigns are produced on (assignment) and also the order, in which the campaigns of different products are produced on each individual machine (sequencing), are to be determined by the solver. For technical reasons, there is a minimal length for each campaign, which depends on the machine chosen and on the product produced. Problems of campaign planning are also dealt with in [3,7,15].

Figure 1 gives an overview of the product flow of this application: Production consists of two stages with intermediate storage in tanks. Both production stages produce 24 hours per day without interruption, but there may exist planned shutdowns as well as breakdowns.

The first stage consists of a few large machines with minimal campaign lengths, sequence-dependent changeover times and sequence-dependent changeover costs. About seven intermediate products are produced on these machines in disjunctive mode. These machines are very restrictive in terms of the changeovers allowed, some products have just two or three possible successors. Production is continuous, both raw material consumption and production output are linear in the campaign duration. Campaigns on these machines have typical sizes of several days up to one or two weeks.

The intermediate storage consists of a small number of tanks of different size. There may be just one product in a tank at any time (thus, a tank must be empty before another product can be stored), but no setup time or cost occur. Different campaigns of the same product may charge into different or into the same tank, one

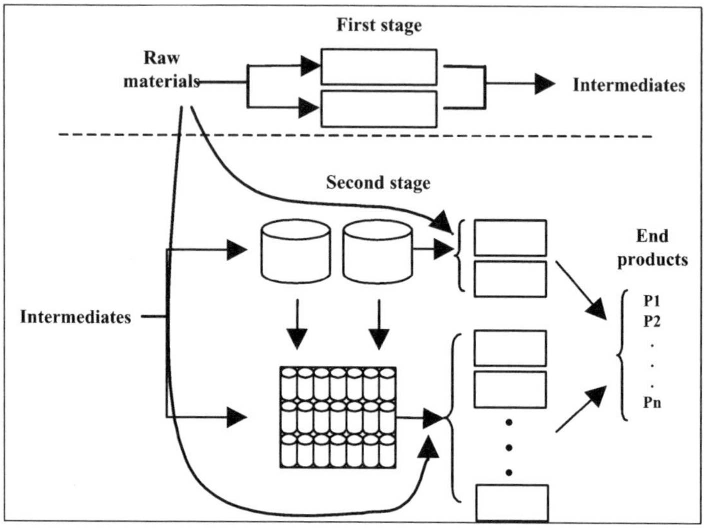

Fig. 1. Production scheme

campaign may charge into different or just one tank. Both is also true for consuming campaigns, but not every tank may feed every end product machine.

In the second stage, around 100 end products are produced on about 10 machines, also in continuous production. Machine dependent minimal campaign lengths have to be considered. Each end product requires one intermediate product and (very small amounts of) additional raw materials. Production rates depend on the product and the machine chosen. Some products may not run on certain machines. Again, there are sequence-dependent changeover times to be considered. Campaigns on the end product machines are typically much shorter, from far less than one day up to several days.

Demands are defined for end products and intermediates. Some demands are related to customers with given due dates. Additionally, we have an average demand for each product per week. Production should start at the latest when inventory level reaches minimal or safety stock, however, short-term disregard of safety stock is possible. There is no maximal stock level for end products.

A special feature of this problem are the *changeover products*. Whenever a changeover between two different products occurs on one of the machines of the first stage, a certain amount of a changeover product is produced. Think of it as a mixture of the two intermediate products involved in the changeover. The changeover product is, in a certain dilution, suitable for the production of some (low quality) end products. In order not to use up tank volume for the changeover product, the task on the end product machines using the changeover product are to be scheduled immediately after the changeover. Which end product has to be produced depends on which changeover is performed.

The overall objectives are to minimize late production (orders not satisfied at their due dates, or stock level less than specified safety level), early production leading to inventory costs (production starts before safety stock level is reached) and (sequence-dependent) changeover costs on the end product machines.

While from a technical point of view, a separation of the problem into parts each suited to be treated by a single dedicated solver is desirable, an analysis of the interdependencies of the three aspects (*lot sizing*, *assignment*, *sequencing*) in this case showed clearly that by starting with any subset of the aspects, we will most probably get a solution which will be infeasible or at least very poor with regard to the other aspects. Instead, a single all-in-one approach had to be pursued.

Last but not least, a practical constraint had to be considered. Because the system is supposed to work in an environment very close to production, solution times of more than ten minutes would not be tolerated by the user. The reader will not be surprised that a full solution in the sense of proven optimality cannot be reached in this time. However, the user is quite happy with a quick good feasible solution to this problem.

4 Modeling: hybrid MIP/CP formulation

Before the hybrid model as it is described here was developed, several trials were made to stick to easier formulations. At first, a pure CP model approach was investigated. It was technically correct and produced feasible solutions, but solution times were so large that we never managed to solve a problem data set of full size. A pure MIP model was never tested because the complicated changeover restrictions can, to the best of our knowledge, not be modeled efficiently with MIP techniques, at least not if a hundred products are involved. An algorithm similar to [11] was also tried, but it suffered from the fact that the corresponding CP model was very large and therefore very slow, and cuts were produced too late. This lead to the idea of cuts being derived as early as possible in the solution process. A MIP and two different CP models were developed. The MIP model acts as the master process, which calls a small CP model every three nodes. This small CP model entails just a few important constraints very efficiently, and is able to cut off infeasible trials of the MIP solver very early. The MIP model itself is a time-bucketed model and is responsible for the lot-sizing and the assignment component of the problem. It is solved using a sort of gliding-window-heuristic. Immediately after a time bucket of the MIP model is solved (all binary variables belonging to the period are feasible), the full CP model is generated and started. The CP model is responsible for the sequencing of that particular period. The result of this run is either a (valid) inequality, or, if the CP model could be solved, information about changeovers and product dependencies, which can be applied back in the MIP model and will influence the solution of the next period.

This new approach has the advantage that any infeasibility of the CP problem can be located very accurately, so that the resulting MIP model constraint can be very efficient. The two models are presented in the sequel.

4.1 Data and notation

First, we give an overview of all indices, variables and data used in the mixed MIP-CP model formulation. Capital letters denote constants, script capitals denote sets. Small characters denote variables and indices.

Note that we use the concept of time dependency in both models, but the length of time slices is very different. Time slices in the MIP model have a duration of two to four weeks, while time slices in the CP model have length of ten minutes. The function $DUR(t)$ returns the duration of MIP model period t in days. The duration of period $NT + 1$ is set to equal the sum of the lengths of all other periods. We also define a function $\mathrm{sol}(x)$, which gives the value of variable x in the current relaxation. (This is needed in the definition of the cuts.)

Indices

name	*cardinality*	*description*
$i \in \mathcal{I}$	100+7+100	raw materials, intermediates, end products
$k \in \mathcal{K}$	12	machines
$j \in \mathcal{J}$	200	jobs, tasks (per period)
$t \in \{1, .., NT + 1\}$	6–8	time slices (MIP model)
$o \in \mathcal{O}$	3000	task types (recipes)
$c \in \mathcal{C}$	3	customer categories

MIP model variables

name	*type*	*description*
x_{kt}^i	semi-continuous	production
s_t^i	cont.	amount stocked
y_{kt}^i	0/1	setup
r_{ct}^i	cont.	backlog
rx_{kt}	cont.	backlog of production quantity
rs_t^i	cont.	infeasibility breaker for stock constraints

CP model variables

name	*type*	*description*
$start_j$	integer	start of task j
end_j	integer	end of task j
$succ_j$	integer	successor of task j
$mach_j$	integer	machine task j runs on
$first_k$	integer	first task on machine k

Recipe data: Recipes are always dedicated to a particular output product and a particular machine, and describe the type and quantity of raw materials and intermediates to produce a certain amount of a product or intermediate.

Product data: Demands for products come in two types: customer orders and average demand. While customer orders must by all means be delivered on time, it is acceptable that regular demand is delayed if necessary.

BC_{ic} = backlogging cost per day
IC_i = inventory cost per day
$RCPT_{it}$ = external receipts, including initial stock
DEM_{ict} = demand
$SMIN_i$ = lower bound on stock level per product
$SMAX_i$ = upper bound on stock level per product (intermediates)
RT_i = replenishment time (for raw materials)

Machine data

$MCL_{koo'}$ = cleaning time (sequence-dependent)
$P0_k$ = last product produced before the start of the planning horizon
$MCAP_{kt}$ = capacity in time units per machine and time slice
$IR_{kii'}$ = input rate (quantity of i needed to produce one unit of i')
ST_{ki} = minimal setup time
SC_{ki} = minimal setup costs
LB_{ki} = minimal campaign length
UB_{ki} = maximal campaign length
CR_{ki} = capacity utilisation rate (production coefficient)

4.2 Mathematical formulation of the MIP model

The MIP model is a very standard *capacitated dynamic lot-sizing problem.* ($<$ Big-bucket, capacitated, with setup times $>$ according to the classification proposed in [2].) “Big-bucket” means that time buckets are long in relation to production campaigns, therefore we assume that production campaigns (and setups) do not extend into the next period. On the other hand, buckets must be short enough that the fact that we allow for just one setup per product, machine and period does not take away too much flexibility. A third (and the least justified) assumption is that all data are deterministic: Demand forecast will surely materialize, machines will not break down, production output will always accord with specification.

The model is presented here in some detail, so that the reader gets an impression of the complexity of the problem at hand. We leave out the constraints modeling the use of tanks, because this is very technical and does not give additional insight into the idea of hybrid solution.

4.2.1 Material balance

The first set of constraints consists of the material balance equations for each product $i \in I$. Material balance is only constrained for those time slices that cover the period before the replenishment time of the material. The idea is that, if we order the material now, it will be delivered before the replenishment time, so the actual stock is irrelevant after that period. This applies mostly to raw materials, RT_i is very large for products without external source. Note that receipts, demand and backlog applies to all three material types (raw material, intermediate, end product). (A customer for a raw material or a supplier for an end product will usually be another factory of the same company.) Constraint (5) is used to prevent using backlogged intermediates for production of finished products.

$$\forall i \in \mathcal{I},\ t \in \{2, ..., NT\},\ RT_i \geq t :$$

$$\begin{aligned} s^i_{t-1} + \sum_{c \in \mathcal{C}} r^i_{ct} + RCPT_{it} + \sum_{k \in \mathcal{K}} x^i_{kt} = \\ s^i_t + \sum_{c \in \mathcal{C}} r^i_{ct-1} + \sum_{c \in \mathcal{C}} DEM_{ict} + \sum_{\substack{i' \in \mathcal{I} \\ i' \neq i}} \sum_{k \in \mathcal{K}} IR_{kii'} x^{i'}_{kt} \end{aligned} \tag{1}$$

For the first period ($t = 1$), $\sum_{c \in \mathcal{C}} r^i_{ct-1}$ has to be omitted.

We define an additional time slice $NT + 1$ of arbitrary length, which can be used to produce all the backlogged material in case the total demand cannot be fulfilled in the time horizon. The constraint for this period reads:

$$\forall i \in \mathcal{I},\ RT_i \geq NT + 1 :$$

$$\begin{aligned} s^i_{NT} + \sum_{c \in \mathcal{C}} r^i_{cNT+1} + \sum_{k \in \mathcal{K}} x^i_{kNT+1} = \\ s^i_{NT+1} + \sum_{c \in \mathcal{C}} r^i_{cNT} + \sum_{\substack{i' \in \mathcal{I} \\ i' \neq i}} \sum_{k \in \mathcal{K}} IR_{kii'} x^{i'}_{kNT+1} \end{aligned} \tag{2}$$

In case the backlog is not caused by lack of production capacity, but raw material, the additional time slice does not help, so there might still be some backlog at the end.

4.2.2 Machine usage

Machine time is used up by production and setup. (A *setup* is the process of preparing the machine for the next product. It differs from the *changeover* insofar that it disregards the preceding product.) The capacity constraint is *not* enforced in the additional time slice (giving it arbitrary length). A special variable is the "production quantity backlog" rx_{kt}. This variable does not occur in any other constraint, including the objective function. Therefore, the MIP solver will always drive this variable to zero. We will use it to communicate the solution of the CP problem into the MIP model. The data of the MIP model are prepared in such a way that the MIP model solution will always underestimate the time needed for changeovers. When the CP solver has finished successfully, the real duration of the schedule on all machines of the given period is known. The difference between the time really needed and the guess of the MIP model solution (which is significant in some cases) is subtracted from the length of the successive period. Technically, this is done by adding a local cut on the variable rx_{kt} (instead of reducing the right-hand-side value $MCAP_{kt}$). This has the advantage that the bound is removed on backtracking automatically by the cut manager of XPRESS-MP.

$$\forall k \in \mathcal{K},\ t \in \{1, ..., NT\}: \qquad (3)$$
$$\sum_{i \in \mathcal{I}} CR_{ki} x^i_{kt} + \sum_{i \in \mathcal{I}} ST_{ki} y^i_{kt} + rx_{kt} \leq MCAP_{kt}$$

4.2.3 Link setups to production

The following constraints link the setup variables to the production variables, i.e., production of a product is possible only after the product has been set up.

$$\forall k \in \mathcal{K},\ i \in \mathcal{I},\ t \in \{1, ..., NT+1\}: \qquad (4)$$
$$LB^i_{kt} \cdot y^i_{kt} \leq x^i_{kt} \leq UB^i_{kt} \cdot y^i_{kt}$$

4.2.4 Upper bounds on backlogs

Because we use different backlog variables for different customer groups, care must be taken that the optimizer does not use the "wrong" variable, if a particular customer group cannot be satisfied.

$$\forall c \in \mathcal{C},\ i \in \mathcal{I},\ t \in \{2, ..., NT\}: \qquad (5)$$
$$r^i_{ct} \leq r^i_{ct-1} + DEM_{ict}$$

For the first period ($t = 1$), r^i_{ct-1} has to be omitted.

4.2.5 Inventory bounds

For every product, the stock must lie between the minimal (safety) and maximal stock.

$$\forall i \in \mathcal{I},\ t \in \{1, ..., NT\}:$$
$$s^i_t \leq SMAX_i \qquad (6)$$
$$s^i_t + rs^i_t \geq SMIN_i$$

4.2.6 Objective function

The objective of this model is to minimize the sum of setup costs, stock holding costs and backlogging costs.

$$\min \sum_{i \in \mathcal{I}} \sum_{t=1}^{NT+1} \left(\sum_{k \in \mathcal{K}} SC_{ki} \cdot y_{kt}^{i} + DUR(t) \cdot \left(\sum_{c \in \mathcal{C}} BC_{ic} \cdot r_{ct}^{i} + BC_{i2} \cdot rs_t^i + IC_i \cdot s_t^i \right) \right) \tag{7}$$

Note that we assume that all stock and backlogging costs grow linearly with the quantity. This does not need to be the case in general, especially not for backlogging costs. Customer group 2 is the group of most important customers, the term $BC_{i2} \cdot rs_t^i$ means that for those customers it is allowed (if all other means fail) to use safety stock. It was thought of including an additional "communication variable" into the objective function in order to be able to communicate changeover costs computed in the CP model to the MIP model (see 4.2). This has not been implemented, because an exhaustive search in the tree is not possible due to restrictions in computation time. Thus it is not essential to know the exact value of the objective.

4.2.7 Enumeration strategy

Branching in the MIP model is on integer (binary) variables. When using powerful MIP solvers like XPRESS-MP (or CPLEX), it is usually not worthwile to try to improve the default enumeration strategy. In this case, however, since cuts are being added frequently to the MIP model, the system cannot derive as much information from the relaxation as it usually can. Therefore, it had to be strengthened by defining variable selection criteria.

The first requirement is that the priority of all integer (or binary) variables belonging to period t must be strictly higher than the priority of variables belonging to periods $t+1 \ldots NT+1$, since we want to solve the periods in their natural order. The second, less obvious requirement is that variables inside the individual periods should be selected according to their position in the production process: first stage variables prior to second stage variables, and for each stage, a group of important (big) machines prior to a group of less important (smaller) machines.

Even more important is the control of the backjumping mechanism. Since the current solution is frequently made infeasible by adding a new cut, care must be taken that the solver does not jump back "too high", trying a new combination which is so different from the previous one that the cut just derived does not cut off anything. Instead, the solver should retry at a level where the information of the cut can be exploited optimally. Usually, this is the first tree node that belongs to "this" period.

4.3 Small CP model

Every few nodes of the Branch & Bound tree, a small CP problem is started which contains only restrictions concerned with changeovers on the first level machines and the necessity of producing changeover products at the second stage when a changeover is done at the first stage. The model takes as input the assignment as far as it is already decided by the MIP solver, i.e. for every binary y^i_{kt}-variable set to 1, a *task* of size 1 is created. It is checked (using full enumeration) whether it is principally possible to sequence these tasks. Consider the following example: There are 10 setup variables $\{1, .., 10\}$ on machine one and 10 $\{11, ..., 20\}$ on machine two, and for every product $n \in \{1, .., 8\}$ only $n + 2$ is a possible successor, and for every changeover $n \to n + 2$ on machine one we have to produce $n + 10$ on machine two. There are three possible outcomes of the model:

1. The assignment currently active in the MIP model is feasible. (In our example, we might have 2 – 4 – 6 – 12 – 14.)
2. The assignment currently active in the MIP model is not feasible, but can be made feasible by producing at least n additional products out of set $\{i_m\}$ on certain machines $\{k_l\}$. (In our example, we might have 2 – 6 – 12. Then $n = 2$ and $\{i_m\} = \{4, 14\}$.) Then a local constraint

$$\{i_m\} \subset \mathcal{I},\ \{k_l\} \subset \mathcal{K},\ t \in \{1, ..., NT + 1\} : \\ \sum_{i \in \{i_m\}} \sum_{k \in \{k_l\}} y^i_{kt} \geq n \tag{8}$$

   is added to the MIP model formulation. This contraint has to be removed on backtracking, since it is not a globally valid cut.
3. The assignment currently active in the MIP model is not feasible, and cannot be made feasible. (In our example, we might have 2 – 5.) Then we cut off the assignment:

$$k \in \mathcal{K},\ t \in \{1, ..., NT + 1\} : \\ \sum_{\substack{i \in \mathcal{I} \\ \mathrm{sol}\left(y^i_{kt}\right)=1}} y^i_{kt} \leq \sum_{i \in \mathcal{I}} \mathrm{sol}\left(y^i_{kt}\right) - 1 \tag{9}$$

   The runtime of the small model is just fractions of seconds.

4.4 Full CP model

Unfortunately, there is no uniform way to describe CP problems as it is in the MIP world. We give a verbal description of the objects involved, and of the principles of the constraints. The CP model was built using the CHIP C++ system [6], [www.cosytec.fr]. The constraints used are *primitive* constraints and relations, and the following global constraints: `Cumulative_trapeze`, `Diffn`, `Cycle`, `Element`.

4.4.1 Objects

Tasks

A task j is the production of a specified quantity of a certain product. It has two variables $start_j$ and end_j for start and end time (in ten-minute-intervals). The duration of the task $dur_j = end_j - start_j$ is not yet known, since it is computed by dividing the (MIP-fixed) quantity by the output rate of the (CP-chosen) machine.

As soon as a partial solution is found in the MIP model (a solution where all binary variables for a given period are feasible), the CP model is called to do the full sequencing for that period. The CP model may also partly rearrange the assignment. For preparing the input data of the CP model, the following steps are performed:

1. Copy the last task of the previous period into the task list for the CP model, this will be the first task in the actual period. This task may not be included in the inventory constraints, because it was already considered in the previous period. There might also be some task which the user of the system has fixed already. These tasks have to be scheduled without any modification.
2. For all production activities x^i_{kt} on a machine in the actual time slice, a task for that product of the respective size is created. To allow for some flexibility, not only the machine selected by the MIP solver, but also some other *similar* machines are allowed for this task. (The similarity relation on the machines is provided by the user. Different aspects, e.g. throughput of the machines, preferred products or connectivity can be considered here.)
   The latest ending times of the tasks (upper bound on end_j) are derived from the inventory profile of the corresponding product in such a way that the stock level never falls below zero or the minimal stock defined.
   The earliest starting time of the task (lower bound on $start_j$) is always set to equal the beginning of the given time slice.

About 200 production tasks have to be scheduled each time. Each production task has a cleaning task directly following it. The length of the cleaning task depends on the product sequence.

Machines

Machines are unitary renewable resources used to perform tasks. Each is characterized by a sequence-dependent cleaning time $MCL_{koo'}$. Every machine is assigned a pointer $first_k$ that points to the first task on this machine. The first task in a given period is the same as the last task of the previous period.

Breaks

Break objects are very similar to task objects. They are, however, irrelevant for the consideration of the sequence-dependent changeover times.

4.4.2 Constraints

4.4.2.1 Task duration. The duration $end_j - start_j$ of a (not-prescheduled) task is linked to the machine choice variable $mach_j$. The different possible durations are computed from the size of the task and the output rates of the machines suitable for that task.

4.4.2.2 Machine nonoverlap. Each machine performs at most one job at any time, also taking into account cleaning times. Although the capacity limits of the machines are considered in the MIP model as well, it may happen that the changeover restrictions lead to additional idle times on particular machines. If by this the load on machines $\{k_l\}$ becomes too high, the following constraint is added to the MIP model formulation, stating that production on this group must be at least 10 tons less than it actually is. The idea of using machine groups instead of individual machines here as well as in other constraints is to leave more flexibility in the MIP model. The infeasibility is detected at one particular machine. It results, however, from the interaction of several constraints and machines. Therefore it would "mislead" the MIP model if a constraint was stated just for this particular machine. In $\{k_l\}$ we group together "similar" (as defined by the user) machines.

$$\{k_l\} \subset \mathcal{K},\ t \in \{1, ..., NT+1\}:$$

$$\sum_{i \in \mathcal{I}} \sum_{k \in \{k_l\}} x^i_{kt} \leq \sum_{i \in \mathcal{I}} \sum_{k \in \{k_l\}} \operatorname{sol}\left(x^i_{kt}\right) - 10 \tag{10}$$

It may also happen that a particular set of products may not be sequenced on a particular group of machines respecting all the complicated changeover constraints, without the small CP model having noticed it. This is due to the fact that the size of tasks is not considered there. Then, very similar to (9), we add a cut stating that at least one product more out of $\{i_m\}$ has to be scheduled on machines $\{k_l\}$, or one less.

$$\{i_m\} \subset \mathcal{I},\ \{k_l\} \subset \mathcal{K},\ t \in \{1, ..., NT+1\}:$$

$$\sum_{i \in \mathcal{I}} \sum_{\substack{k \in \{k_l\} \\ \operatorname{sol}\left(y^i_{kt}\right)=1}} y^i_{kt} - \sum_{i \in \{i_m\}} \sum_{\substack{k \in \{k_l\} \\ \operatorname{sol}\left(y^i_{kt}\right)=0}} y^i_{kt} \leq \sum_{i \in \mathcal{I}} \sum_{k \in \{k_l\}} \operatorname{sol}\left(y^i_{kt}\right) - 1 \tag{11}$$

4.4.2.3 Flow balance. The inventory control (6) of the MIP model is weak in a sense, since it looks only at the time slice boundaries. In the CP model, the material balance of end products is considered implicitly by the determination of ending times of tasks, but the flow balance of raw materials and intermediates through the network must be maintained. The flow balance constraint can never fail because of a lack of product, since the quantity needed in total must have been made available by the material balance constraint (1) of the MIP model, and feasibility can always be obtained by postponing the consuming task. If the producer-consumer constraint fails because of excess inventory of product i^*, a constraint is added to the MIP model formulation, which states that one must produce less of i^*, or consume more:

$$i^* \in \mathcal{I},\ t \in \{1, ..., NT+1\} :$$

$$\sum_{k\in\mathcal{K}} x_{kt}^{i^*} - \sum_{k\in\mathcal{K}} \sum_{\substack{i\in\mathcal{I} \\ IR_{ki^*i}>0}} IR_{ki^*i} \cdot x_{kt}^{i} \le \sum_{k\in\mathcal{K}} \mathrm{sol}\left(x_{kt}^{i^*}\right) - \sum_{k\in\mathcal{K}} \sum_{\substack{i\in\mathcal{I} \\ IR_{ki^*i}>0}} IR_{ki^*i} \cdot \mathrm{sol}\left(x_{kt}^{i}\right) - 10 \tag{12}$$

Only for those products and periods, for which the replenishment time exceeds the beginning of the actual period, the flow balance constraints are needed. This makes the CP models for later periods significantly faster than those for the earlier periods, because the inventory constraints are the most time-consuming part of the CP model. Note that it is not sufficient to employ the widely-used "rectangular block" scheme to model the production/consumption of a task, where the input is consumed at once at the beginning, and the output produced at once at the end of a task, because of the long runtimes of campaigns.

4.4.2.4 Storage nonoverlap. There may be at most one product in a given tank at any time. If the corresponding constraints fail, one can also add a cut to the MIP model.

4.4.2.5 Objective function. The overall objective function is considered only implicitly in the heuristic of the enumeration strategy. Within the CP model we use the different objective of minimizing makespan. This is because CP solvers "do not like" objective functions consisting of large sums - they cannot deduce anything, and propagation is poor. Using makespan is more efficient, and it is reasonable, since we are looking at just one period, and try to schedule just the tasks the MIP solver has proposed for this period. By minimizing the makespan we simultaneously minimize the total changeover time, and this in turn generally reduces (not necessarily minimizes) changeover costs. (In (3) we argued that changeover costs can be seen as the sum of two terms, one of which is proportional to changeover time.)

4.4.3 Enumeration strategy and solving

The enumeration strategy is not only responsible for finding a solution, but also the changeover product constraint is modeled procedurally, i.e., the enumeration is done in such a way that the constraint is always fulfilled. First, the machines are assigned to the tasks of the first production step, and at the same time these tasks are sequenced. Then the same procedure is carried out for the second production step.

The basic idea is to try to select the most urgent task and schedule it as early as possible, but this has to be done in such a way that there remains a feasible sequence for the tasks not yet scheduled, and, for every changeover at the first stage,

a corresponding task consuming the changeover product at the second stage has to be scheduled at the right time. Experience shows that propagation of this CP model is rather weak in some cases, and complete depth-first search is not able to cope with "fatal mistakes" that have been made high up in the tree. So we cut off some branches using the following heuristic. If the enumeration is not able to overcome a certain level of the tree, we move backwards up the tree, cutting off all branches until a decision is reached the changing of which makes the overcoming of that level possible.
If a schedule cannot be found for a set of machines, a cut to the MIP model formulation is added which should redistribute the current load between machine groups $\{k_l\}$ and $\{k_m\}$: Either produce less on $\{k_l\}$,or more on $\{k_m\}$. Again we use the "similarity" relation to form the two groups:

$$\{k_l\} \subset \mathcal{K},\ \{k_m\} \subset \mathcal{K},\ t \in \{1, ..., NT+1\}:$$

$$\sum_{i \in \mathcal{I}} \sum_{k \in \{k_l\}} x^i_{kt} - \sum_{i \in \mathcal{I}} \sum_{k \in \{k_m\}} x^i_{kt} \leq \sum_{i \in \mathcal{I}} \sum_{k \in \{k_l\}} \mathrm{sol}\left(x^i_{kt}\right) - \sum_{i \in \mathcal{I}} \sum_{k \in \{k_m\}} \mathrm{sol}\left(x^i_{kt}\right) - 1 \tag{13}$$

4.4.4 Application of the CP model result in the MIP model

If the CP model turns out to be infeasible, the constraint (cut) described is added to the matrix to cut off this and similar solutions. Some constraints are not valid model cuts in the sense that they cut off infeasible solutions only; instead they might also cut off feasible solutions. This applies to constraints (10) and (12). The reason for doing so is the fact that the CP returns just an information of infeasibility, without any indication how big the infeasibility actually is. If we restricted ourselves to using strictly this information, this would mean we can exclude just a single point of the solution space, since a solution with the binary variables identical and the continuous variables slightly perturbed might be feasible. Practically, this would result in excessive solution times. Therefore, and since an exhaustive search of the tree is not possible anyway, it is better to cut "deeper" and try a completely different combination than to stay in the same part of the tree forever.

If the CP model is feasible, the communication variable rx_{kt_0+1} is set to the correct value (which is the excess time needed for scheduling all the products proposed on machine k in the MIP model time slice t_0, we subtract this from the duration of the succeeding period $t_0 + 1$), and the production variables $x^i_{kt_0}$ are fixed to their present values. The latter is necessary to maintain the synchronisation of the MIP model and the CP model for the periods computed already. A change of the (continuous) production quantities later on might make the schedule found for the earlier periods invalid, even if the binary variables are not changed.

When the last period has been scheduled successfully, the schedules found for the different periods are "glued together" by copying the partial solutions into one table.

5 Implementation and results

5.1 Implementation

The application has been programmed in C++, using Dash's XPRESS-MP library functions for the MIP model part and the CHIP C++ library for the CP model part. With a typical data set, we get a MIP problem with about 25000 rows and 30000 variables, 3500 of which are binary and 600 semi-continuous. (For information on the size of data sets, refer to the section "Data and Notation".) The first full CP model has several hundred variables and up to about 100 global constraints. The CP models for later periods are significantly smaller, since the inventory constraints for raw materials are not needed anymore.

We tested about 20 data sets which were obtained directly from the production plant.

The quality of the solutions is hard to estimate. The integrality gap reported by the MIP solver is several hundred percent. However, since the MIP model does not contain all constraints, the lower bound provided by the relaxation of the MIP must be rather weak.

Concerning solution times, we did not observe significant variations. The solution of the linear relaxation of the MIP takes 4 to 8 seconds, the branch and bound procedure solves about 100 nodes per minute. The first full CP model is called after about 100 nodes. The set-up of this model needs 1 to 3 seconds, the solution 0 to 10 seconds, depending on the period. The first full solution will be available after about five minutes. Sometimes, a second full solution is found some minutes later, which is not significantly better than the first. (All solution times refer to a Pentium III 500 MHz PC with 384 MB memory.) Letting the algorithm run longer did not lead to significant further improvements, however, this was not tested thoroughly.

These results gave rise to the following solution heuristic. The algorithm is run until either a second solution is found, or the time limit of ten minutes is exhausted (and at least one solution could be found, which was the case for all data sets).

5.2 Results

The algorithm was installed as part of a production planning and scheduling system on the users's PC. The main user interface is an interactive gantt chart, where tasks are represented as horizontal bars. The typical workflow of the production planner is like as follows. When she starts the system, she views the plan that was valid at the end of the previous session, together with the new product demands downloaded from the SAP R/3 system, which are not covered by this plan. She may make first arrangements herself and/or fix any part of the plan, which means the algorithm is not allowed to change it. Usually the coming week will be fixed, because production management does not like sudden changes in the near future. Afterwards, the algorithm described in this paper is started, which delivers a plan in at most 10 minutes. Although, from a mathematical point of view, this plan is feasible in terms of the constraints and reasonably good in terms of the planning goals, it will usually not be directly implementable from a practical point of view.

This is due to the fact that in an environment so close to production there are some additional constraints and information every day, which are not general enough to include them into the planning system. Therefore, the aim of installing such a system should not be to make the human scheduler superfluous, because this can usually not be achieved; instead, the system should support her to make better plans in less time. So the planner will make some adjustment, fix further parts, start the algorithm again, and so on, until she is satisfied with the plan, at which time it will be transfered back into the SAP R/3 system. The necessity to call the planning system several times more or less interactively is the reason why the solution time is limited to 10 minutes.

The gain of the company consists mainly in increased reactivity. Without the system, manual planning took several days and was therefore performed once a month only. With the planning system, a plan can be generated in less than one hour. This enables the company to react to sudden changes, like machine breakdowns or urgent orders more efficiently, and to increase customer satisfaction.

6 Conclusion and outlook

We presented an industrial application of a combined MIP/CP algorithm, which is able to find a good feasible solution to a problem otherwise intractable in reasonable time. The key to efficiency of the system was the careful assignment of problem aspects to the two solvers involved, and the design of solver communication. Nevertheless, we would like to see improvements in the following directions:

1. Try to gather better information on the quality of the resulting schedule by providing lower bounds.
2. Try to use more information of the relaxation in the decisions of the CP algorithm.

Hopefully, our approach will provide some motivation for further research on hybrid MIP/CP algorithms. Both the aspects of interaction of models and interaction of solvers deserve attention.

References

1. Aggoun A, Beldiceanu N (1993) Extending CHIP in order to solve complex scheduling problems. Journal of Mathematical and Computer Modelling 17(7): 57–73
2. Belvaux G, Wolsey LA (1999) Lot-sizing problems: modelling issues and a specialised branch-and-cut system bc-prod. CORE Discussion Paper 9849 09/1998. Université Catholique de Louvain, Louvain-la-Neuve, Belgium
3. Berning G, Brandenburg M, Gürsoy K, Mehta V, Tölle FJ (2002) An integrated system solution for supply chain optimization in the chemical process industry. OR Spectrum 24: 371–401
4. Bockmayr A, Kasper T (1998) Branch-and-infer: A unifying framework for integer and finite domain constraint programming. Informs Journal of Computing 10(3): 287–300

5. Bockmayr A, Pisaruk N (2001) Solving assembly line balancing problems by combining IP and CP. Proceedings of the 6th Annual Workshop of the ERCIM Group on Constraints, Prague (http://arXiv.org/abs/cs.DM/0106002)
6. Dincbas M, Van Hentenryck P, Simonis H, Aggoun A, Graf T, Berthier F (1998) The constraint logic programming language CHIP. In: Proceedings of International Conference on Fifth Generation Computer Systems (FGCS-88), Tokyo, pp 693–702
7. Grunow M, Günther HO, Lehmann M (2002) Campaign planning for multi-stage batch processes in the chemical industry. OR Spectrum 24: 281–314
8. Heipcke S (1999) Combined modeling and problem solving in mathematical programming and constraint programming. Doctoral Thesis, University of Buckingham
9. Hooker JN, Kim HJ, Ottosson G (1998) A declarative modeling framework that integrates solution methods. Annals of Operations Research 104: 141–161
10. Hooker JN, Ottosson G, Thorsteinsson ES, Kim HJ (1999) On integrating constraint propagation and linear programming for combinatorial optimization. Proceedings of the Sixteenth National Conference on Artificial Intelligence (AAAI-99), The AAAI Press/The MIP Press, pp 136–141
11. Jain V, Grossmann IE (2001) Algorithms for hybrid MILP/CP models for a class of optimization problems. INFORMS Journal of Computing 13: 258–276
12. Kallrath J (2002) Planning and scheduling in the process industry. OR Spectrum 24: 219–250
13. Marriot K, Stuckey PJ (1998) Programming with constraints. MIT Press, New York
14. Nemhauser GI, Wolsey LA (1988) Integer and combinatorial optimization. Wiley, New York
15. Neumann K, Schwindt C, Trautmann N (2002) Advanced production scheduling for batch plants in process industries. OR Spectrum 24: 251–279

Makespan estimation and order acceptance in batch process industries when processing times are uncertain

Cristina V. Ivanescu, Jan C. Fransoo, and J. Will M. Bertrand

Department of Technology Management, Technische Universiteit Eindhoven,
P.O. Box 513, 5600 MB Eindhoven, The Netherlands
(e-mail: {c.v.ivanescu,j.c.fransoo,j.w.m.bertrand}@tm.tue.nl)

Abstract. Batch process industries are characterized by complex precedence relationships between operations, which renders the estimation of an acceptable workload very difficult. A detailed schedule based model can be used for this purpose, but for large problems this may require a prohibitive large amount of computation time. We propose a regression based model to estimate the makespan of a set of jobs. We extend earlier work based on deterministic processing times by considering Erlang-distributed processing times in our model. This regression-based model is used to support customer order acceptance. Three order acceptance policies are compared by means of simulation experiments: a scheduling policy, a workload policy and a regression policy. The results indicate that the performance of the regression policy can compete with the performance of the scheduling policy in situations with high variety in the job mix and high uncertainty in the processing times.

Key words: Batch process industries – Regression analysis – Order acceptance

1 Introduction

The batch process industries covers areas such as food, specialty chemicals, pharmaceuticals and brewing. The industry is characterized by the production of a diverse range of final products starting from a few basic raw materials, often in liquid or powder form, which are processed according to production recipes through a number of processing steps in batches of varying size to produce the final product. We can distinguish two basic types of process equipment: multiproduct which is used to produce different variants of a basic product type following the same routing (cf. a flow shop), and multipurpose which allows diverse processing tasks to be carried out by a particular equipment unit (cf. a job shop) [15]. In this paper, we focus on batch process industries using multipurpose equipment.

Correspondence to: C.V. Ivanescu

Batch process industries produce a variety of different products that follow different routings through a production department. Differences exist between products with regard to the number and duration of the required processing steps. Intermediate products may be unstable, which means that a product needs to be processed further without delay. These no-wait restrictions and the large variety of products with different routings cause complex scheduling problems. For an extensive description of the planning situation and planning problems we refer to [10].

This paper builds on earlier research of Raaymakers et al. [13, 12, 11]. The models developed in that research estimate the makespan (i.e., the completion time of the last job) of a job set containing jobs with deterministic processing times of the processing steps using a number of aggregate job set and resource configuration characteristics. These estimation models are linear regression models, and are compared to estimation models based on the work content of the jobs, and exact makespan prediction models where a detailed schedule is constructed. Let us define the *ex ante (ex post)* makespan as the makespan obtained when a schedule based on expected (actual) processing times is constructed. In a deterministic situation, the ex ante makespan of a constructed detailed schedule is identical to the ex post makespan, and is therefore the best possible estimate. Although the performance of regression models is surprisingly good in a deterministic situation [11], there still is a considerable difference in performance between the regression-based estimation models and the fully detailed schedule model. In most real life production situations, however, processing times are uncertain and an aggregate statistical estimation model might perform equally well or even better than a detailed deterministic scheduling model to estimate the ex post makespan. Indications of the quality of aggregate estimations in complex situations have been provided by Ten Kate [16].

In this paper, we will examine the performance of aggregate makespan estimation models, such as workload-based models and regression-based models, and compare them to using detailed schedules to estimate the makespan in situations when processing times are uncertain. The probability distribution functions of the processing times are based on observations of data from a real-life batch process plant. Generating a wide variety of scheduling problems in different settings, using computer simulation, we construct estimation models by means of linear regression. Then, under another set of problem settings, we conduct extensive simulation experiments to assess the performance of the aggregate estimation models and compare them to the predictive quality of the detailed schedule models when used to support customer order acceptance decisions. The performance is measured by the service level, the job set lateness and the acceptance rate. We conclude that, under relatively high levels of uncertainty in the processing times in combination with high levels of variability in job structure characteristics, the regression policy performs equally well as the scheduling policy. However, for situations with low and moderate uncertainty, the detailed schedule based model performs better than the regression based model, at the expense of a longer computational time.

Apart from the work by Raaymakers et al. cited above, the available literature on makespan estimation is limited, although some attention has been paid to the related problem of flowtime estimation. Flowtime estimation is generally used for determining achievable due dates for customer orders. Upon arrival of a customer order, the flowtime is estimated for the job related to that customer. For an overview of flowtime estimation research we refer to Cheng and Gupta [3].

The remainder of this paper is organized as follows. Section 2 presents the basic assumptions underlying the model. Section 3 describes how the makespan estimation models are constructed. Section 4 is devoted to the description of the order acceptance policies investigated in this paper. Section 5 presents the simulation experiments we conducted and the results of the simulation study. The final section contains the conclusions and directions for further research.

2 Description of the problem

We consider a batch chemical production department consisting of N resources of M different resource types. Resources of the same type are identical. The production department performs production tasks which are jobs (j) each consisting of a specific number of processing steps (s_j). These processing steps may have an overlap in time. The overlap is due to the fact that two resources are needed simultaneously. The product being processed is generally a fluid that needs containerization in order to be stored. If there are two consecutive processing steps, these could be decoupled by storing the product in a silo or bin, but this is often not possible due to the lack of stability of the product. As a consequence, the product is transferred from one resource to the next (e.g., from a reactor to a distillation unit). The product will then occupy both units. This overlap of processing steps is modeled as the time delay (δ_{ij}) between the start time of the processing step i ($i = 1, 2, ..., s_j$) relative to the start time of the job j.

We study the problem in a hierarchical production control structure consisting of two levels:

1. a planner who accepts or rejects orders that are requested by the market for delivery at the end of a specified planning period, and
2. a scheduler who allocates the processing steps of a job (the accepted order) to specific resources and determines the exact sequence and timing of the (planned) execution of the processing steps on the resources.

This type of hierarchical structure is quite common, both in industrial and theoretical settings (see, e.g. [1,4,10]). In this paper we consider a single planning period setting, where the period starts with an empty system and at the end of the period the system must be empty again.

Orders arrive at the planner randomly. The planner can decide whether or not to accept a certain order. All orders have the same due date and have to be completed within one planning period. Each order consists of exactly one job with a specified structure of no-wait processing steps. At the start of each planning period, the planner releases a job set (J) consisting of all accepted orders allocated to that period to be scheduled. The following assumptions regarding the job set and resources are made:

- All jobs in the job set are known at the start of the planning period, i.e. the processing steps, timing and sequencing constraints on the processing steps, the probability distribution of the processing times; only integer values of the processing times are considered.
- Resources are available from the start of the planning period and without interruptions.
- No precedence relations exist between jobs.
- Each processing step has to be performed without preemption on exactly one resource of a specific resource type.
- More than one processing step of a job may require a resource of the same resource type. These processing steps have to be performed on different resources of that type if they overlap.

Processing times are uncertain. We define the *effective processing time* (p_{ij}) of the i-th processing step of a particular job j as the elapsed time between the start and the completion of the processing step. This elapsed time consists of the actual processing time (which may include time for set-up, inspection, rework), and any interruptions (such as resource or operator unavailability, machine failure, and preventive maintenance). Actual production data obtained from a batch chemical processing department in industry showed that the distribution of processing times is close to being Erlang-distributed. The effective processing times (p_{ij}) in the problem considered in this paper are therefore determined by drawing a random variable from an Erlang distribution with mean $E[p_{ij}]$ and Erlang shape parameter k. Thus, processing time uncertainty decreases as k increases. Note that the Erlang shape parameter k is the same for each processing step i of every job j.

The problem to be solved at the second level of the hierarchical production control structure is a stochastic scheduling problem. The stochastic scheduling literature is substantial and usually addresses special classes of stochastic scheduling problems with one or two machines (or multiple parallel machines)[9]. To the best of our knowledge, optimal policies for multiprocessor stochastic job-shop problems with no-wait restrictions have not been developed. Therefore, similar to the work of [5], in this paper, we chose the following approach to solve the problem. The scheduler simplifies the stochastic scheduling problem to a deterministic problem and constructs a schedule (S) using the *expected processing times* of the processing steps $(E[p_{ij}])$. For constructing the deterministic schedule, a simulated annealing algorithm is used which was developed for no-wait job shops [14]. Let $C_{\max}^{\text{ex ante}}(S)$ denote the makespan of the schedule S under deterministic processing times. Then, the resulting schedule is applied to the original stochastic problem.

The scheduler has complete knowledge about the effective processing time of a particular processing step only after that processing step has been completed. Non-feasibility problems may occur if the effective processing times differ from the expected processing times and therefore, rescheduling may be needed. One of the difficulties in no-wait job-shop scheduling is that changes in the job sequence on one resource are likely to affect several other resources. This is caused by the no-wait restrictions between processing steps of a job. If the effective processing times are not known at the moment the schedule is determined, it is impossible to satisfy all no-wait restrictions throughout the entire planning period, because the effective processing times gradually become known. We define the *resource-feasible schedule* as the schedule for which, at the same time, at most one processing step is scheduled on a resource and a minimum possible number of no-wait restrictions are violated, given the status of information at the time of (re)scheduling. When rescheduling, it is often desirable for the new schedule to be as similar to the original schedule as possible. Moreover, the rescheduling process should not be too time consuming. In this paper, the rescheduling procedure used is a "right-shift" control policy [7]. This simple control policy entails a "right-shifting" of the schedule in order to restore the resource-feasibility while always maintaining the original sequence, and can be used for locally revising the schedule in real time. A description of the rescheduling procedure and an example is included in Appendix A. Let $C_{\max}(S)$ denote the ex post makespan of S after rescheduling.

In this hierarchical planning and control structure, the planner has to be able to estimate whether the resulting job set for a specific period will be achievable when allocating jobs to planning periods taking into consideration the uncertainty of the processing times. A job set is achievable in a planning period if the estimated makespan of the job set is less than the length of that period. The planner thus needs to be able to estimate the ex post makespan of the job set, taking into account that the processing times are uncertain.

In the next section, we will develop an aggregate model for this order acceptance and makespan estimation problem, and, in Section 5, compare this model to a number of alternatives.

3 Building the makespan estimation model

The makespan of a job set is influenced by its workload. More specifically, the workload on the bottleneck resource puts a lower bound to the makespan. A lower bound on the makespan is obtained by dividing the workload on the bottleneck resource by the number of resources of that type. We may round this up, because all processing times are integer values and no pre-emption is allowed. Because of job interactions (timing and no-wait sequencing constraints), the minimal makespan for which a feasible schedule is realized will often exceed this lower bound. Different job sets may have different levels of job interaction. The job interaction is measured by the interaction margin, which is defined as the relative difference between the ex post makespan ($C_{\max}(S)$), obtained by simulated annealing [14] and the rescheduling procedure described in Section 2, and a single resource lower bound on the makespan (LB), obtained according to Carlier [2]. The interaction margin (I) is defined as follows:

$$I = \frac{C_{\max}(S) - LB}{LB} \tag{1}$$

Raaymakers et al. [12] have identified a number of aggregate resource and job set characteristics that influence the amount of job interaction for a batch processing situation in which the processing times are deterministic. Using linear regression, they reach a fairly good level of estimation quality of the actual makespan. In this section, we will construct and evaluate models for estimating the makespan when processing times are uncertain.

3.1 Constructing the interaction estimation models

Following the earlier work of Raaymakers et al. [12], we will use a linear regression model to estimate the interaction margin, I. In order to determine the variables that need to be included in the regression model, and the coefficients of these regressors, we need to construct a set of problems to use as a starting point. To construct this problem set, we first determine the core of the problem set by generating a large number of job sets. A job set and resource configuration combination is referred to as a problem instance. For reasons of self-containedness, and in order to allow us to explain how we incorporated the stochasticity into the problem generation process, we will briefly describe the method of generating the problem instances. However, for full details we refer to [12].

The notation for the experimental factors and variables used in this paper is defined as:

n_{jobs}	=	the number of jobs in the job set;
$F_{E[p]}$	=	the distribution function of the expected processing time;
μ_s	=	the average number of processing steps;
μ_g	=	the average overlap;
$cv^2_{E[p]}$	=	the squared coefficient of variation of the expected processing times;
$\rho_{\max}$	=	the workload balance;
cv^2_p	=	the squared coefficient of variation of the effective processing time.

Detailed definitions of these variables are given in Appendix B.

For each problem instance, a job set is generated randomly and has to be completed on 10 resources consisting of 5 resource types, with 2 identical resources of each type. The factors included in the generation process are the number of jobs in the job set, n_{jobs}, the number of processing steps per job, $s_j(j = 1, 2, ...n_{jobs})$, the average overlap, μ_g, and the distribution function of the expected processing time, $F_{E[p]}$. All the experimental factors are varied on three levels. In Table 1 the values of these factors are presented.

A full factorial design is used to generate the core of the problem set. In total, 81 combinations of different factors levels are possible. However, if the number of processing steps per job is equal to 1, overlap between processing steps cannot occur. Hence, these combinations are removed from the experimental design. For each of the remaining 63 combinations of the different levels of experimental factor values, five replications have been made. This results in a total of 315 problem instances, that form the core of the problem set. For each problem instance, a schedule S_u ($u = 1, 2, ..., 315$) is constructed by using a simulated annealing algorithm for no-wait job shops [14] and the corresponding makespan $C^{\text{ex ante}}_{\max}(S_u)$ is computed.

The core of the problem set employed in this paper is similar to the problem set used by Raaymakers et al. [12]. This enables us to single out the effect of stochastic processing times on the performance of the regression-based approach. In the next step in our simulations we modelled different levels of uncertainty in processing times by varying the shape parameter of the Erlang distribution on six levels: 2, 3, 4, 5, 6 and 10. The industrial data we studied showed Erlang shape parameters between 2 and 6. In order to be able to assess the influence of this parameter statistically, we considered one more value than we encountered in the industrial data. For each of the 315 problems and each of the uncertainty levels, 10 replications have been made. This resulted in a problem set with a total of 18, 900 problem instances. Let $C_{\max,k,v}(S_u)$ denote the ex post makespan of S_u after disruptions, for each Erlang parameter shape k and each replication $v = 1, 2, ..., 10$.

The interaction variables used in the model construction step are the average number of processing steps, μ_s, the average overlap, μ_g, the squared coefficient of variation of the expected processing times, $cv^2_{E[p]}$, the workload balance, $\rho_{\max}$, the number of jobs in the job set, n_{jobs} and the squared coefficient of variation of the effective processing time, cv^2_p. The first four variables are identical to the ones used in the deterministic estimation models [12]. The fifth variable has been added during the construction process since it was found to be significant. The sixth variable is computed in the following way:

$$cv^2_p = \frac{(b-a)^2}{3\cdot(b+a)^2} + \frac{4\cdot(b^2+a\cdot b+a^2)}{3\cdot k\cdot(b+a)^2} \tag{2}$$

where the expected processing times are uniformly distributed on $[a, b]$ and k is the shape parameter of the Erlang distribution. The derivation of cv^2_p is included in Appendix C, along with the density function and the first two moments of the effective processing times.

We performed an analysis of variance in order to determine the contribution of the interaction variables to the explanation of the variability in the interaction margin. The results showed that both the main effects and the two-way interactions between the six variables have a significant contribution to explaining the variation in the interaction margin. Therefore, both main effects and two-way interactions of these variables have been used to build estimation models that estimate the interaction margin for specific job sets. The estimation models were generated by means of multiple linear regression techniques. Before performing the regression analysis the data was checked on orthogonality of the regressor variables. These

Table 1. Factor levels for the experiments

Factor	-1	0	1
n_{jobs}	25	50	75
s_j	1	5	10
μ_g	0	0.5	1
$F_{E[p]}$	constant expected processing times, all equal to 25 time units	$U(15, 35)$	$U(1, 49)$

results showed that there is a considerable correlation between some of the variables, indicating multicollinearity risks.

First, an estimation model (A) was constructed that only contains the main effects of the interaction variables. Backward regression was used to eliminate variables that did not show a significant contribution. Second, estimation models were constructed that included two-way interactions. The number of possible two-way interactions is large (15) and hence, the number of regressor variables is 21. Stepwise regression was used to determine the estimation models. Not all the regression models found by stepwise regression were suitable because of the multicollinearity problems. We excluded from further analysis the models for which the multicollinearity appeared as a problem.

The estimation models are presented in Table 2. Given are the name of the model, the number of regressor variables, the regressor variables, the adjusted R^2, and the square root of the residual mean square (RMS). The latter two statistics are measures of fit of the regression model.

Table 2. Estimation models

	No	Regressor variables	adj. R^2	RMS
A	6	$\mu_s, \mu_g, cv^2_{E[p]}, cv^2_p, \rho_{\max}, n_{jobs}$	0.74	0.25
B	1	μ_s	0.51	0.35
C	2	μ_s, μ_g	0.61	0.30
D	3	$\mu_s, \mu_g, cv^2_p \cdot \rho_{\max}$	0.70	0.27
E	4	$\mu_s, \mu_g, cv^2_p \cdot \rho_{\max}, cv^2_{E[p]}$	0.74	0.25
F	5	$\mu_s, \mu_g, cv^2_p \cdot \rho_{\max}, cv^2_{E[p]}, cv^2_p \cdot n_{jobs}$	0.75	0.25

Apart from models B and C, the performance measures looked quite good, so we only eliminated from further analysis the models B and C. Table 3 shows the regression parameters' least squares estimates.

Table 3. Unstandardized coefficients for the regression models

Regressors	Model			
	A	D	E	F
$Intercept$	−0.503	0.424	0.380	0.388
μ_s	0.066	0.086	0.082	0.081
μ_g	−0.402	−0.404	−0.406	−0.405
$cv^2_{E[p]}$	−1.003		−1.140	−1.134
cv^2_p	1.088			
$\rho_{\max}$	1.225			
n_{jobs}	−0.002			
$cv^2_p \cdot \rho_{\max}$		0.698	1.274	1.490
$cv^2_p \cdot n_j$				−0.003

3.2 Interaction estimation model validation

Model validation is directed towards determining if the model will function successfully in its intended operating environment. The most effective method of validating a regression model with respect to its predictive performance is to collect fresh data and directly compare the model predictions against it [8]. Therefore, the validation of the estimation models was based on their estimation quality for a number of newly generated problem instances under a variety of parameter settings. The model validation consisted of six test sets. Each test set consisted of 50 job sets that were generated respectively with:

1. similar characteristics as the original data set
2. variations in the number of processing steps and in the overlap
3. variations in the number of jobs in the job set
4. variations in the average processing time
5. an unbalanced workload
6. a combination of the variations in the previous test sets

To determine the interaction margin for the job sets in the test sets we proceeded in the same way as was done for the original data set. For each of the 50 job sets and each of the uncertainty levels (the shape parameter was varied now on the following five levels: 2, 4, 5, 7, and 10), 5 replications have been made. This resulted in a total of 1250 problem instances per test set.

The quality of the estimation models was evaluated using the mean estimation error (ME) and the standard deviation of the estimation error (SDE). Additionally, the percentage of variability in the new data explained by the model (R^2_{pred}) is given in order to be compared to the R^2 of the building model. R^2_{pred} is computed as follows:

$$R^2_{pred} = 1 - \frac{\sum_i (I_i - \hat{I}_i)^2}{\sum_i (I_i - \bar{I})^2} \tag{3}$$

where, for a problem instance i, I_i is the actual interaction margin according to the detailed schedule, $\hat{I}_i$ is the interaction margin estimated according to the estimation model and $\bar{I}$ is the average of the actual interaction margins.

The results of the evaluation are presented in Table 4. Comparison of the prediction performance of the regression models on the newly generated data provides us with a basis for final model selection.

Variation in the average processing time, in the number of processing steps and variation in the overlap, variation in the number of jobs and workload balance do not seem to have a large impact on the estimation quality. Therefore we may conclude that the estimation models are highly robust with respect to these variations in the problem instance.

3.3 Makespan estimation models

The estimation models developed and tested in the previous sections are used to estimate the interaction margin of a given job set. This estimated interaction margin

Table 4

(a) Experimental results for the interaction margin, test sets 1–2

	Test set 1			Test set 2		
Model	ME	SDE	R^2_{pred}	ME	SDE	R^2_{pred}
A	0.031	0.147	0.86	0.049	0.130	0.83
D	0.053	0.190	0.75	0.084	0.174	0.67
E	0.049	0.148	0.85	0.078	0.132	0.79
F	0.054	0.147	0.84	0.082	0.131	0.78

(b) Experimental results for the interaction margin, test sets 3–4

	Test set 3			Test set 4		
Model	ME	SDE	R^2_{pred}	ME	SDE	R^2_{pred}
A	0.054	0.156	0.82	0.031	0.135	0.88
D	0.039	0.200	0.73	0.062	0.176	0.78
E	0.040	0.158	0.83	0.048	0.138	0.86
F	0.066	0.163	0.80	0.053	0.137	0.86

(c) Experimental results for the interaction margin, test sets 5–6

	Test set 5			Test set 6		
Model	ME	SDE	R^2_{pred}	ME	SDE	R^2_{pred}
A	−0.002	0.123	0.90	0.022	0.125	0.87
D	−0.118	0.170	0.72	−0.065	0.171	0.73
E	−0.090	0.135	0.83	−0.040	0.138	0.84
F	−0.072	0.134	0.85	−0.025	0.136	0.85

may be used to estimate the makespan of the job set. From a production point of view, the makespan is more interesting than the interaction margin because the estimated makespan indicates whether a job set can be completed within a given planning period.

In the previous section we saw that accurate estimates of the interaction margin can be given based on a number of aggregate job set characteristics. Following the earlier work of Raaymakers et al. [12], we may therefore expect that accurate makespan estimates can be realized using the lower bound on the makespan and the estimated interaction margin. Thus, the estimated makespan is obtained according to the following formula:

$$\hat{C}_{\max}(S) = (1 + \hat{I}) \cdot LB \tag{4}$$

where $\hat{I}$ is the interaction margin estimate according to one of the interaction margin estimation models. The hats are used to indicate the estimate of a variable.

We first checked the quality of the makespan estimation models for the initial data set used at the construction of the interaction margin estimation models. Two characteristics of importance are investigated when the "quality" of a makespan estimate is considered: accuracy and precision. Accuracy refers to how close, on the average, the individual estimates are from their true values. This is the same quantity as the expected value of the estimation errors (ε), which are defined as the difference between the actual (ex post) makespan and the estimated makespan:

$$\varepsilon = C_{\max}(S) - \hat{C}_{\max}(S) \tag{5}$$

Precision refers to the variability of the estimation errors. These two characteristics are quantified by the mean estimation error (ME) and the standard deviation of the estimation error (SDE). Column 2 and 3 from Table 5 give the mean estimation error and the standard deviation of the estimation error for the makespan estimation models determined according to (4).

Table 5. Quality of the makespan estimation models, construction data

	Before correction		After correction	
Model	ME	SDE	ME	SDE
A	−4.00	267.10	0.43	265.33
D	−8.23	288.35	−3.81	280.86
E	−5.77	266.37	−1.34	266.06
F	−16.63	265.53	−12.20	265.47

The results in Table 5 show that the accuracy of the models can be improved such that the estimation errors will achieve a zero mean. Thus, we introduced a correction factor (α). The model can now be written as follows:

$$\hat{C}^{\alpha}_{\max}(S) = \hat{C}_{\max}(S) + \alpha \cdot LB \tag{6}$$

The correction factor is determined, for each Erlang shape parameter k, in the following way:

$$\alpha = \frac{E[\varepsilon]}{\overline{LB}} \tag{7}$$

where $\overline{LB}$ is the average lower bound. Both are computed from the initial data set used at the construction of the interaction margin estimation model.

Column 4 and 5 from Table 5, reveal that by adding the correction factor α not only the mean estimation error is closer to zero, but also the variability in the estimation error is decreased. Therefore, we selected the corrected models for further use.

We further evaluated the quality of the corrected makespan estimation models for the same test sets as for the interaction margin estimation models. Table 6 gives the mean estimation error (ME), the standard deviation of the estimation error (SDE) and the percentage of variability in the new data explained by the model (R^2_{pred}).

A comparison of the estimation quality for different test sets shows that some models have a better estimation quality than others. The models A and F have a similar estimation quality. Since the model A is less complex than F, we selected A to be the model that will be used further on.

Table 6

(a) Experimental results for the makespan estimation models, test sets 1–2

Model	Test set 1			Test set 2		
	ME	SDE	R^2_{pred}	ME	SDE	R^2_{pred}
A	**44.36**	**115.46**	**0.98**	**52.72**	**96.09**	**0.97**
D	59.97	146.94	0.96	76.54	125.99	0.95
E	58.79	118.73	0.98	72.50	101.44	0.96
F	62.92	119.11	0.97	75.81	101.84	0.96

(b) Experimental results for the makespan estimation models, test sets 3–4

Model	Test set 3			Test set 4		
	ME	SDE	R^2_{pred}	ME	SDE	R^2_{pred}
A	**89.28**	**168.62**	**0.99**	**52.24**	**168.29**	**0.99**
D	49.07	188.80	0.99	84.92	208.28	0.98
E	50.04	148.66	0.99	70.57	175.46	0.99
F	105.69	175.72	0.98	75.62	176.89	0.99

(c) Experimental results for the makespan estimation models, test sets 5–6

Model	Test set 5			Test set 6		
	ME	SDE	R^2_{pred}	ME	SDE	R^2_{pred}
A	**13.97**	**118.50**	**0.98**	**49.47**	**124.14**	**0.99**
D	−87.22	158.52	0.95	−48.94	150.07	0.98
E	−60.32	129.29	0.97	−27.10	127.71	0.99
F	−44.73	128.24	0.97	−6.55	125.81	0.99

4 Modelling the order acceptance policies

Order acceptance policies discussed in the literature are based on either using the overall workload or on constructing a detailed schedule (see, e.g. [17, 16]). Raaymakers et al. [11] introduced a third policy, the makespan estimation policy, which is based on a regression model of specific job set and resource configuration characteristics. This policy falls in-between the two other policies with respect to the level of detail of the information used. They obtained that for situations with high demand and job mix variety, in a setting with deterministic processing times, the makespan estimation policy clearly outperformed an acceptance policy based on workload only. Under deterministic circumstances, their scheduling-based acceptance policy performed however considerably better (be it at the expense of very long computation times). Therefore, it makes sense to investigate the performance that can be obtained by using a regression based model for accepting orders if the uncertainty is present.

In this paper three order acceptance policies are compared: a workload-based policy, a regression-based policy and a detailed scheduling-based policy. As mentioned above, the planning and control structure is a hierarchical production control structure consisting of two levels. At the first level, the planner accepts or rejects orders. An order, which consists of exactly one job, is accepted only if, according to the policy used, sufficient capacity is expected to be available to complete the resulting job set. Orders that fail this test are rejected and leave the system. At the second level, a complete schedule of all accepted orders is constructed. The same simulated annealing algorithm for no-wait job shops [14] is used for all three policies.

The policies differ in the way the acceptance of an order is evaluated. Under the workload policy, jobs are accepted as long as the workload per resource does not exceed the available capacity per resource and the total workload for the entire job set (for all resources) does not exceed a specified maximum workload. Under the regression-based policy, that will be called the *regression policy* in the remainder of this paper, jobs are accepted as long as the estimated makespan of the resulting job set does not exceed the period length. Under the detailed scheduling policy a job is accepted if a schedule can be constructed for the resulting job set such that the deterministic makespan plus an amount of slack, which is empirically determined, does not exceed the period length.

In the remainder of this section a detailed description of the order acceptance policies is given. A description of the simulation experiments that we conducted to assess the performance of these acceptance policies is given in the next section.

4.1 Scheduling policy

The scheduling policy is based on constructing a new detailed schedule after every order arrival. An order is accepted if a schedule of the resulting job set can be constructed with the following condition fulfilled:

$$(1+\gamma_k)\cdot C_{\max}^{\textit{ex ante}}(S) \leq T \tag{8}$$

where

γ_k	=	slack factor for the Erlang shape parameter k, $0 \leq \gamma_k < 1$;
T	=	period length.

We cannot expect to find optimal solutions in reasonable time because job shop scheduling problems with no-wait restrictions are NP-hard [6]. Therefore, we have chosen to construct the schedules by using a simulated annealing algorithm for no-wait job shops [14]. Raaymakers and Hoogeveen [14] have investigated the performance of this algorithm and conclude that it performs reasonably well. Note that the schedule algorithm used at the first level (the order acceptance level) is the same as the one used at the second level (production scheduling level) in the hierarchical production control structure described in Section 2.

In a stochastic situation, the ex ante makespan is no longer the best estimate possible of the ex post makespan. Therefore, the ex post makespan is estimated making use of the ex ante makespan and an empirically determined slack in order to cope with the uncertainty in the processing times. The slack factor (γ_k) is determined from the same initial data set used at the construction of the interaction estimation model in the following way:

$$\gamma_k = \frac{1}{315 \times 10} \cdot \sum_{u=1}^{315} \sum_{v=1}^{10} \frac{C_{\max,k,v}(S_u) - C_{\max}^{ex\ ante}(S_u)}{C_{\max}^{ex\ ante}(S_u)} \tag{9}$$

4.2 Workload policy

The rule used to accept and allocate orders under the workload policy is similar to the one used by Raaymakers et al. [11]. Consequently, a job set is considered achievable if the following conditions are met:

$$\sum_{j \in J} \sum_{i=1}^{s_j} \frac{E[p_{ij}]}{(1-\tau_k) \cdot N} \leq T \tag{10}$$

and

$$\forall r : \sum_{j \in J} \sum_{i \in P_r} \frac{E[p_{ij}]}{N_r} \leq T \tag{11}$$

where

J	=	job set to be evaluated;
τ_k	=	slack factor for Erlang shape parameter k, $0 \leq \tau_k < 1$;
N_r	=	total number of resources of type r;
P_r	=	set of processing steps that need to be executed at resource r.

The slack factor (τ_k) determines the maximum total workload allowed and takes into consideration the processing uncertainty. As in the case of scheduling policy, τ_k is computed from the same initial data set used at the construction of the interaction estimation model as follows:

$$\tau_k = 1 - \frac{1}{315 \times 10} \cdot \sum_{u=1}^{315} \sum_{v=1}^{10} \frac{\sum_{j \in J} \sum_{i=1}^{s_j} p_{ij}}{N \cdot C_{\max,k,v}(S_u)} \tag{12}$$

4.3 Regression policy

In Section 3 we developed a model to estimate the interaction margin of a job set by means of multiple linear regression techniques. By using a lower bound on the makespan and the estimated interaction margin we further estimate the makespan of a job set (see equation (6)).

The regression model is used dynamically, i.e., it is used each time an order arrives to investigate the makespan consequences of accepting this order in addition to the orders already accepted. Under the regression policy, orders may be accepted as long as the estimated makespan of the job set $\hat{C}_{\max}^{\alpha}$, computed according to (6), remains smaller than the period length:

$$\hat{C}_{\max}^{\alpha}(S) \leq T \tag{13}$$

5 Simulation experiments

We conducted simulation experiments to compare the performance of the described order acceptance policies when the processing times are uncertain. The performance is measured by the service level (SVL), the job set lateness (JSL), the acceptance rate (ACR) and the feasibility performance. We chose the service level as a performance indicator because it indicates the reliability of the due dates agreed with the customers. The service level is defined as the percentage of accepted orders that are completed before their due dates.

We chose the lateness of the job set as a performance indicator because this performance measure indicates the difference between the estimation and the realized makespan. The lateness of the job set is defined as follows:

$$JSL = (C_{\max}(S) - MS)^{+} \tag{14}$$

where MS is the makespan estimate computed according to each policy, i.e.

$$MS = \begin{cases} (1+\gamma_k) \cdot C_{\max}^{\textit{ex ante}}(S) & \text{, under the scheduling policy} \\ T & \text{, under the workload policy} \\ \hat{C}_{\max}^{\alpha}(S) & \text{, under the regression policy} \end{cases} \tag{15}$$

The third performance measure, the acceptance rate, is defined as the percentage of orders that are accepted within a planning period.

We define the last performance measure, the feasibility performance, as 1 − the fraction of processing steps that violate the no-wait restrictions through the entire planning period. Due to the no-wait restrictions between the processing steps of a job and due to the fact that the effective processing times differ from the expected processing times, unavoidable non-feasibility problems will occur. In real life situations a smaller number of violations of the no-wait constraints is preferred.

5.1 Experimental design

The general setting of the simulation experiments is the same one as described in the previous sections. Customer orders arrive with a given due date, which is equal to the end of the planning period. The orders keep on arriving until the stopping conditions of the acceptance policies are met.

We used a simulation run length of one planning period. The length of the planning period is chosen such that the job set consists of a realistic number of jobs. The empirical study of Raaymakers et al. [11] showed that a job set of 40 to 50 jobs is realistic for this type of industrial process. In line with that study, the length of the planning period has been fixed at 1000 time units. We performed 26 replications of the planning period.

The performance of the order acceptance policies may be affected by the demand/capacity ratio (φ), job mix variety (χ) and workload balance (ψ). Therefore, two levels are considered for each of these experimental factors. We chose the two levels in such a way that the difference between them allows us to properly assess the influence of each of these factors on the performance of the order acceptance policies. At the high level, the average demand requirements for capacity are equal to the total available capacity per planning period. At the low level, the average demand requirements for capacity are equal to 70% of the total available capacity per planning period. As has been shown by Raaymakers et al. [11], due to the no-wait structure of processing steps in each job, capacity utilization in this type of industry is at most between 50 and 60%. Thus, both demand levels investigated represent situations where demand effectively exceeds available capacity.

Each order consists of exactly one job with a specified structure of no-wait processing steps. The job characteristics have been generated randomly on arrival of the order. Hence, each job arriving at the system may be different. Thc parameters that determine the job mix variety are the number of processing steps per job and the distribution of the expected processing times. In order to carry out a realistic simulation study, the values of the parameters used have an order of magnitude similar to the one observed in reality. Moreover, the number of processing steps per job cannot exceed the number of resources, because each processing step of a job has to be performed on a different resource if the processing steps are overlapping. Nevertheless, jobs in a job set may differ in the number of processing steps. In the situation with high job mix variety, i.e. less homogeneous, the number of processing steps per job is uniformly distributed between 1 and 10 and the expected processing time is uniformly distributed between 1 and 49. In the situation with low job mix variety, i.e. more homogeneous, the number of processing steps per job is uniformly distributed between 4 and 7, and the expected processing time is uniformly distributed between 20 and 30. Note that in both situations the average number of processing steps and the average processing time is the same. In generating the jobs, each processing step is allocated at random to a resource type. In the situation with high workload balance, the allocation probability is the same for each resource type. In the situation with low workload balance, the allocation probability is different for each resource type. On average 30, 25, 20, 15 and 10% of the processing steps were allocated to the five different resource types respectively.

The same random number seeds are used for each combination of experimental factor values in order to obtain identical order arrivals for the different policies.

Table 7. Levels of the experimental factors

	L	H
demand/capacity ratio(φ)	0.7	1.0
Job mix variety (χ)	$s_j \sim U(4,7)$ $E[p_{ij}] \sim U(20,30)$	$s_j \sim U(1,10)$ $E[p_{ij}] \sim U(1,49)$
Workload balance (ψ)	30, 25, 20, 15 and 10% of demand requirements for resource type 1 to 5	20% of demand requirements for each resource type

The performance of the order acceptance policies may be also affected by the level of the uncertainty in the processing times. The uncertainty was introduced in the same way as in the previous sections. The effective processing times (p_{ij}) are determined by drawing a random variable from an Erlang distribution with mean $E[p_{ij}]$ and an Erlang shape parameter k. Three levels of uncertainty were considered. At the low uncertainty level the shape parameter is equal to 10, at the moderate uncertainty level the shape parameter is equal to 5 and at the high uncertainty level the shape parameter is equal to 2. 50 replications have been made for each of the uncertainty levels.

The values of the slack/correction factors for the three levels of uncertainty considered are presented in Table 8.

Table 8. Values of the slack/correction factors

k	γ_k	τ_k	α
2	$0.5, 0.55, 0.6$	0.59	0.03
5	0.3	0.52	-0.01
10	0.2	0.48	-0.05

5.2 Experimental results

The workload policy and regression policy are based on aggregate models of the production system, while the scheduling policy is based on constructing detailed schedules every time a new order arrives. In all three policies we used slack/correction parameter settings in order to make them comparable.

To compare the order acceptance policies we measured the feasibility performance, service level, the lateness of the job set and the acceptance rate. We observed that concerning the feasibility performance, it ranges from 0.83 to 0.86 and that all three order acceptance policies have a similar behavior. There is no significant difference between the three policies in the different scenarios considered and different levels of uncertainty.

The main results of our experiments are summarized in Table 9, which reports the average service level, the average lateness of the job set and the average acceptance rate for different scenarios under different levels of uncertainty. An ANOVA showed that all main effects (four experimental factors and three policies) have a statistically significant contribution towards the value of all these three performance measures. Below, we will discuss the relevant differences between the policies under the various scenarios.

Let us first focus on the performance of the workload policy as compared to the performance of the other two approaches. The results in Table 9 show that both the regression policy and the scheduling policy clearly outperform the workload policy with respect to both service level and job set lateness performance, in all the scenarios considered and under different uncertainty levels. The difference

Table 9. Simulation results (a) Erlang shape parameter $k = 2$

Scenarios				Policy		
φ	χ	ψ		Scheduling[a]	Workload	Regression
H	H	H	SVL	0.95	0.87	0.95
			JSL	47.91	132.99	56.96
			ACR	0.51	0.44	0.47
H	H	L	SVL	0.96	0.85	0.96
			JSL	32.25	166.46	47.11
			ACR	0.47	0.44	0.45
H	L	H	SVL	0.97	0.87	0.92
			JSL	18.09	114.95	66.47
			ACR	0.40	0.42	0.40
H	L	L	SVL	0.99	0.82	0.94
			JSL	8.83	176.74	41.27
			ACR	0.35	0.42	0.37
L	H	H	SVL	0.97	0.87	0.96
			JSL	33.12	138.33	46.41
			ACR	0.62	0.62	0.60
L	H	L	SVL	0.97	0.85	0.96
			JSL	26.06	162.54	42.29
			ACR	0.60	0.62	0.58
L	L	H	SVL	0.98	0.88	0.91
			JSL	12.81	99.45	68.91
			ACR	0.55	0.61	0.58
L	L	L	SVL	0.99	0.82	0.94
			JSL	6.70	168.39	44.89
			ACR	0.50	0.61	0.54

[a] $\gamma_k = 0.55$. We present in the Appendix D the results for $\gamma_k = 0.5$ and $\gamma_k = 0.6$

between the workload policy and the regression policy is smaller in the case of more homogeneous job mix variety (χ) and a balanced workload (ψ). This may be explained by the fact that in the case of low job mix variety, the amount of job interaction is expected to be more homogeneous, nearly constant. An order acceptance based on a fixed allowance may perform as good as the regression policy, which is based on a variable allowance. Namely, the makespan estimation model uses a variable allowance depending on the job set characteristics, whereas, the workload policy uses a fixed allowance, the slack factor. This observation is in accordance with the results reported in [11]. They performed a simulation study on a similar data setting, in a deterministic case. We further observe that the difference between the regression policy and the workload policy is larger as the uncertainty in the processing times increases, for all the considered scenarios.

Table 9 (continued)

(b) Erlang shape parameter $k = 5$

Scenarios				Policy		
φ	χ	ψ		Scheduling	Workload	Regression
H	H	H	SVL	0.96	0.88	0.92
			JSL	31.42	119.92	91.81
			ACR	0.57	0.51	0.57
H	H	L	SVL	0.97	0.86	0.92
			JSL	27.28	153.03	89.12
			ACR	0.55	0.51	0.54
H	L	H	SVL	0.98	0.89	0.90
			JSL	11.35	89.60	80.21
			ACR	0.48	0.49	0.48
H	L	L	SVL	0.98	0.80	0.91
			JSL	10.23	179.56	75.82
			ACR	0.44	0.49	0.45
L	H	H	SVL	0.98	0.87	0.92
			JSL	17.96	126.39	79.64
			ACR	0.70	0.71	0.72
L	H	L	SVL	0.98	0.85	0.92
			JSL	17.61	149.10	76.66
			ACR	0.68	0.71	0.69
L	L	H	SVL	0.99	0.88	0.90
			JSL	10.22	95.04	86.07
			ACR	0.66	0.71	0.71
L	L	L	SVL	0.99	0.82	0.91
			JSL	6.73	165.55	69.62
			ACR	0.60	0.71	0.65

Due to the rather poor performance of the workload policy, we will now further discuss only the differences between the scheduling policy and the regression policy.

In the deterministic case [11], the schedule constructed upon order acceptance is an exact representation of the future state of the production system since no production disturbances are considered. Therefore, the scheduling policy performs best here. One may expect that in a highly stochastic environment (i.e., high uncertainty in the processing times) the performance of the scheduling policy will be affected heavily whereas, the regression policy may be expected to be more robust since it is based on aggregate information. This is confirmed, in the case of high job mix variety (χ), by the results in Table 9. However, the performance of the scheduling policy is better than the performance of the regression policy for the other combinations of the experimental factors. One can see that in the case of high job mix variety and low and moderate uncertainty in the processing times the

Table 9 (continued)

(c) Erlang shape parameter $k = 10$

Scenarios				Policy		
φ	χ	ψ		Scheduling	Workload	Regression
H	H	H	SVL	0.97	0.89	0.91
			JSL	19.34	100.19	105.62
			ACR	0.63	0.55	0.62
H	H	L	SVL	0.98	0.86	0.90
			JSL	15.55	132.19	109.58
			ACR	0.59	0.55	0.60
H	L	H	SVL	0.99	0.88	0.90
			JSL	5.45	92.32	81.76
			ACR	0.51	0.53	0.53
H	L	L	SVL	0.99	0.80	0.88
			JSL	4.70	143.95	91.11
			ACR	0.47	0.53	0.49
L	H	H	SVL	0.98	0.89	0.92
			JSL	12.56	90.11	87.02
			ACR	0.76	0.75	0.78
L	H	L	SVL	0.99	0.85	0.91
			JSL	9.96	133.65	96.93
			ACR	0.74	0.75	0.76
L	L	H	SVL	0.99	0.89	0.90
			JSL	4.21	89.87	88.64
			ACR	0.72	0.77	0.77
L	L	L	SVL	1.00	0.81	0.89
			JSL	2.59	143.77	89.42
			ACR	0.66	0.77	0.71

difference in performance is much smaller. We further observe that as uncertainty increases, the performance of the regression policy increases, while the performance of the scheduling policy decreases. Apparently, the regression policy copes very well with higher levels of uncertainty. The results also show that the performance sensitivity of the scheduling policy to the uncertainty becomes greater in a high demand/capacity ratio (φ) case than is in the case with the low φ. Finally, one should keep in mind that the scheduling policy uses full detailed information which may not be always available at the order acceptance level, whereas the regression policy uses aggregate information.

Additional experiments with different slack/correction factors show that while the absolute performance of all policies increases with an increase in the slack/ correction factors, the relative differences between the policies remain similar. The absolute performance of the policies is however very sensitive to the slack/correction factor.

The acceptance rate is low for all three policies, especially in case of high uncertainty in the processing times. Table 9 further reveals that for all three policies there is a significant difference between the situation with high demand/capacity ratio and low demand/capacity ratio (φ). This difference is reduced as the processing times uncertainty increases.

We conclude this section with a comment regarding the computational requirements for the various order acceptance policies. The average CPU time on a Pentium 1.7 GHz is about 12 seconds to evaluate a single order under the scheduling policy, whereas it takes negligible time under the regression policy or workload policy.

6 Conclusions and directions for further research

Previous research [11] showed that a regression model performs reasonably well when used to accept orders in a setting with deterministic processing times. In this paper, we investigated the performance that can be obtained by using such a model for accepting or rejecting orders in the presence of production uncertainty. Three different order acceptance policies, which differ with respect to the level of detail of information used, are compared:

- scheduling policy (order acceptance based on detailed schedule information);
- workload policy (order acceptance based on aggregate workload profiles);
- regression policy (order acceptance based on a limited number of aggregate characteristics of the job set).

The performance of these three policies have been compared with respect to service level, lateness of the job set, acceptance rate and feasibility performance by means of simulation experiments. The data setting for the experiments was based on previous research [11].

The scheduling policy accepts orders based on a detailed schedule that has to be constructed each time an order arrives. In a stochastic situation, this schedule is only an estimate of what can be realized later by the production system. Therefore, a slack factor is introduced when the schedules are constructed upon order acceptance, in order to cope with the uncertainty in the processing times. The results show that, in general, the scheduling policy also performs best in a situation with uncertain processing times. However, due to the great scheduling complexity and high interrelations between jobs in multipurpose batch chemical shops [14], acceptance decisions based on detailed schedules are time consuming.

The other two acceptance policies investigated in this paper are based on aggregate information. The main advantage of the aggregate policies is that they are quick. Most customers may have a confirmation of the order fast. Experiments indicate that in all the scenarios investigated and under different uncertainty levels the regression policy clearly outperforms the workload policy.

Given the close performance of the scheduling and regression policies, other considerations rather than the actual performance may determine the choice which of the two methods to use in a particular situation. First, if all data are available and reliable, and the computational time is not important, it is clear that the scheduling policy should be used for order acceptance. In the case where only general job characteristics are available at the time of order acceptance, it might be worth considering the regression policy, due to its good performance with limited information. Under a multiperiod setting, in which orders need to be allocated optimally to a particular planning period, it may also be better to use the regression policy, since in that case multiple schedule alternatives (under different allocations of jobs) need to be computed, leading to long response times when using the scheduling policy for order acceptance. Finally, it is not clear what the optimal choice would be if the policies are tested dynamically over multiple periods. Further research is needed in this area.

The actual set of orders that are accepted will differ in general for different order acceptance policies. Analysis of the detailed experimental output suggests that the scheduling policy shows a particular selectiveness by not accepting one or two "difficult" jobs that the regression policy does accept. Further research is required to assess this properly. Moreover, an interesting question is whether the selective acceptance is commercially feasible.

A lot of past data is needed, also for the workload policy and scheduling policy, in order to tune the slack factors. The results show that under relatively high levels of uncertainty in processing times and high levels of variability in the characteristics of the job set, the value of using this average past performance information decreases. Further research is needed to better understand this effect. Our current results are not conclusive with regard to this aspect.

Appendix A

The scheduler monitors the schedule progress and the schedule is reviewed such that any new information with regard to the effective processing time is taken into consideration. If at one review moment t, the effective processing time of a particular processing step is larger than the expected processing time, an infeasibility may occur. The right-shift rescheduling procedure implies to delay processing if necessary while keeping the processing order of the initial schedule. Therefore, all the processing steps that have a scheduled start time larger than or equal to t are delayed by one time unit until the full information about the effective processing time of that particular processing step is available to the scheduler. On the other hand, if the effective processing time is smaller than the expected one, then the start time of all remaining jobs may be decreased while maintaining the no-wait restrictions.

An example problem with 4 jobs to be processed on 5 machines was generated for illustrating the proposed rescheduling procedure. The job parameters are showed in Table 10.

The Gantt chart of the schedule obtained by using simulated annealing algorithm is shown in Figure 1. Numbers inside the blocks represent the job number and the associated processing steps (e.g. 4.1 means the first processing step of job 4).

Table 10. Job characteristics

Job	Processing step	Expected processing time	Time delay
1	1	25	0
	2	23	12
	3	26	24
2	1	9	0
	2	38	4
	3	35	23
3	1	14	0
	2	29	7
	3	22	22
4	1	44	0
	2	29	22
	3	2	36

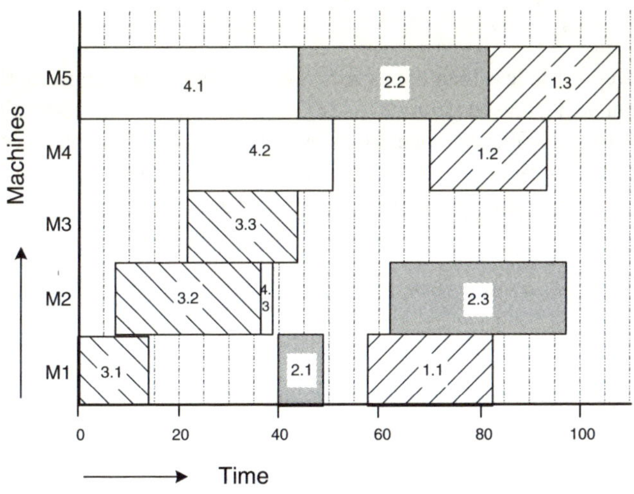

Fig. 1. Initial schedule for the chosen example

The right-shift procedure was applied to the example problem and the result is shown in Figure 2.

Figure 2 shows the ex post schedule for the considered example. The first non-feasibility problem occurs when on machine 2, the effective processing time of processing step 3.2 is longer than the expected processing time. At time T=36, because the completion time of processing step 3.2 was changed, the start time of all the other processing steps that had a scheduled start time larger than or equal to 36 were delayed until the complete information about the effective processing time of processing step 3.2 is known. At time T=38, the effective processing time

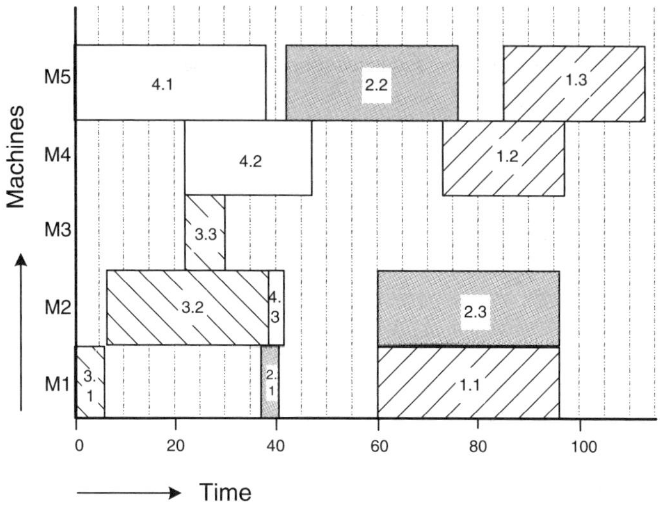

Fig. 2. Ex post schedule for the chosen example

of processing step 4.1 is less than the expected processing time, therefore, the start time of job 2 can be decreased (from 40 to 38) such that the no-wait restrictions are not violated. At time T=39, the processing step 3.2 is completed and the processing step 4.3 can start. But by delaying the start time of the processing step 4.3 (from 36 to 39) the no-wait restrictions within job 4 is violated.

Appendix B

Definitions of aggregate resource and job set characteristics.

Notation:

n_{jobs}	=	number of jobs in the job set;
N	=	number of resources;
M	=	number of resource types;
n_m	=	number of resources of resource type m;
L_m	=	workload on a single resource of type m, $m=1,2,...,M$;
LB	=	lower bound on the makespan;
s_j	=	number of processing steps of job j;
δ_{ij}	=	time delay of step i of job j;
$E[p_{ij}]$	=	expected processing time of step i of job j;

1. Average number of processing steps:

$$\mu_s = \frac{1}{n_{jobs}} \sum_{j=1}^{n_{jobs}} s_j \tag{16}$$

2. Average overlap:

$$\mu_g = \frac{1}{n_{jobs}} \sum_{j=1}^{n_{jobs}} g_j \tag{17}$$

where

$$g_j = \frac{1}{s_j - 1} \sum_{i=2}^{s_j} \left(1 - \frac{\delta_{ij} - \delta_{i-1,j}}{E[p_{i-1,j}]}\right) \tag{18}$$

3. Squared coefficient of variation of the expected processing times:

$$cv^2_{E[p]} = \frac{\sigma^2_{E[p]}}{\mu^2_{E[p]}} \tag{19}$$

where

$$\sigma^2_{E[p]} = \frac{1}{S-1} \sum_{j=1}^{n_{jobs}} \sum_{i=1}^{s_j} (E[p_{ij}] - \mu_{E[p]})^2, \tag{20}$$

$$S = \sum_{j=1}^{n_{jobs}} s_j \tag{21}$$

and

$$\mu_{E[p]} = \frac{1}{S} \sum_{j=1}^{n_{jobs}} \sum_{i=1}^{s_j} E[p_{ij}] \tag{22}$$

4. Workload balance, represented by the maximum utilization if the makespan were equal to the lower bound:

$$\rho_{\max} = \frac{\bar{L}}{LB} \tag{23}$$

where

$$\bar{L} = \frac{1}{N} \sum_{m=1}^{M} \sum_{n=1}^{n_m} L_n \tag{24}$$

and

$$N = \sum_{m=1}^{M} n_m$$

Appendix C

The derivation of the squared coefficient of variation of the effective processing times.

B1. No variation in the expected processing times

The expected processing times are constant. Let denote $E[p_{ij}] = c$. The effective processing times are determined by drawing a random variable from an Erlang distribution with mean c and shape parameter k. So, the squared coefficient of variation for the effective processing times is:

$$cv_p^2 = \frac{1}{k} \tag{25}$$

according to the properties of an Erlang distributed random variable.

B2. Variation in the expected processing times, $E[p_{ij}]$.

The effective processing times are determined by drawing from a random variable X, Erlang distributed with mean $E[p_{ij}]$ and shape parameter k. We also know that $E[p_{ij}]$ is at its turn a random variable uniformly distributed $U(a,b)$. In the following we compute the density function and the first two moments of the effective processing time random variable. Having this known we are able to finally compute the squared coefficient of variation used in the regression model.

$$f(t) = \frac{d}{dt}P(X \leq t) = \int_a^b \frac{d}{dt}P\left(X \leq t \mid \frac{k}{\mu} = z\right)\frac{1}{b-a}dz =$$

$$= \frac{1}{b-a}\int_a^b \frac{k}{z}\frac{\left(\frac{k}{z}t\right)^{k-1}}{(k-1)!}e^{-\frac{k}{z}t}dz \tag{26}$$

Replacing $\frac{k}{z}$ with ω in (26) we have:

$$f(t) = \frac{k}{b-a}\int_{\frac{k}{b}}^{\frac{k}{a}} t\frac{(\omega t)^{k-2}}{(k-1)!}e^{-\omega t}d\omega \tag{27}$$

Through the substitution $\omega t = s$ we arrive further at the following formula for the density function of the effective processing time random variable.

$$f(t) = \frac{k}{b-a}\frac{1}{(k-1)!}\int_{\frac{kt}{b}}^{\frac{kt}{a}} s^{k-2}e^{-s}ds \tag{28}$$

The first two moments are now derived straightforward :

$$E[X] = \int_0^\infty t f(t)dt = \int_0^\infty t \left(\frac{k}{b-a} \frac{1}{(k-1)!} \int_{\frac{kt}{b}}^{\frac{kt}{a}} s^{k-2} e^{-s} ds \right) dt =$$

$$= \int_0^\infty \left(\int_{\frac{as}{k}}^{\frac{bs}{k}} t dt \right) \frac{k}{b-a} \frac{1}{(k-1)!} s^{k-2} e^{-s} ds =$$

$$= \int_0^\infty \frac{(b^2-a^2)s^2}{2k^2} \frac{k}{b-a} \frac{1}{(k-1)!} s^{k-2} e^{-s} ds = \frac{b+a}{2} \int_0^\infty \frac{s^k}{k!} e^{-s} ds = \frac{b+a}{2} \tag{29}$$

knowing that $\int_0^\infty \frac{s^k}{k!} e^{-s} ds = 1$ (see properties of the gamma function).

The second moment is computed according to the well known formula:

$$Var[X] = E[X^2] - (E[X])^2 \tag{30}$$

where

$$E[X^2] = \int_0^\infty t^2 f(t)dt = \int_0^\infty t^2 \left(\frac{k}{b-a} \frac{1}{(k-1)!} \int_{\frac{kt}{b}}^{\frac{kt}{a}} s^{k-2} e^{-s} ds \right) dt =$$

$$= \int_0^\infty \left(\int_{\frac{as}{k}}^{\frac{bs}{k}} t^2 dt \right) \frac{k}{b-a} \frac{1}{(k-1)!} s^{k-2} e^{-s} ds =$$

$$= \int_0^\infty \frac{(b^3-a^3)s^3}{3k^3} \frac{k}{b-a} \frac{1}{(k-1)} s^{k-2} e^{-s} ds = \frac{b^2+ab+a^2}{3} \frac{k+1}{k} \int_0^\infty \frac{s^k}{k} e^{-s} ds =$$

$$= \frac{b^2+ab+a^2}{3} \frac{k+1}{k} \tag{31}$$

So, we have then the following formula for the second moment:

$$Var[X] = \frac{b^2+ab+a^2}{3}\frac{k+1}{k} - \left(\frac{b+a}{2}\right)^2 = \frac{(b-a)^2}{12} + \frac{b^2+ab+a^2}{3k} \quad (32)$$

Thus, the squared coefficient of variation is:

$$cv^2 = \frac{Var[X]}{(E[X])^2} = \frac{(b-a)^2}{3\cdot(b+a)^2} + \frac{4\cdot(b^2+a\cdot b+a^2)}{3\cdot k\cdot(b+a)^2} \quad (33)$$

Appendix D

The experimental results for the scheduling policy, in case of high uncertainty level, for different slack factors are presented in Table 11.

Table 11. Experimental results for $k = 2$ and different slack factors

Scenarios				Policy	
φ	χ	ψ		$Scheduling$ ($\gamma_k = 0.5$)	$Scheduling$ ($\gamma_k = 0.6$)
H	H	H	SVL	0.94	0.97
			JSL	74.05	30.95
H	H	L	SVL	0.95	0.98
			JSL	60.29	24.81
H	L	H	SVL	0.96	0.99
			JSL	35.47	8.52
H	L	L	SVL	0.97	0.99
			JSL	21.89	4.60
L	H	H	SVL	0.95	0.98
			JSL	52.52	20.92
L	H	L	SVL	0.96	0.98
			JSL	43.46	17.61
L	L	H	SVL	0.96	0.99
			JSL	29.34	7.30
L	L	L	SVL	0.98	0.99
			JSL	16.30	3.74

References

1. Bertrand JWM, Wortmann JC, Wijngaard J (1990) Production control: a structural and design oriented approach. Elsevier, Amsterdam
2. Carlier J (1987) Scheduling jobs with release dates and tails on identical machines to minimize the makespan. European Journal of Operational Research 29: 298–306

3. Cheng TCE, Gupta MC (1989) Survey of scheduling research involving due date determination decisions. European Journal of Operational Research 38: 156–166
4. Hopp WJ, Spearman ML (2000) Factory physics: foundations of manufacturing management, 2nd edn. Irwin McGraw-Hill, New York
5. Lawrence SR, Sewell EC (1997) Heuristic, optimal, static and dynamic schedules when processing times are uncertain. Journal of Operations Management 15: 71–82
6. Lenstra JK, Rinnooy Kan AHG, Brucker P (1977) Complexity of machine scheduling problems. Annals of Discrete Mathematics 1: 343–362
7. Leon VJ, Wu SD, Storer RH (1994) Robustness measures and robust scheduling for job shops. IIE Transactions 26: 32–43
8. Montgomery DC, Peck EA (1992) Introduction to linear regression analysis. Wiley, New York
9. Pinedo M (1995) Scheduling: theory, algorithms and systems. Prentice Hall, New Jersey
10. Raaymakers WHM, Bertrand JWM, Fransoo JC (2000) The performance of workload rules for order acceptance in batch chemical manufacturing. Journal of Intelligent Manufacturing 11: 217–228
11. Raaymakers WHM, Bertrand JWM, Fransoo JC (2000) Using aggregate estimation models for order acceptance in a decentralized production control structure for batch chemical manufacturing. IIE Transactions 32: 989–998
12. Raaymakers WHM, Bertrand JWM, Fransoo JC (2001) Makespan estimation in batch process industries using aggregate resource and job set characteristics. International Journal of Production Economics 70: 145–161
13. Raaymakers WHM, Fransoo JC (2000) Identification of aggregate resource and job set characteristics for predicting job set makespan in batch process industries. International Journal of Production Economics 68: 137–149
14. Raaymakers WHM, Hoogeveen JA (2000) Scheduling no-wait job shops by simulated annealing. European Journal of Operational Research 126: 131–151
15. Reklaitis GV (1996) Overview of scheduling and planning of batch process operations. In: Reklaitis GV, Sunol AK, Rippin DWT, Hortacsu O (eds) Batch processing systems engineering, pp 660–705. Springer, Berlin Heidelberg New York
16. Ten Kate HA (1994) Towards a better understanding of order acceptance. International Journal of Production Economics 37: 139–152
17. Wester FAW, Wijngaard J, Zijm WHM (1992) Order acceptance strategies in a production-to-order environment with setup times and due-dates. International Journal of Production Research 30: 1313–1326

A multi-agent system for chemical supply chain simulation and management support

Rodolfo García-Flores and Xue Zhong Wang

Department of Chemical Engineering, The University of Leeds, Leeds LS2 9JT, UK
(e-mail: x.z.wang@leeds.ac.uk)

Abstract. Modern chemical production is customer-driven and the desired delivery time for the products is often shorter than their campaign length. In addition, the raw materials supplying time is often long. These features make it desirable to provide tools to support collaborative supply chain decision making, preferably over the Internet, and where there are conflicts, compromise decisions can be quickly reached and the effects of the decisions can be quantitatively simulated. This paper describes such a multi-agent system (MAS) that can be used to simulate the dynamic behaviour and support the management of chemical supply chains over the Internet. Geographically distributed retailers, logistics, warehouses, plants and raw material suppliers are modelled as an open and re-configurable network of co-operative agents, each performing one or more supply chain functions. Communication between agents is made through the common agent communication language KQML (knowledge query message language). At the simulation layer, the MAS allows distributed simulation of the chain behaviour dynamically, so that compromise decisions can be rapidly and quantitatively evaluated. Because in a chemical supply chain the scheduling of the plant often dominates the chain performance, an optimum scheduling system for batch plants is integrated into the MAS. The functions of the system are illustrated by reference to a case study for the supply and manufacture using a multi-purpose batch plant of paints and coatings.

Key words: Supply chain management – Supply chain simulation – Software agents – Paints and coatings

---

Financial support for the first author from CONACyT is acknowledged. The project has benefited considerably from discussions with Dr G.E. Goltz at the Keyworth Research Institute of the University of Leeds and Dr. T. F. Burges at the School of Economic and Business Studies at the University of Leeds.

Correspondence to: X.Z. Wang

1 Introduction

Chemical supply chains are integrated networks where a number of business entities, i.e., suppliers, manufacturers, distributors and retailers, work together to acquire raw materials, convert them into final chemical products, and deliver the products to the retailers and final customers. It is estimated that the supply chain can represent as much as 60-80% of a typical chemical manufacturer's costs, and that a 10% reduction in supply chain costs can bring a 40–50% improvement in before-tax profits (Gibson, 1998). It has also been reported that one of Du Pont's polymers businesses reduced working capital tied up in inventory from £160,000,000 to £95,000,000 (Kafoglis, 1999). Compared with other industrial sectors such as the car manufacturing industry, the chemical process industry (CPI) is thought to have lagged behind in addressing supply chain issues. Benson (1998) estimated that the typical costs of operating the supply chain in the chemical industry are about 14% of sales, compared with the world-class figure of about 6% in other industrial sectors, therefore there is a gap of 8% for improvement. Several researchers have pointed out that CPI supply chains have some specific features compared with other industrial sectors (Bodington and Shobrys, 1999), for example, the chain is often longer and the finished products co-exist with by-products in the market. The most distinguishable feature is considered to be on the engineering side, i.e., the chemical plant, because its equipment connection and configuration as well as scheduling and control schemes are not flexible and agile enough to meet the dynamic demand of the supply chain.

The overall objective of this paper is to describe a distributed and co-operative system for support of supply chain management over the Internet, and integrate the scheduling and control of the plant into the dynamics of a supply chain. The system will be used to develop a prototype system for the supply and manufacture of paints and coatings using a multi-purpose batch plant. Using the prototype system, issues related to distributed software agent design and implementation, including legacy software wrapping, conflict resolution through negotiation and optimal problem solving using well-established OR methods will also be explored.

The rest of the paper is organised as follows. In Section 2, a brief review of computer aided supply chain management will be given, in order to clarify the contribution of the present work. Section 3 presents the system architecture and implementation of the multi-agent supply chain system. Section 4 describes the structure of the supply chain and the production process for paints and coatings. Section 5 presents the simulation considerations for each agent. Sections 6 and 7 discuss the use of the system for communication and negotiation for conflict resolution, which are followed by discussion of some illustrative simulation results in Section 8 and concluding remarks in Section 9.

2 Previous work on supply chain studies

Previous research on computer aided supply chain management has mainly adopted three approaches. These are methods based on *control theory*, *operational research* methods such as mathematical programming, business games and statistical analysis, and *simulation* approaches.

2.1 Previous work based on control theory and operations research

Supply chain studies based on control theory make an analogy between supply chains and control systems. Towill (1997) used feedback control block diagrams to represent the supply chain, and difference equations to describe chain dynamic behaviour. These studies suggest improving chain performance through modifications of the chain such as through removal of one layer of the chain, information integration, reduction in lead-time, and modification of ordering rules as well as combinations of these. In the chemical industry, Perea-Lopez et al. (2001) also studied the control theory based methods. They proposed a model representing balances of inventories, balances of orders, shipping rates and initial and boundary conditions. Their framework was applied to a company that produces three types of polymers and supplies a distribution network composed of warehouses, distribution centres and retailers. Each site is considered as a "tank" that stores material or information. Bose and Pekny (2000) proposed the use of model predictive control for planning and scheduling of a generic supply chain. They also studied the effect of varying co-ordination structures and other parameters on the overall customer service level.

Supply chain optimisation can be roughly clustered into three groups. The first group is concerned with multi-echelon inventory systems and focuses research on cost and service optimisation of warehousing policies at different levels of a supply chain. In turn, these may use a decentralised view (when ordering is triggered solely by inventory position at each stock-point, or *installation stock policy*) or a centralised view. The latter is called *echelon stock policy* and in this case ordering is triggered by the echelon inventory position, that is, the sum of all stock in transit to this stock-point plus its physical stock plus that in transit to or on hand at its downstream stock-points minus back orders at its end stock-points. The literature related to multi-echelon optimisation is abundant nowadays mainly due to the increased ease of exchange of product information between supply chain components (Sherbrooke, 1968). A literature review of optimal multi-echelon problems from a service level perspective was given by Diks et al. (1996). Most work dealt with one distribution centre and N retailers (e.g. Axsater and Zhang, 1999; Cachon and Zipkin, 1999; Seo et al., 2002; Svoronos and Zipkin, 1988). Tempelmeier (2000) developed a procedure for estimating the probability distribution of the order waiting time in a discrete time periodic (s, S)-inventory system, which is particularly useful for determination of inventory location that serves downstream nodes (e.g. production processes, regional warehouses or customers) in a supply chain. Among these papers, very few deal with optimal solution methods for systems of more than two-echelons, apart from the work of Diks and de Kok (1998, 1999) who developed a model for divergent multi-echelon systems and tested it using a decomposition approach. Van der Heijden et al. (1999) presented a method for multi-echelon divergent systems with stochastic lead times, which computes the order-up-to level and the allocation fractions required to achieve given target fill rates.

The second group includes mainly those that use classic operations research methods and consider more than one of the following aspects of supply chain management: plant design, production scheduling, logistics of distribution and inventory management (e.g., Chandra and Fisher, 1994; Haq et al., 1991; Van Roy, 1989). General literature reviews have been made by Thomas and Griffin (1996) and Maloni and Benton (1997). The review by Vidal and Goetschalckx (1997) paid special attention to mixed-integer programming models of production-distribution problems. Baita et al. (1998) reviewed dynamic routing and inventory problems. The majority of works on chemical industries' supply chain optimisation also belong to this group. Bok et al. (2000) proposed a multi-period optimisation model that considers inventory profiles, process operating modes and product sales. Zhou et al. (2000) used goal programming to optimise a continuous chemical plant with a multi-objective function. Gjerdrum et al. (2001) modelled multi-enterprise supply chains assuming fair profit distribution as a mixed-integer non-linear program. Timpe and Kallrath (2000) modelled a chemical supply chain considering production and distribution using a MILP approach. Kallrath (2002) presented a comprehensive review of literature on planning and scheduling of batch and continuous process plants. Other works studied sub-optimal solutions using commonly applied policies, as for example Agrawal and Cohen (2001), who analysed the effect of a fair share policy in a production-inventory problem over service level. An interesting method for coordinating information and materials flows in a supply chain using optimisation models was presented by Haehling von Lanzenauer and Pilz-Glombik (2002) and applied to a modified version of MIT's Beer Distribution Game. It demonstrated enormous potential for performance improvement using analytical decision support over human decision making. In chemical supply chains, optimisation using mathematical programming is probably the most widely studied approach. However, this approach also has some limitations. For example, as indicated by Kafoglis (1999), it is technically insufficient in handling a high volume of what-if scenarios, and it is very difficult to solve a problem where more than two management issues are considered. Based on the study of networked batch plants with interdependent production schedules, multi-stage production at multi-purpose facilities and chain production, Berning et al. (2002) also addressed the insufficiency of pure mathematical models for scheduling and proposed an integrated framework that consists of three layers, i.e. an optimiser for scheduling solution, a mechanism for collaborative planning among the involved plants, and a tool for manual updates and scheme changes.

The third group includes mainly statistical analysis-based methods, where at least one of the variables is unknown and is assumed to follow a particular probability distribution. Business game models can also be assigned to this group. An example of statistical approaches is the work of Applequist et al. (2000), who presented a method for evaluating supply chain projects where significant risk exists using statistical methods. Corbett (2000) presented a two-echelon study considering incentive conflicts and information asymmetries. Gupta et al. (2000) considered demand uncertainty in a production-inventory problem and converted stochastic features of the problem into a chance constraint programming problem. Cohen and

Lee (1989) used stochastic models to represent production, inventory and distribution in the supply chain.

2.2 Previous work based on simulation and agents

Simulation approaches experiment with an executable model of the system. In most cases these models are created to support one of the previously described approaches. Research in this area dates back to 1958 with Forrester's seminal work on industrial dynamics (Forrester, 1958). We will centre our review on the application of agents as simulation tools to solve operational research problems. Agent technology has been already applied to several operations research problems. For example, Pendharkar (1999) explored design issues of a multi-agent system that uses genetic learning of job scheduling strategies. Knotts et al. (2000) solved a series of published project scheduling problems using agents. They compared results when using sets of reactive and deliberative agents and concluded that deliberative agents perform much better than purely reactive agents. In recognition of the geographical nature of a supply chain, most recent work on simulation of supply chain behaviour has centred on developing distributed systems (Parunak et al., 1999), and some used agent technology (Gjerdrum et al., 2000; García-Flores and Wang, 2000). The method models the dynamics of individual components as well as the emergent interacting behaviour. Since it mimics the natural structure of a supply chain and interacting mechanism, it has the advantage of being able to be easily reconfigured when the chain structure is reorganised.

Other researchers (Fox et al., 2000) have also studied agents as a technique to develop co-operative supply chain systems. Their work incorporates the three levels of decision making: *strategic*, e.g. where to allocate production, or what is the best sourcing strategy, *tactical*, that is, forecasting, scheduling, ordering of short lead materials and *operational*, that refers to inventory deployment and detailed scheduling. In a parallel study, Chen et al. (1999) studied the negotiation methods using agents in supply chain management. Swaminathan et al. (1998) presented a sophisticated framework of a supply chain modelling library which is composed of agents (classified as production and transportation), its control elements (classified as flow, inventory, demand, supply and information) and their interaction protocols. In their work, three message classes were implemented including money, information and materials. These components are exemplified after analysing different real supply chains.

Compared with the many efforts on control theory and optimisation based methods for supply chain support, the work on agent based systems is still very limited, especially those dedicated to the chemical industries. The publications on agent based supply chain systems have either focused on co-operative decision support or distributed simulation, none has simultaneously addressed both. The current study is aimed at developing a multi-agent based system for co-operative decision support over the Internet as well as distributed simulation of chain behaviour, for chemical industries' supply chains. Agent based technique has a number of distinctive features (Jennings and Wooldridge, 1999; Han et al., 2000; Batres et al., 1997; McGreavy et al., 1995), making it attractive for developing this project. First, agents constitute a very effective technique for designing distributed supply chain systems over the Internet. This is very important considering the fact the chain components are actually geographically distributed. In addition, agents mimic the supply chain structure, i.e., systems for individual chain components can be developed and maintained independently and the overall system behaviour and decision making is through interactions of the subsystems. Such a system can be easily maintained. It also offers the advantage of being re-configurable when necessary. Furthermore, it is easy to develop such framework as an open system, i.e., new tools can be easily incorporated into it without major changes in the rest of the structure.

3 The multi-agent system and its implementation

3.1 Agent structure for supply chain modelling

A software agent is an encapsulated computer program that possesses some properties, most importantly, communication for the purpose of co-operation and negotiation, learning to improve performance over time, and autonomy implying that agents can act pro-actively over their environment rather than passively waiting for commands (Jennings and Wooldridge, 1999). These properties make agents different to traditional standalone software systems. Software agents have proved to be a useful technique in designing distributed and co-operative systems in many industrial and business sectors, including telecommunication, air traffic control, traffic and transportation management, entertainment and medical care (Jennings et al., 1998). They have also been used for developing distributed systems that can collaboratively solve domain problems over the Internet. Agent-based systems are difficult to build and debug from scratch. Therefore it is important to use agent building tools, such as IBM Aglets Software Development Kit (also known as Aglets Workbench), JAFMAS (Jave-based Agent Framework for Multi-agent Systems) (Chauhan, 1997), and JATlite (Java Agent Template Lite).

It is important to note that the agent paradigm fits better the real structure of a supply chain, where different entities are in charge of their own responsibilities. As members of a loosely coupled network, these entities work together to solve a problem (optimising supply chain operation) that is beyond the individual capabilities or knowledge of each component. Of course, tackling this specific problem in this way does not conflict with other aspects of supply chain management where a centralised database is desirable, as for example project management or customer information repositories for *e*-commerce.

Peng et al. (1999) studied the use of agents for designing supply chain management systems and concluded that there are three requirements that are essential:

1) *A common agent communication language (ACL) and protocol.* This is the language the agents use to exchange messages. In this work, we used the *Knowledge Query and Manipulation Language* (KQML). KQML is both a message format and a message-handling protocol to support run-time knowledge sharing among agents. KQML can be used as a language for an application program to interact with an intelligent system or for two or more intelligent systems to share knowledge in support of co-operative problem solving. KQML focuses on an extensible set of

Table 1. Examples of supply chain agent communication performatives

Performatives	Sender	Receiver	Information
Start	External agent	Purchasing	–
Order	Purchasing	RMS	–amount ordered, –due times
Advertise	RMS	Purchasing	–RM unit prices
Delivery	RMS	Plant	–delivery times
Forward	Purchasing	Logistics	–RM unit prices
forward	Plant	Warehouses	–
advertise	Logistics	Retailers	–P unit prices
Delivery	Warehouses	Retailers	–delivery times
Order	Retailers	Logistics	–amount ordered, –due times
Order	Logistics	Warehouses	–transportation plan, –amounts ordered, –due times
Order	Warehouses	Plant	–amount ordered, –due times
Delivery	Plant	Warehouses	–delivery times, –Plant operating cost, –RM demand
report–w	Warehouses	Logistics	–Warehouse op. Costs, –Plant operating cost –RM demand, –transportation costs
pay–budget	Logistics	Purchases	–
pay–budget	Purchases	Plant	–
Order	Plant	Purchases	–amount ordered, –due times
report–p	Purchases	Logistics	–
authorise	Logistics	Purchases	–
Order	Purchasing	RMS	–amount ordered, –due times

performatives, which define the permissible operations that agents may attempt on each other's knowledge base. A performative is a predicate with some parameters.

The performatives comprise a substrate to develop higher-level models of inter-agent interaction such as contract nets and negotiation. It is useful for programs to communicate attitudes about information, such as querying, stating, believing, requiring, achieving, subscribing and offering. Table 1 shows some examples of performatives used in the current system.

KQML messages are indifferent from the format of the information itself, thus KQML expressions will often contain sub-expressions, i.e. so-called "content" languages. The only restrictions to KQML messages are that (1) expressions must be viewed as entries in a knowledge base, and (2) sentences must be encoded as ASCII strings. We have used some of the reserved performatives (Labrou and Finin, 1997) and created some new ones.

2) A *common format* for the contents of communications is necessary. Agents should be able to access a set of neutral commands. Neutrality implies that the

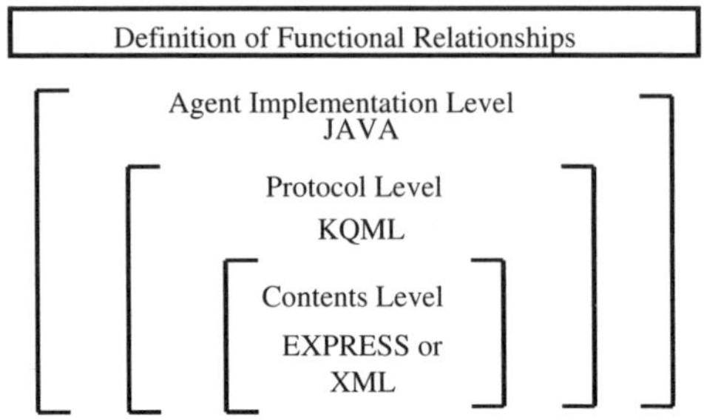

Fig. 1. Levels of implementation

contents may be used for different purposes. STEP (the STandard Exchange of Product model data) and XML (eXtensible Mark Language) are investigated as the content languages.

3) The agents also need to use *a shared ontology.* Ontology is a formal description of entities and their properties, relationships, constraints and behaviour. Ontology is a "model of the world". Agents must share the same conceptual model of the domain problem in order to understand the concepts and get effective communication, and this model of the environment is represented by the data models supported by the content language.

3.2 Agent implementation tools

The language chosen to implement the system is Java. The main reason for using Java is that it is Internet oriented and platform independent. The most important Java library for the development of this project is JATLite, a package of programs developed in Java by the Centre for Design Research of the University of Stanford. It allows users to quickly create new-layered software agents that communicate robustly over the Internet and exchange information with other agents on other computers where they are running. JATLite provides a set of Java templates and a Java agent infrastructure that makes it easy to build systems in a common way. The template for building agents uses a common high-level language and protocol and provides users with numerous pre-defined Java classes that facilitate agent construction. Figure 1 shows the language implementation levels. A useful feature of JATLite is that it can "wrap" existing and often standalone programs by providing them with a front-end that allows them automatically to communicate with other programs, sending and receiving messages and files.

The second level of implementation in our simulation uses KQML as the protocol language. KQML is the "shell" that provides information about the sender and receiver agents, the subject of the message and additional information. Finally, the innermost layer in the agent communication framework is represented by a neutral file exchange format, for which we use EXPRESS or XML.

4 The supply chain and the manufacturing process

The supply chain used as a case study is composed of six classes of agents: retailers, logistics, warehouses, purchasing departments, plants and raw material suppliers (RMS). They can also be grouped as *functional* and *information* agents. Functional agents are those through which there is a material flow (retailers, warehouses, plants and RMS), whereas information agents refer to those that support the activities of functional agents through taking decisions or facilitating certain operations (logistics and purchasing departments). Functional agents are subject to optimisation by OR methods and carry inventories of raw materials (if they are located in the supply side of the chain, i.e. RMS and plants) or finished products (if they are located in the demand side of the chain, i.e. warehouses and retailers) and are assigned a physical location. There is no material flow through information agents, so no physical locations are necessary for them. Figure 2 depicts the class relationships. The supply chain examined in this study consists of two warehouses that distribute the finished products to eight retailers spread in four countries, i.e. United Kingdom, Ireland, France and Belgium. The warehouses get the finished products from a single plant located in the UK.

The production plant is based in the UK and requires supply of three groups of raw materials (resin, paint components and packaging materials). These are supplied respectively by three RMS that are also located in the UK.

The simulation is conducted at different time scales, as can be seen in Figure 3. In the contents level, the processes that occur at the smallest time units (hours-days) are managed at the plant by the scheduling optimisation software. At the protocol level, each time step in the simulation corresponds to planning horizons of

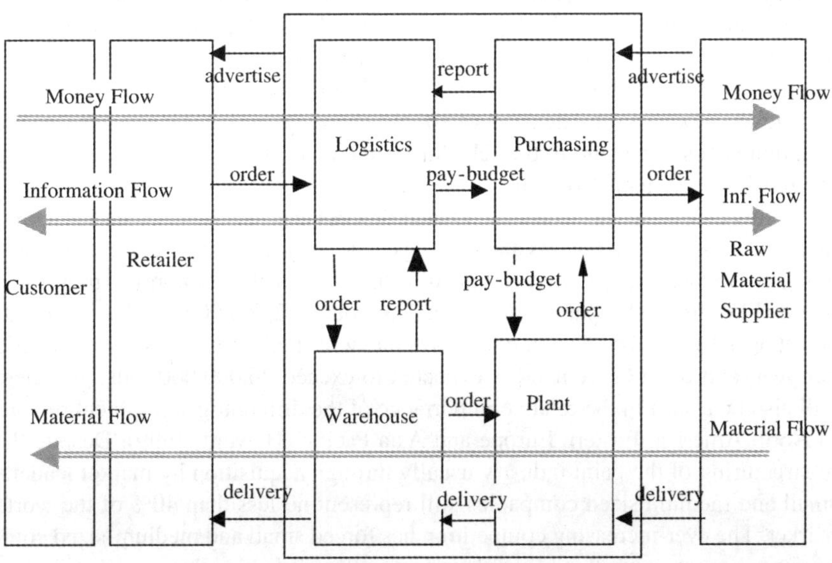

Fig. 2. The agent classes and their relationship

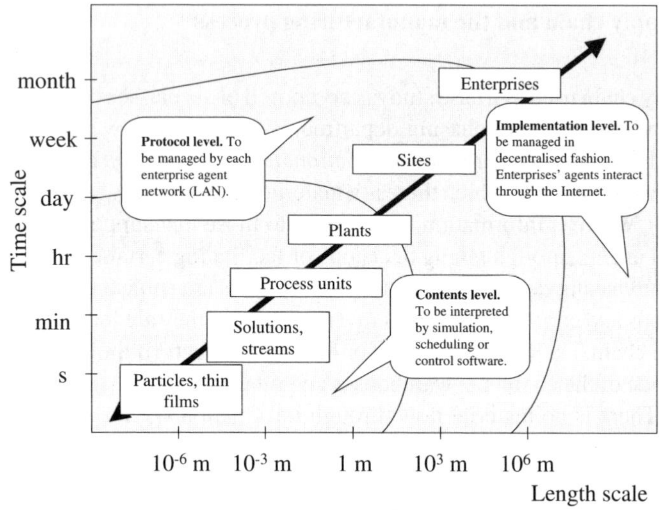

Fig. 3. Integrated chemical supply chain at different levels of implementation

three months, or *periods*, on which each agent in the system must forecast its own demand, and place orders according to the forecast. The implementation level is represented by the overall performance of the supply chain in a simulation horizon of years. The negotiation process between chain components occurs in the protocol level, and its results affect both the output of the software involved in the contents level and the implementation level.

A distinctive feature of chemicals supply chains is that the plant, which is also called the engineering side of the chain, plays a more dominant role than other chain components. The desired delivery time for the products is often shorter than the production campaign length, and the raw materials supplying time is often long (Janicke, 2000). In addition, the plants are often not flexible and agile enough to be able to respond to the demand of supply chain dynamics. As a result, special attention needs to be paid to the scheduling and control of the plant in addressing chemical supply chain problems.

In this case study, we consider a batch process and its supply chain for manufacturing and supplying paints and coatings. Due to the diversity of end customers, this is an industry that is tightly linked to the state of the economy. The present worldwide demand for paints is estimated at around 22,500,000 tons/year and valued at over £65 billion. The global growth rate will stabilise between 2% and 2.5% per year, and by 2005 demand is expected to exceed 26,000,000 tons, and there will also be a shift in the relative importance of the different geographical regions to North America, Eastern Europe and Asia Pacific (Howard, 1999). Despite the re-structuring of the paint industry usually through acquisition by market leaders, small and medium sized companies still represent no less than 40% of the world market. The ever-increasing competition has forced small and medium-sized companies to become more flexible, and modernise the production lines and distribution networks.

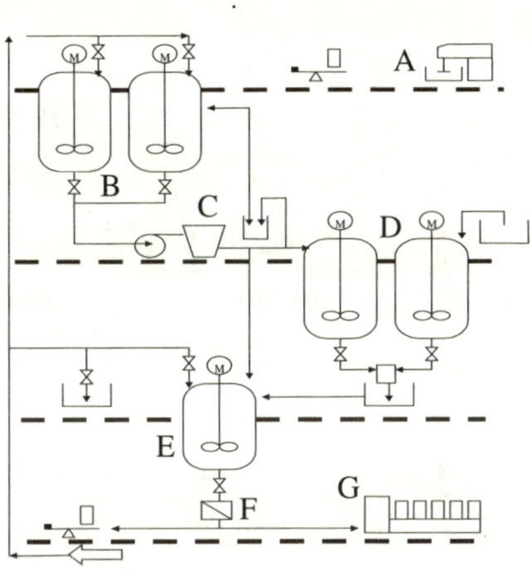

Fig. 4. The batch paint production process. (A – dispersers, B – dissolvers, C – rotary mill, D – paste mixers, E – adjusters, F – Filter, G – packing machine)

Paints and coatings production is performed on *multi-purpose* batch plants, where products can follow different routes through the plant and two or more products may be produced at any one time. Multi-purpose plants are more flexible than *multi-product* plants where all products are produced following the same route, one at a time, the plant is designed for a fixed set of products, and campaigns are carried out for each product in turn.

One of our objectives in developing the system is that it can be used to study the impact of supply chain dynamics on plant operation and scheduling. With this consideration in mind, we assume that the only non-significant delay between agent replenishments occurs when the plant delivers to the warehouses. This delay is equal to the time it takes to the plant to finish the production campaign. At this stage of implementation, no delays are assumed to happen in our model for other agent interfaces.

A generic description of the paint and coating production process can be found in Stoye and Freitag (1998) and Figure 4 shows a simplified flow diagram. This is a typical multi-purpose batch plant where the production stages are arranged vertically. In the upper level, raw materials are weighed and charged into the dispersers (the equipment denoted as A on the flowsheet). These concentrated mixtures pass then to the dissolvers (B) to produce a more diluted mixture. This resin is then pumped and dispersed using an agitator mill (C). After this it goes to the paste mixers (D), where additives including pigments, stabilisers, accelerators, plasticizers and extenders are added. The product goes then to an adjuster (E), where the fine properties required for each specific batch are tuned-up. The paint is finally filtered to remove the impurities and packaged using the packaging machine G.

To describe the production procedure, a state task network (STN) is used. The STN is a directed graph that describes the sequence of activities, represented by rectangles and the states represented by circles that have to undergo in order to get the desired products (Kondili et al., 1993; Papageorgiou and Pantelides, 1996).

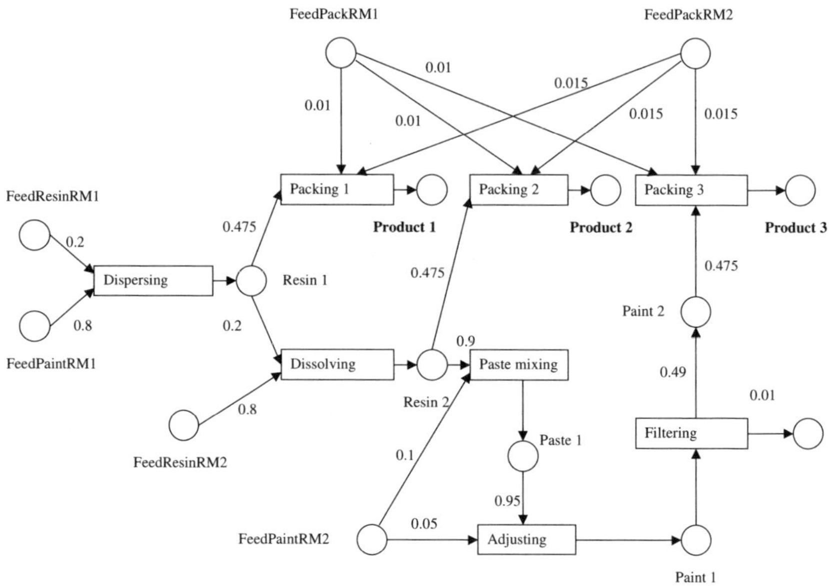

Fig. 5. State task network for paint production process

The numbers attached to the arrows indicate the composition (i.e., the fraction of each state) that enters the following activity. The STN network for the paints and coatings production is shown in Figure 5.

The recipe depicted in this Figure shows that there are three products produced at different stages that can be sold after being packaged. These products are obtained from six different raw materials, each belonging to any of three groups (resin, paint and pack) and named accordingly FeedResinRM1, FeedResinRM2, FeedPaintRM1, FeedPaintRM2, FeedPackRM1 and FeedPackRM2.

The STN network describes only the recipe for production, but does not give information on the equipment used. Different vessels may be used for the same task, or a piece of equipment can be employed to perform more than one processing activity. In addition, there are also storage vessels for intermediate products and finished products. All the equipment considered in the example is listed in Table 2.

The scheduling problem that the plant agent solves is a common operational research problem that may be stated as follows. Given the mode of operation, the product orders, the product recipes, the number and capacities of the various types of existing equipment and the list of equipment types allowed for assignment to each task, determine the order in which tasks use equipment and resources, and the detailed timing of the execution of all tasks so as to minimise the delay in fulfilling orders. To perform this calculation, the commercial package gBSS® optimisation software for multipurpose plants developed at Process Systems Enterprise Ltd. from Imperial College, London, was used to calculate the plant optimal schedule. The details of the mathematical problem formulation can be found in Kondili et al. (1993) and Shah et al. (1993) and therefore will not be described here. However, the

Table 2. Equipment used for production in the case study

Equipment	Unit No.	Capacity (ton)	Function
B100	1	10	Dispersing 1
B110	2	10	Dispersing 1
D200	3	40	Dissolving 1, Paste mixing 1
D210	4	40	Dissolving 1, Paste mixing 1
P300	5	60	Dissolving 1, Paste mixing 1
P310	6	60	Dissolving 1, Paste mixing 1
A400	7	100	Adjusting 1
A410	8	100	Adjusting 1
F500	9	Continuous	Filtering 1
PK600	10	Continuous	Packaging 1, Packaging 2, Packaging 3

terms of the objective function for minimising production make-span are outlined and the values of the parameters used in optimisation are given.

For this problem, the objective function to be minimised is as follows:

F = *Contribution from each order + Cost of feedstocks - Cost of storage - Costs of utilities and tasks + penalties*

When the objective is the minimisation of delay in fulfilling orders, a *nominal* and a *hard* deadline must be specified for each order. Any order delivered after the nominal deadline incurs a penalty, which increases with the difference between the actual delivery time and the nominal deadline. If the solution results in all orders being fulfilled by their nominal deadlines, no delay penalty will be incurred. In any case, all orders must be fulfilled by the hard deadline.

When more than one product needs to be produced in a single run, it may be the case that delays in the production of some items are tolerated more than delays in others, so it becomes necessary to assign priorities. When used to minimise delays, gBSS attempts to minimise the total weighed delay in fulfilling orders, where the delay of each order is weighed by its priority. A priority is defined together with the form of the penalty function, which may be linear or square. A very high priority will insure that no delay is incurred unless it is absolutely unavoidable. Orders' priorities must be specified, together with a penalty form F_p. The penalty added to the objective function F is:

IF actual delivery time $\leq$ nominal deadline
THEN penalty = 0
IF nominal deadline $<$ actual delivery time $\leq$ hard deadline
THEN penalty = amount delayed * priority * F_p (actual delivery time – nominal deadline)

The penalty form chosen for the simulation (see Sect. 5) is linear and the priority set for all orders is 300. Nominal and hard deadlines were specified as 48 hrs and 144 hrs respectively.

The following assumptions are made for the formulation of the mathematical optimisation model:

(1) The time variable is discretised and the minimum period is 1 hour.

(2) No cleaning operations are performed after each job.

(3) A list of identical units within a stage, as can be found in table 2.

(4) Batch sizes are variable.

(5) There is unlimited storage capacity for intermediates.

(6) The minimal campaign length is set to 48 hours due the shape of the objective function, as has been explained.

In addition, it is assumed that the plant agent performs its inventory management activities in a similar way as the warehouses and retailers do.

5 Simulation considerations

5.1 The inventory policies

Four inventory policies have been implemented, including the *order-point and order-quantity* (s, Q) *system*, *order-point and order-up-to-level* (s,S) *system*, *periodic review and order-up-to-level* (R, S) *system*, as well as (R, s, S) *system* (Silver et al., 1998). The *order-point, order-quantity* (s, Q) *system* is a continuous review system where a fixed quantity Q is ordered whenever the inventory position (i.e., items on hand plus items on order) drops to the reorder point s or lower. The *order-point, order-up-to-level* (s,S) *system*, a continuous review system where an order is placed when the inventory position drops below s, but the amount ordered is variable and is enough as to raise the inventory position to the order-up-to-level S. The *periodic review, order-up-to-level* (R, S) *system* is also known as replenishment cycle system. Every R units of time, enough is ordered to raise the inventory position to level S. Because it is periodic, it offers advantages when coordinated inventory management of related items is necessary. The (R, s, S) *system* is a combination of the (s,S) and (R,S) systems. When using it, the inventory position must be checked every R time units. If it is below the minimum s, an order is placed so to reach the maximum S, but if it is above s, nothing is done until the decision has to be repeated in the next review period.

5.2 Retailers

The retailer agent fulfils the customers' demands on wood paints and coatings. Retailers use a "customer" function that sets boundary marks in demand by defining a monthly rate of demand that the retailers know up to the present moment and whose value they have to forecast for the next period. Retailers strive to fulfil the demand defined by the "customer" function, although their success depends on the warehouses' ability to supply and the success of the negotiation procedure when stock is scarce. The demand forecast is currently made using a weighed averages method. The reason is that this method can easily be adjusted to change the relative importance of the demand in previous months to forecast future demand. Operations research forecasting methods like moving averages or exponential smoothing may be implemented easily at a later stage. With this forecast and the inventory model, the orders are placed to the company (Logistics). Each retailer shares a fraction of the monthly demand through a weight constant.

5.3 Logistics

Logistics receives orders from all the retailers. These will be used together with other information about the physical locations of retailers and warehouses and the transportation cost per unit of finished product per unit of distance. Thus, logistics solves a classical transportation problem where the warehouses have equal capacity to fulfil the retailer's orders.

5.4 Warehouses

Warehouses are distribution centres for the finished products. They place orders to the plant according to their inventory policies.

5.5 Purchasing

The purchasing agent is in charge of ordering raw materials according to the information supplied by the plant.

5.6 Raw material suppliers (RMSs)

Each RMS provides a family of raw materials including resin, paint and package materials. These agents state a price based on the function that the raw material producers provide, and pass it to the company (purchasing agent).

6 Agent communication

The communication action between agents can be explained as follows. After the agent initialisation, an external agent sends a signal to the purchasing agent so to place an order to the raw material suppliers (beginning of period n). This in turn delivers the required raw materials to the plant and simultaneously place an *advertise* performative with the prices (i.e., the prices at period n+1) of the raw materials to purchasing. In turn, the plant *forwards* to the warehouses the materials that they ordered in period n-1 so they can deliver to the retailers according to the optimal transportation plan calculated by logistics in period n-1 for deliveries in the current period, and purchasing forwards to logistics so it can advertise prices for period n+1 to the retailers. The retailers then calculate their expected requirements using their inventory levels and forecasting methods and place their *orders* to logistics.

Logistics calculates the optimum transportation schedules for period n+1 given the transportation costs per distance unit, retailers and warehouses locations and supply abilities of each warehouse and proposes this to the warehouses. If the warehouses accept, they place in turn their *orders* to the plant. In contrast, if they reject Logistics' proposal, the negotiation process continues as described in the next Section. In any case, the negotiation process ends when the warehouses place their orders to the plant. Once the Plant receives the Warehouses' orders, it "produces" the materials that will be delivered at period n+1 by running the wrapped optimisation software gBSS, which calculates as a mixed-integer nonlinear program the production schedule that minimises production make-span. Once this has been done, the plant *tells* the warehouses that finished product has been delivered and places its own raw material requirements to purchasing, always according to its inventory levels. These raw materials will be received by the plant at n+1. Purchasing orders again to the raw materials suppliers and the cycle starts again.

As already mentioned, a neutral format for the message contents is necessary. We used the EXPRESS language and XML. One criticism of STEP is that STEP focuses more on data storage, for example, most tools developed are for transfer data between database systems, but less on exchange, particularly over the Internet. XML is now regarded as a potentially more powerful tool for data exchange.

The system starts with the Purchasing agent sending an *order* to the raw material suppliers. If the warehousing policy for the plant is continuous review with S = 1000, s = 500 for FeedResinRM1 and 4000, 2000 for FeedResinRM2, so that inventory is filled up to S in each case. If the plant's stock of FeedResinRM1 runs low at 10 months and FeedResinRM2 runs low at 11 months, the *order* is placed to the raw material supplier 1 (RMS1) with the following performative,

```
(order
   :sender Purchasing
   :receiver RMS1
   :language KQML
   :content ((K 3)(amountsOrdered00(508.49))
             (dueTimes00(10.0))
            (fullPrice00(50849.00))(amountsOrdered01
                                       (2000.46))
            (dueTimes01  (11.0))(fullPrice01(200046.00))))
```

The first piece of information in the body of the message, (K 3), means that the order is being sent in the third round of message exchanges, that is, the third period or planning horizon. The remaining contents are self-explanatory. There are equivalent orders to other RMSs.

A RMS fulfils the order with a delivery of material to the plant (i.e., the RMS communicates about the delivery with a *tell* performative), and at the same time sends an *advertise* to the Purchasing agent of the plant, containing the prices of raw materials for the next period. The buying cost for the raw material supplier are 100 £/ton of raw material, but as there is a gain of 0.2 for the supplier and a transportation cost from the RMS at Hull to the plant at Derby to be considered (0.5 £/ton Km), the price raises (100(1.2) + 0.5(139.8) = £ 189.9). The performatives sent are as follows,

```
(advertise
   :sender RMS1
   :receiver Purchasing
   :language KQML
   :content ((K 4)(unitPrice00 189.91)
             (unitPrice01 189.91)))
(tell
   :sender RMS1
   :receiver Plant
   :language KQML
   :content ((K 4)(amountDelivered00(508.49))
             (deliveryTimes00(10.0))
             (amountDelivered01 (2000.46))
           . (deliveryTimes01 (11.0))))
```

Purchasing simply *forward*s the prices to Logistics, as the latter is the responsible agent for managing budget. The plant then sends an empty *forward* to the warehouses to inform them that they can deliver the material requested by the retailers.

Logistics *advertise*s to the retailers the unit prices of its products for this period, considering gain, transportation costs from plant to warehouses and from warehouses to retailers, production costs and number of retailers in an analogous way as RMS did with purchasing, using the following performatives,

```
(advertise
   :sender Logistics
   :receiver Retailer0
   :language KQML
   :content ((K 4)(unitPrice00 162.39)
             (unitPrice01 162.39)
             (unitPrice02 162.39)))
```

Each warehouse delivers (i.e. *tell*s about the deliveries) to the retailers just as the RMS did with the plant. The amounts ordered depend on each retailer warehousing policy parameters. We have one message like this for each warehouse for each retailer.

```
(tell
   :sender Warehouse0
   :receiver Retailer0
   :language KQML
   :content ((K 4)(amountDelivered00
             (112.50 112.50))
             (deliveryTimes00 (10.25 11.75))
             (amountDelivered01 (112.50 112.50))
```

```
                (deliveryTimes01 (10.25 11.75))
                (amountDelivered02 (103.12))
                (deliveryTimes02 (10.75))))
```

The retailers *order* to Logistics. The total monthly demand is 300 ton/month of products 1 and 2, and 150 ton/month of product 3. As there are eight retailers, the monthly demand of each is 37.5 ton/month. The warehousing policy parameters are $S = 300$, $s = 200$, so the order placed is of around 100 ton on each order. The amount ordered comes from what the retailer forecast the monthly consumption of the customers for the next period (a customer is simply the function that defines what the monthly rate of demand is going to be) through a weighed averages method. Following is one of these messages sent by each retailer to Logistics.

```
(order
   :sender Retailer0
   :receiver Logistics
   :language KQML
   :content ((K 4)(amountOrdered00 (112.50 112.50))
             (dueTimes00 (13.25 14.75))
             (fullPrice00 (18269.32 18269.32))
             (amountOrdered01 (112.50 112.50))
             (dueTimes01 (13.25 14.75))
             (fullPrice01 (18269.32 18269.32))
             (amountOrdered02 (103.12))
             (dueTimes02 (13.5))
             (fullPrice02 (16746.88))))
```

With this information, logistics sends the orders of each retailer of each product to the warehouses and solves the transportation problem as explained.

The warehouses *order* to the plant independently, but if several orders arrive at the plant at the same time, it will result in a longer delay in production. This is a sample *order* message,

```
(order
   :sender Warehouse0
   :receiver Plant
   :language KQML
   :content ((K 4)(orderP00
             ((337.50 337.50 337.50)
             (12.75 13.25 14.5)))
             (orderP01 ((337.50 337.50 337.50)
             (12.75 13.25 14.5)))
             (orderP02 ((309.37)(12.75)))))
```

After gBSS finds a minimum make-span schedule, the plant sends the finished product to the warehouses and informs about the delivery through a *tell*, assuming that the delay between order placement and order reception equals the production delay.

```
(tell
   :sender Plant
   :receiver Warehouse1
   :language KQML
   :content ((K 4)(deliveryP10
              ((337.5 337.50 337.50)
              (12.25 13.75 15.0)))
              (deliveryP11 ((337.5 337.5 337.50)
              (12.25 13.75 15.0)))
              (deliveryP12 ((360.93)(14.5)))))
```

The warehouses *report* their performance to Logistics. Afterwards, Logistics pays to purchasing the plant operation costs, transportation costs and raw material price for the next period. Purchasing sends its part of the budget to the plant. The plant will inform Purchasing the raw materials needed for production in the next period. On receiving the message from the plant, Purchasing informs (*reports*) logistics that is ready to purchase raw materials for the next period. Logistics *authorises* purchasing to proceed to buy material for the plant. Purchasing places orders to the raw material suppliers and the cycle of performatives is repeated until the time reaches the pre-set number of periods for simulation.

7 Conflict resolution through negotiation

Three stages are involved in the agent negotiation process for conflict resolution: *proposal*, *counterproposal* and *termination*. A proposal is a solution to a problem proposed by an agent, based on its own knowledge and interests. A counter-proposal is the action taken by the receiving agent of amending the original proposal and sending it back to the originating agent. Proposals and counterproposals can be carried out for a number of times until they reach the state of termination, i.e. the end of a negotiation process. The state of termination may mean that a compromise decision has been reached, or that the negotiation failed. When a number of agents are involved, a co-ordinator is needed which observes the process and makes strategic decisions.

A negotiation process can be conveniently represented by a state transition diagram (Fig. 6). A transition diagram is a collection of circles (states) connected through labelled arcs (transitions). These labels give information about the required input to produce a change of state and/or output. One of the circles has a pointer (the *initial state*) and at least one of the states is represented as a double circle to represent a stop (an *accept state*). In our application, the circles in the transition diagram represent states of an agent when it receives different sequences of performatives from other agents or the environment. The possible sequences are represented as arcs.

The transition diagram in Figure 6 represents a negotiation protocol where the logistics agent acts as a co-ordinating agent involved in conflict resolution with two warehouses. The purpose of negotiation is to reach an agreement on whether the

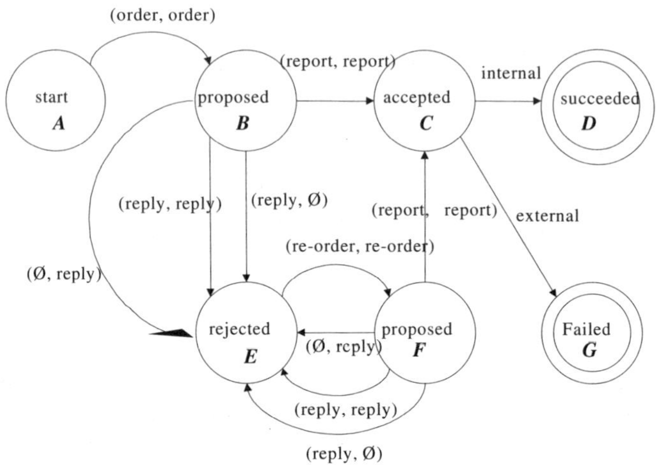

Fig. 6. Transition diagram for anegotiation protocols

warehouses are able to fulfil the Retailers' demand as requested by Logistics. As explained in Section 6, the warehouses have to cope at period n with deliveries that fulfil retailers' demands made in the previous period before receiving their own replenishment for the current period. So it may happen that a large order is received by a warehouse after the reorder point has been reached (assuming a continuous review (s, S) policy). The warehouse knows in this case that its own inventory level is too low, so it can inform logistics that it is not able to deliver as this agent had first requested according to the original optimal transportation schedule. This message is interpreted by the logistics agent as a request for a different solution, and therefore it is a petition to set a different constraint in the transportation problem. The affected constraint is the relative capacity of each warehouse to fulfil orders, set originally to equal capacities. Logistics assigns 1% less capacity to the "grumbling" warehouse until an agreement is reached. If this solution allows acceptable lowest inventory levels during the current planning horizon for both warehouses (that is, until replenishment arrives), the negotiation succeeds. The negotiation process continues until it succeeds or until there is no way to fulfil the demand without incoming shortage, in which case the negotiation fails.

The first word of the pair in brackets that labels a transition in Figure 6 represents the performative sent by the first warehouse and the second label represents the answer of the second warehouse to logistics. The symbol $\emptyset$ represents no answer. The labels internal and external represent the way demand is fulfilled. If negotiation concludes successfully, (node ***D***) the supply chain does not require outsourcing. Otherwise external supply is necessary. The negotiation process starts with the logistics agent sending orders to warehouses 0 and 1 (node ***A*** to ***B***). If the orders are accepted by both warehouses, they carry on with their duties, prepare reports and the negotiation is terminated successfully (node ***B*** to ***C*** to ***D***). If the orders are rejected by at least one of the warehouses (***B*** to ***E***), the negotiation starts (***E*** to ***F***). If the inventory level in any warehouse drops below its specified reorder point, it will inform the logistics. Logistics will calculate a new solution with a modified constraint and will send a new proposal. The negotiation will continue until reaching a state of termination (***C*** to either ***D*** or ***G***), with a success or failure negotiation result.

The conflict resolution mechanism does not guarantee an optimum solution be found each time, a dynamic simulation can help to evaluate the impact of a compromise decision on the dynamic behaviour of the supply chain.

8 llustrative results of distributed simulation

Simulation of supply chain dynamic behaviour is particularly useful when a compromise decision reached through negotiation needs to be evaluated. The simulation can also be used to study the impact of chain dynamics and uncertainty in demand and supply on the operating and control policies of chain components, such as inventory control policies as well as plant scheduling strategies. In contrast to traditional methods of modelling the whole supply chain as a group of algebraic and differential equations, the current system simulates the supply chain behaviour as a network of distributed and interacting agents, where each can be developed and maintained independently and has its own operations. In this section we present some illustrative results of simulation runs.

8.1 Comparison of different warehouse inventory policies

In this section, we compare results when using continuous review (s, S) and periodic review (R, s, S) inventory policies. The reviewing period is set to one month. The purpose is to demonstrate that various management and control strategies can be evaluated using the distributed simulation system. The customer's rate of demand is considered to be constant at 200 ton/month for products 1 and 2 and 100 ton/month for product 3.

It was found that under continuous review policy there are more withdrawals between orders than in periodic review. This is because the retailers in the previous link of the chain are also able to reorder every month only. Undershoots are also larger when working under periodic review. These differences also have an impact on the production pattern of the plant. Figure 7 shows the production patterns of the plant operating under periodic review with periods of one and two months respectively. Clearly under continuous review policy the plant needs to schedule the production more frequently but the amount of product processed is less in each run. The frequency that a process unit is used increases from continuous review of one month to periodic review periodic review of two months, but the differences in used capacity are not so big.

Of course, reducing the revision period when using periodic review increases the retailers' and warehouses' ordering frequencies and also the intervals between orders in the plant. The periodic review policy is reduced to continuous review when the revision period approaches zero.

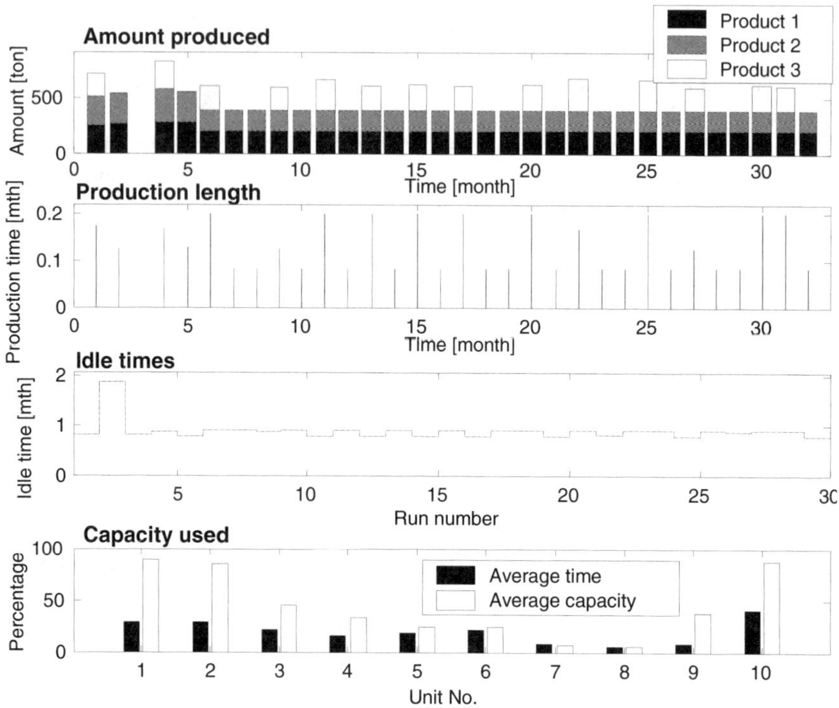

Fig. 7. Production pattern under periodic policy (period = 1 month)

8.2 *Effect of negotiation*

Optimal inventory parameters should minimise the probability of stock-out while keeping the stored amount to a minimum. As explained earlier, when a stock-out happens in a warehouse, the negotiation process is triggered. The results presented here were produced by combining a low reorder point (s is set to 70 tons) with a sudden increase in demand from 200 to 300 tons/month for products 1 and 2 and from 100 to 150 for product 3.

Figures 8 and 9 show the warehouses' inventory levels and the plant production regime when negotiation takes place. The net effect is a hectic delivery schedule for warehouse 0 between months 18 and 21, that is when the increased demand reaches the company's echelon (that is, the warehouses and the plant). No stock-out is incurred by warehouse one due to the negotiation carried out. The effects were also observed for the plant which was under great pressure to satisfy demand for all products between months 18 and 19. There were also longer production lead times that were calculated to optimality and increased transportation cost (Fig. 10). In a first instance, the transportation costs for the "shock period" are considerably larger than at any other time. These costs in following periods are larger than in the time before the increase; negotiation is still necessary after the step increase in demand. A re-tuning of warehousing policy parameters becomes necessary at

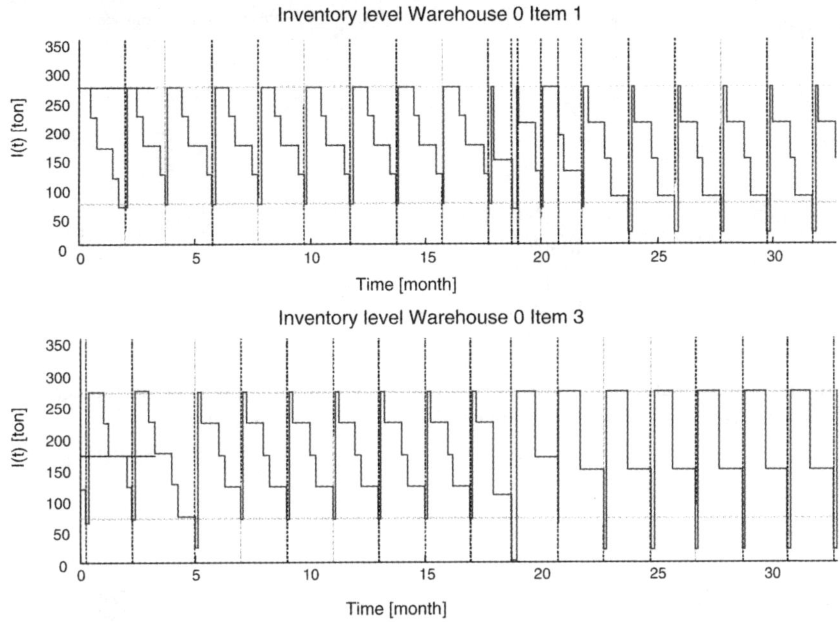

Fig. 8. Inventory plots for warehouse 0 (above) and warehouse 1 (below) when negotiation occurs (reorder point = 70 tons, step increase fromm 200 to 300 tons/month)

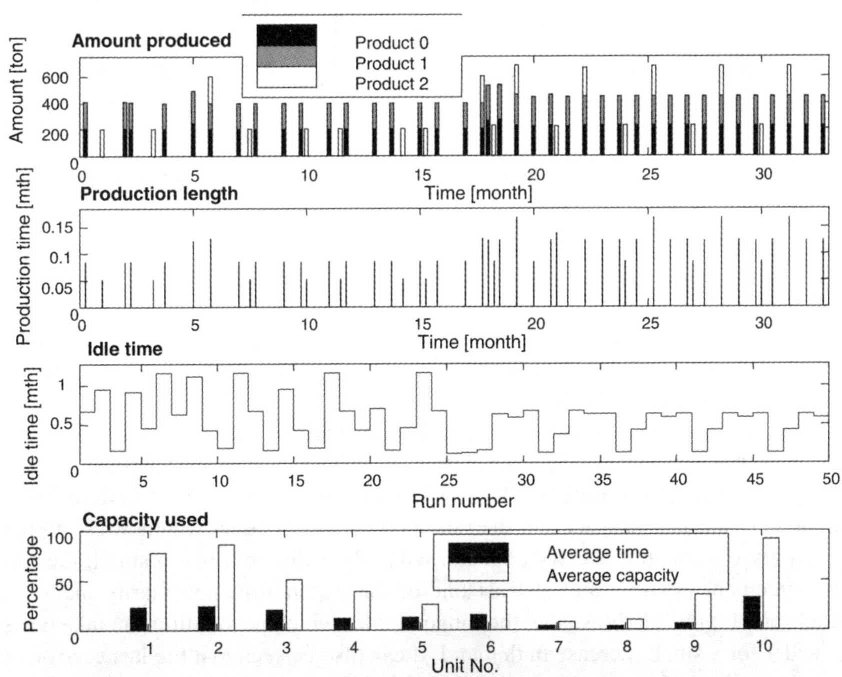

Fig. 9. Plant utilisation record for step from 200 to 300 tons/month (oproducts 0 and 1) and 100-150 tons/month (product 2) using negotiations, r= 70 tons

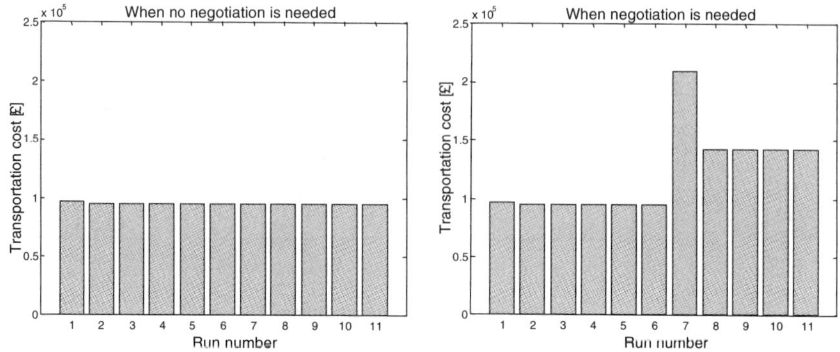

Fig. 10. Transportation costs with no step (left) and when negotiation takes place, step 200-300 tons/month for product 1 and reorder point set to 70 tons (right)

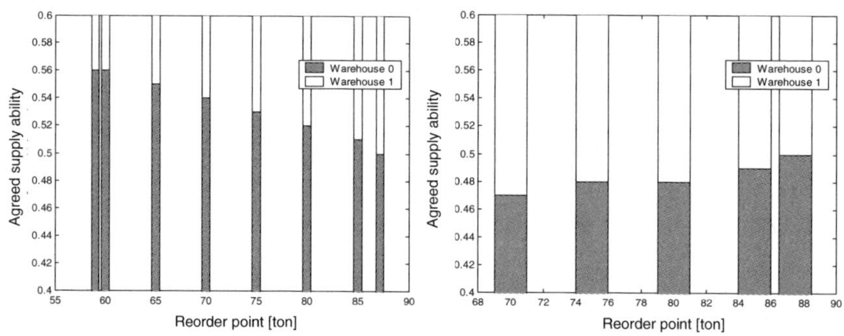

Fig. 11. Agreed supply abilities of each warehouse as a result of negotiation with a step of 200–250 tons/month (*left*); 200–300 tons (*right*)

this point to avoid all the inconveniences caused by plant overload, stock-outs and increased transportation costs.

Figure 11 shows the final agreed supply amount for each warehouse after negotiation has succeeded when the minimum inventory level s (identical in both warehouses) is set to different values. When the value of s exceeds a certain value, no negotiation is necessary and the initial constraint of equal warehouses' ability to supply is enough to satisfy demand; when the value of s is too small, there is no agreement possible and at least one of the warehouses necessarily incurs in stock-out. Figure 11 shows that the range of s for which negotiation can take place is wider for a small increase in demand. It can also be seen that the last cession of supply ability before negotiations fail is larger for the small increase ($|0.5\text{-}0.56|$) than for the large increase ($|0.5\text{-}0.47|$), so it can be said that small increases in demand allow the system to show more flexibility than large perturbations.

9 Concluding remarks

The architectural and functional characteristics of a multi-agent system for chemicals supply chain simulation and management support has been described. The agent-based architecture provides a natural paradigm for implementing distributed and co-operative systems over the Internet. Geographically distributed supply chain components including retailers, warehouses, chemical plants, logistics and raw material suppliers are modelled as a network of agents, each with its internal models and algorithms and being able to communicate with others. This is a flexible and open structure because it can be easily maintained and reconfigured when chain structure is reorganised.

It has also been shown that supply chain behaviour can be analysed using the distributed simulation system. The results have shown a sensible behaviour of the plant production patterns. The system has also demonstrated the potential to be used to study how to improve the plant production scheduling and control policies in order to cope with the uncertainty of chain dynamics. Using the prototype, it has been shown that re-tuning of warehousing policy parameters became necessary after a steep increase on demand in order to avoid inconveniences produced by plant overload, stock-outs and increased transportation costs. However, a specific feature of agent systems (automated negotiation) managed to avoid stock-outs and keep service level standards at the most difficult moment combining varied OR techniques.

Study on supply chain dynamics may also have an impact on the way a manufacturing plant is designed. Traditionally the plants are not designed flexible and agile enough to respond to changes in the market environment. Plants that are part of an ever-changing economic environment should be designed to be re-configurable within hours rather than days and even weeks in order to satisfy ever-changing demand.

References

Applequist GE, Pekny JF, Reklaitis GV (2000) Risk and uncertainty in managing chemical manufacturing supply chains. Computers and Chemical Engineering 24: 2211–2222

Agrawal N, Cohen M (2001) Optimal material control in an assembly system with component commonality. Naval Research Logistics 48: 409–429

Axsater S, Zhang WF (1999) A joint replenishment policy for multi-echelon inventory control. International Journal of Production Economics 59: 243–250

Baita F, Ukovich W, Pensenti R, Favaretto D (1998) Dynamic routing-and-inventory problems: a review. Transportation Research 32: 585–598

Batres R, Lu ML, Naka Y (1997) An agent-based environment for operational design. Computers and Chemical Engineering 21: S71–S76

Benson R (1998) A wider view of competitiveness. The Chemical Engineer April: 16–17

Berning G, Brandenburg M, Gürsoy K, Mehta V, Tölle FJ (2002) An integrated system solution for supply chain optimisation in the chemical process industry. OR Spectrum 24, No. 4 (to appear)

Bodington CE, Shobrys DE (1999) In: Kane LA (ed) Optimise the supply chain Advanced process control and information systems for the process industries, pp 236–240. Gulf Pub, Houston

Bok JK, Grossmann IE, Park S (2000) Supply chain optimisation in continuous flexible process networks. Industrial and Engineering Chemistry Research 39: 1279–1290

Bose S, Pekny JF (2000) A model predictive framework for planning and scheduling problems: a case study of consumer good supply chain. Computers and Chemical Engineering 24: 329–335

Cachon GP, Zipkin PH (1999) Competitive and cooperative inventory policies in a two-stage supply chain. Management Science 45: 936–953

Chandra P, Fisher ML (1994) Co-ordination of production and distribution planning. European Journal of Operational Research 72: 503–517

Chauhan D (1997) JAFMAS: A Java-based agent framework for multi-agent systems development and implementation. PhD thesis, ECECS Department, University of Cincinnati

Chen Y, Peng Y, Finin T, Labrou Y, Cost S, Chu B, Yao J, Sun R, Wilhelm B (1999) A negotiation based multi-agent system for supply chain management. Working Notes of the Agents '99 Workshop on Agents for Electronic Commerce and Managing the Internet-enabled Supply Chain, Seattle, WA

Cohen MA, Lee HL (1989) Resource deployment analysis of global manufacturing and distribution networks. Journal of Manufacturing Operations Management 2: 81–104

Diks EB, de Kok AG, Lagodimos AG (1996) Multi-echelon systems: a service measure perspective. European Journal of Operational Research 95: 241–263

Diks EB, de Kok AG (1998) Optimal control of a divergent multi-echelon inventory system. European Journal of Operational Research 111: 75–97

Diks EB, de Kok AG (1999) Computational results for the control of a divergent N-echelon inventory system. European Journal of Operational Research 59: 327–336

Forrester J (1958) Industrial dynamics: a major breakthrough for decision makers. Harvard Business Review July-August: 37–66

Fox MS, Barbuceanu M, Teigen R (2000) Agent-oriented supply chain management. International Journal of Flexible Manufacturing Systems 12: 165–188

García-Flores R, Wang XZ, Goltz GE (2000) Agent-based information flow for process industries' supply chain modelling. Computers and Chemical Engineering 24: 1135–1141

Gibson WD (1998) Getting a grip on the supply chain. Chemical engineering November: 35–39

Gjerdrum J, Shah N, Papageorgiou LG (2000) A combined optimisation and agent-based approach for supply chain modelling and performance assessment. Production Planning and Control 12: 81–88

Gjerdrum J, Shah N, Papageorgiou LG (2001) Transfer prices for multienterprise supply chain optimisation. Industrial and Engineering Chemistry Research 40: 1650–1660

Gupta A, Maranas CD, McDonald CM (2000) Mid-term supply chain planning under demand uncertainty: customer demand satisfaction and inventory management. Computers and Chemical Engineering 24: 2613–2621

Haehling von Lanzenauer C, Pilz-Glombik K (2002) Coordinating supply chain decisions: an optimization model. OR Spectrum 24: 59–78

Han SY, Kim YS, Lee TY, Yoon T (2000) A framework of concurrent process engineering with agent-based collaborative design strategies and its application on plant layout problem. Computers and Chemical Engineering 24: 1673–1679

Haq AN, Vrat P, Kanda A (1991) An integrated production-inventory-distribution model for manufacture of urea: a case. International Journal of Production Economics 39: 39–49

Howard J (1999) A colourful prospect. Chemistry and Industry 10: 380–382

Janicke W (2000) Supply chain coordinators in the planning of systems of linked chemical multiproduct plants. Chemie Ingenieur Technik 72: 580–584

Jennings NR, Norman TJ, Faratin P, O'Brien P, Odgers B (1998) ADEPT: an agent-based approach to business process management. ACM SIGMOD Record 27: 32–39

Jennings NR, Wooldridge M (1999) Agent-oriented software engineering. Proceedings of the 12th International Conference on Industrial and Engineering Applications of AI, Cairo, Egypt, pp 4–10

Kafoglis C (1999) Maximize competitive advantage with a supply chain vision. Hydrocarbon Processing February: 47–50

Kallrath J (2002) Planning and scheduling in the process industry. OR Spectrum 24: 219–250

Knotts G, Dror M, Hartman BC (2000) Agent-based project scheduling. IIE Transactions 32: 387–401

Kondili E, Pantelides CC, Sargent RWH (1993) A general algorithm for short-term scheduling of batch operations part I – mathematical formulation. Computers and Chemical Engineering 17: 211–227

Labrou Y, Finin T (1997) A proposal for a new KQML specification. paper TR CS-97-03 http://wwwcseeumbcedu/~jklabrou/publications/tr9703ps

Maloni MJ, Benton WC (1997) Supply chain partnerships: opportunities for operations research. European Journal of Operational Research 101: 419–429

McGreavy C, Wang XZ, Lu ML, Naka Y (1995) A concurrent engineering environment for chemical manufacturing. Concurrent Engineering Research And Applications 3: 281–293

Papageorgiou LG, Pantelides CC, (1996) Optimal campaign planning/scheduling of multipurpose batch/semicontinuous plants. Industrial and Engineering Chemistry Research 35: 488–529

Parunak HV, Savit R, Riolo RL, Clark S J (1999) DASCh: Dynamic analysis of supply chains. DASCh final report CEC-0 115 presented at MABS-98 (Paris)

Pendharkar PC (1999) A computational study on design and performance issues of multiagent intelligent systems for dynamic scheduling systems. Expert Systems with Applications 16: 121–133

Peng Y, Finin T, Labrou Y, Cost RS (1999) An agent-based approach for manufacturing integration - the CIIMPLEX experience. Applied Artificial Intelligence 13: 39–63

Perea-Lopez E, Grossmann IE, Ydtsie BE, Tahmassebi T (2001) Dynamic modelling and decentralised control of supply chains. Industrial and Engineering Chemistry Research 40: 3369–3383

Seo Y, Jung S, Hahm J (2002) Optimal reorder decision utilizing centralised stock information in a two-echelon distribution system. Computers and Operations Research 29: 171–193

Shah N, Pantelides CC, Sargent WH (1993) A general algorithm for short-term scheduling of batch operations – II Computational issues. Computers and Chemical Engineering 2: 229–244

Sherbrooke CS (1968) Metric: a multi-echelon technique for recoverable item control. Operations Research 16: 122–141

Silver EA, Pyke DF, Peterson R (1998) Inventory management and production planning and scheduling, 3 edn. Wiley, New York Chichester Toronto

Stoye D, Freitag W (ed) (1998) Paints, coatings and solvents, 2nd edn. Wiley-VCH, Berlin

Svoronos A, Zipkin P (1988) Estimating the performance of multi-level inventory systems. Operations Research 36: 57–72

Swaminathan JM, Smith SF, Sadeh NM (1998) Modelling supply chain dynamics: a multiagent approach. Decision Sciences 29: 607–632

Tempelmeier H (2000) Inventory service-levels in the customer supply chain. OR Spektrum 22: 361–380

Thomas DJ, Griffin PM (1996) Coordinated supply chain management. European Journal of Operational Research 94: 1–15

Timpe CH, Kallrath J (2000) Optimal planning in large multi-site production networks. European Journal of Operational Research 126: 422–435

Towill DR (1997) The seamless supply chain – the predator's strategic advantage. International Journal of Technology Management 13: 37–56

Van der Heijden M, Diks E, de Kok T (1999) Inventory control in multi-echelon divergent systems with random lead times. OR Spektrum 21: 331–359

Van Roy T (1989) Multi-Level production and distribution planning with transportation fleet optimisation. Management Science 35: 1443–1453

Vidal CJ, Goetschalckx M (1997) Strategic production-distribution models: a critical review with emphasis on global supply chain models. European Journal of Operational Research 98: 1–18

Zhou Z, Cheng S, Hua B (2000) Supply chain optimisation of continuous process industries with sustainability considerations. Computers and Chemical Engineering 24: 1151–1158

An application of mixed-integer linear programming models on the redesign of the supply network of Nutricia Dairy & Drinks Group in Hungary

Francisca H.E. Wouda[1], **Paul van Beek**[2], **Jack G. A. J. van der Vorst**[3], **and Heiko Tacke**[4]

[1] Friesland Dairy&Drinks Group, Rijksweg 66, 2880 Bornem, Belgium
[2] Wageningen University, Operations Research & Logistics Group, Hollandseweg 1, 6706 KN Wageningen, The Netherlands
[3] Wageningen University, Management Research Group, Hollandseweg 1, 6706 KN Wageningen, The Netherlands
[4] ISAGRI Agrarsoftware, Eickerhoeh 1, 58553 Halver, Germany

Abstract. Between 1995 and 1998 Nutricia acquired a number of dairy companies in Hungary. Each of these companies produced a wide variety of products for its regional market. Although alterations had been made to the production system in the last few years, production and transportation costs were still substantial. This paper presents a research study with regard to the optimisation of the supply network of Nutricia Hungary using a mixed-integer linear programming model. Focussing on consolidation and product specialisation of plants the objective was to find the optimal number of plants, their locations and the allocation of the product portfolio to these plants, when minimizing the sum of production and transportation costs. The model is in line with traditional location/allocation models, with a modification concerning inter-transportation of semi-finished products between plants. The production costs used in this model are based on a Green field situation, taking into account new and more advanced technologies available today. The model is used by the Nutricia Dairy and Drinks Group as a decision supporting tool.

Key words: Mixed-integer linear programming – Location/allocation – Economy of scale – Production/distribution – Scenario analysis

Correspondence to: F. H. E. Wouda

1 Introduction

The Nutricia Dairy & Drinks Group (NDDG) has grown continuously since the beginning of its activities in Hungary in 1995. Since then a number of dairy companies with multiple plants has been acquired, each plant producing a wide range of products supplying dairy products to a specific region. In order to improve the logistic system and to reduce the production costs, the number of plants and the product portfolio were rationalised (resulting in three main brands, namely Milli, Pöttyös and Oké) and the distribution system was reconfigured. Nowadays, NDDG produces about 300 dairy products in different packages, different flavours and under different brand names; these are products like yoghurt, cheese, milk powder, fresh and long shelf-life milk, cream, butter and desserts. Although changes have been made to the system in the last few years, total system costs in Hungary are still too high due to a number of inefficiencies. NDDG wanted to evaluate a number of redesign strategies:

- *Regionalisation*: each plant serves the customer with a full range of products within the region.
- *Consolidation*: production is centralized at one plant and serves the whole country.
- *Product specialisation*: each plant is producing a certain product group and serves the whole country (via several distribution centers).
- *Process specialisation*: each plant is specialised in a certain stage of manufacturing, e.g. a plant for milk reception, for preparation and one for packing the product. The customers are served from the packaging plant.

The objective of this research was to evaluate these strategies by identifying the optimal number of plants, their locations and the allocation of the product portfolio to these plants, when minimizing the sum of production and transportation costs.

In literature, a number of models and procedures for the analysis of manufacturing strategies can be found (e.g. Cohen and Lee, 1989; Eppen et al., 1998; Krajewski and Ritzman, 1999). The contribution of this paper is first of all the modification that had to be made for the inter-transportation of semi-finished products between the plants. Secondly, it is an example of the applicability of Operations Research techniques in real life, since the model is actually used by a dairy company.

The remaining part of this paper is structured as follows. First, in Section 2, the case analysis is described. In Section 3 the location/allocation model is formulated. Section 4 describes the results followed by the conclusions and discussion in Section 5.

2 Case analysis and demarcation

Figure 1 presents an overview of the supply network considered in this research. Currently more than 400 farmers situated mostly in the eastern part of Hungary are supplying the 9 plants daily with raw milk. The size of the farms differ from 20 thousand liter to 11 million liter on annual basis, although there is a tendency that the small farms grow larger or disappear. Small farms deliver their milk to so called milk collecting points. The milk is collected by a transport company at these milk collecting points and directly at the larger farms.

The 9 plants produce over 300 products. Some of the plants are already dedicated factories for a certain product group. During production semi-finished products are prepared. These semi-finished products (e.g. cream, whey, buttermilk, permeate) are transported to one of the other plants and processed into finished products. Since most of the products produced are fresh products, the production volume is strongly related to the daily demand of products. For some products like cheese, milk powder and long shelf-life milk an amount of stock is generated. Depending on the demand of certain products the daily inter-transport volumes vary.

The products are forwarded to 17 distribution centres which are serving over 17.000 shops spread all over the country. In almost each county a distribution center is located, taking care of regional distribution to the shops. For some of the big retailer chains transport is done directly from the plant to the central distribution center of this retailer. A part of the production (e.g. cheese, milk powder in bulk and long shelf-life milk) is meant for export and for industrial customers.

If we focus on this part of the supply network, the following cost factors can be identified:

- Milk collection costs: raw milk is collected from the farmers until the milk collection truck has reached its maximum capacity.
- Transportation costs of raw milk: after milk collection the milk is pumped into a bigger truck which will deliver the milk to the plant where it is needed.
- Production costs: the raw milk is processed into finished products and semi-finished products (like cream, whey, permeate and buttermilk).
- Inter-transportation costs: the semi-finished products are transported in bulk to other plants where they are needed.
- Transportation costs of finished products: the finished products are transported to the distribution centers or sometimes directly to the (industrial) customer.
- Warehousing costs within the distribution centers are incurred for order picking and handling.

All costs mentioned above will have impact on the number of plants to be opened, their location(s) and the allocation of products to these location(s). The number of locations is mainly influenced by the production costs; the more economy of scale within production, the lower the production costs per unit. According to Slack et al., (1998) there are two categories of stimuli influencing location decisions: supply-side influences and demand-side influences. In this case, the *supply-side influences* are mainly the transportation costs of raw milk and inter-transportation costs. The *influences on the demand side* are mainly transportation costs of finished products. Finally, the allocation of the product portfolio to plants is mainly determined by the trade-off between transportation costs and production costs.

To model over 400 farmers, 300 products distributed via 17 distribution centers to more than 17.000 shops is very complex. Since this case study is used to find a long term strategy, a number of data aggregations have been made. The first aggregation of data is to consider the milk volumes, production volumes and sales

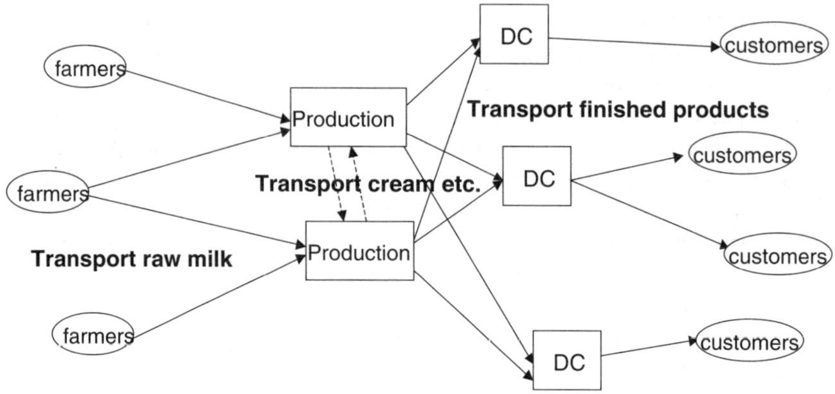

Fig. 1. Supply network of a dairy company

volumes on a yearly basis. No day to day variations are taken into account. The raw milk volume is quite constant during the year. The milk is more concentrated in winter time, smaller volume but a higher protein and fat content than in summer time.

Aggregation of farmers into milk regions

The farmers are grouped into geographical milk regions. The size of a milk region should not exceed a certain diameter or volume. The farmers are grouped according to this rule, resulting in 9 regions and per region a gravity point in milk volume is defined. Within the region the collection costs are calculated. The collection costs depend on the distance between the farms and the annual milk volume of the farms.

Aggregation of products

The main criterion to put up the 300 products into a number of product groups was to combine products which can be produced using the same preparation and packaging equipment, resulting in 13 product groups (Walstra et al., 1998). Figure 2 depicts an overview of the production steps of dairy products. The first number of steps, from raw milk reception to standardisation, is the same for all product groups (see the dashed part of the process description in Fig. 2). The milk is standardised at the right fat content needed for production of a certain product. This can result in a surplus of cream or an additional need of cream if the fat content needed is higher than the current fat content of the raw milk. Additional cream is transported from another plant or surplus cream is transported to another plant. The first common part of the process is analysed separately from the other more specific processing and packaging steps. If a number of product groups is produced in one plant, economies of scale can be gained in the first common part of the process.

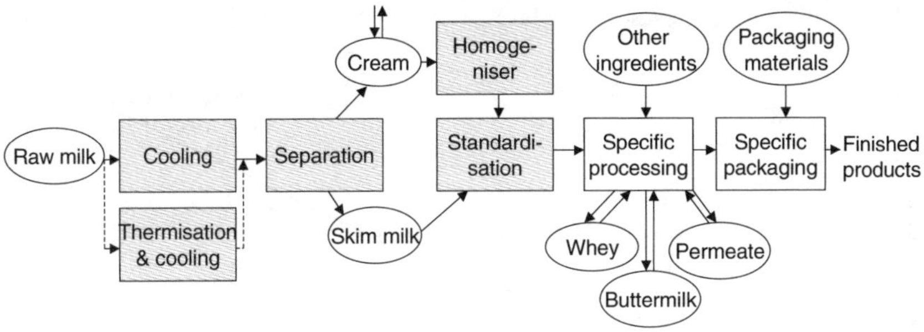

Fig. 2. Production chain of dairy products

Definition of production costs

The costs per product group could be determined in the existing situation, but then it depends on the current capacity of the plant, the product portfolio, the efficiency achieved with the existing equipment etc. If the model determines the strategically optimal location(s) of plants and allocation of product groups, the product portfolio will most probably change. Also the capacity needed for a certain plant, will change. Besides that, it is the question if the existing equipment will be used in the long run. New and improved technologies should be considered. For this case study a method based on "Green field costs" is used. These are the costs associated with a Green field situation; define the capacity needed for a certain product group, invest in the newest and most advanced and proven equipment available today, arrange this into the most efficient layout of the plant and the needed utilities around it. Directly related to the choice of equipment is the need of direct personnel, the energy need and the maintenance cost of the equipment.

For a number of capacities of plants the Green field investments and personnel level in 1, 2 and 3 shifts a day needed (direct and indirect) are determined together with the energy use and maintenance costs of the equipment. This result is presented in Figure 3 for one of the product groups.

It turns out that the relation between total annual costs and annual capacity can be described very accurately by a fixed-charge linear cost function (Hax and Candea, 1984; Vidal and Goetschalckx, 2001). For each product group this linear cost function is determined and used in the model.

Aggregation of distribution centers & shops

In the current situation, finished products are transported from all 9 plants to 17 distribution centers widely spread over Hungary. Most distribution centres serve their own region, but some are right next to the plant and serve their own region as well as other distribution centers and larger customers. Since on the long term the number and location of distribution centres is not stringent, the location and demand from customers (shops) is used. The shops are grouped into 20 geographical regions and per region a gravity point in sales volume is determined. These geographical

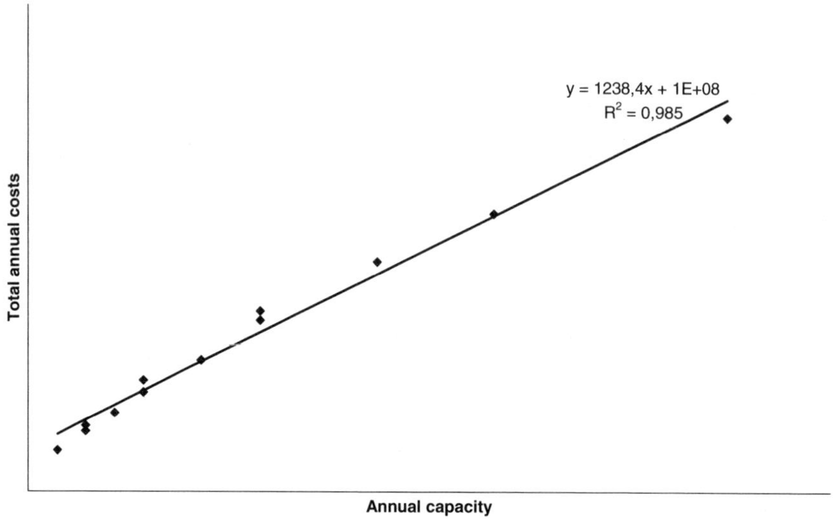

Fig. 3. Total annual costs of a Green field plant related to its annual capacity (the dots refer to the combination of capacity/shift options)

regions are the 19 counties of Hungary and Budapest. The gravity point is the capital of the county, since this is by far the biggest city in most of the counties.

3 MILP formulation of the location/allocation model

The location-allocation problem could be described as:

> *'What would be the optimal number, location and size of the production plants and what product groups should each plant produce, given the market demand and the milk supply, such that the total costs (comprising the production and transportation costs) are minimized.'*

To solve this problem a location/allocation model is build consisting of an objective function and restrictions. The variables and the objective function of the model are described in Section 3.1. The restrictions are described in Section 3.2.

3.1 The variables and the objective function

The variables used in the location/allocation model are defined below:

Indices:

i	milk region (i = 1,2,...,I)
j,h	location of plant (j,h = 1,2,...,J)
k	sales region (k = 1,2,...,K)
p	product group (p = 1,2,...,P)

Variables:

X_{ij}	amount of milk transported from region i to location j (in liters)
$PROD_{jp}$	amount of product p produced at location j (in pallets)
$CREAM_{hj}$	amount of cream transported from location h to location j (in liters)
$CRREQ_j$	net amount of cream required at location j (in liters)
$CRSURP_j$	net amount of cream surplus at location j (in liters)
$SCREAM_j$	surplus of cream in location j after transportation to lack locations (in liters)
$WHEY_{hj}$	amount of whey transported from location h to location j (in liters)
$WHREQ_j$	net amount of whey required at location j (in liters)
$WHPROD_j$	net amount of whey produced at location j (in liters)
$SWHEY_j$	surplus of whey in location j after transportation to lack locations (in liters)
BUT_{hj}	amount of buttermilk transported from location h to location j (in liters)
$BUREQ_j$	net amount of buttermilk required at location j (in liters)
$BUPROD_j$	net amount of buttermilk produced at location j (in liters)
$SBUT_j$	surplus of buttermilk in location j after transportation to lack locations (in liters)
PER_{hj}	amount of permeate transported from location h to location j (in liters)
$PEREQ_j$	net amount of permeate required at location j (in liters)
$PEPROD_j$	net amount of permeate produced at location j (in liters)
$SPER_j$	surplus of permeate in location j after transportation to lack locations (in liters)
Z_{jkp}	amount of product p transported from location j to depot k (in pallets)

Binary variables:

$YPROD_{jp}$	binary decision variable for setting up a production line p at location j
$YMILK_j$	binary decision variable for setting up a milk reception at location j

Coefficients:

mcc_i	milk collection costs in region i (HUF per liter)
tcm_{ij}	milk transportation costs from region i to location j (HUF per liter)
smrc	set-up costs milk reception (HUF)
mrc	variable costs milk reception (HUF per liter)
spc_p	set-up costs production line product p (HUF)
pc_p	variable costs production of product p (HUF per pallet)
tci_{hj}	transportation costs of semi-finished products from location h to j (HUF per liter)
tcf_{jk}	transportation costs finished products from location j to depot k (HUF per pallet)
mav_i	amount of milk available in region i (liters)
d_{kp}	demand of product p in sales region k (pallets)
a_p	amount of skim milk required per pallet of product p (liters)
skm	skim milk percentage of the raw milk
crm	cream percentage of the raw milk
cr_p	amount of cream required per pallet of product p (liters)
kwh_p	amount of whey required or produced per pallet of product p (liters)
bu_p	amount of buttermilk required or produced per pallet of product p (liters)
pe_p	amount of permeate required or produced per pallet of product p (liters)
bigpos	big positive number

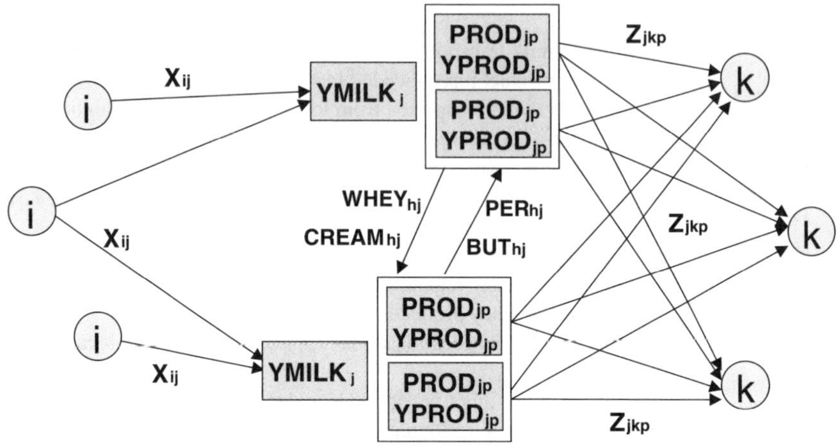

Fig. 4. Supply network with material flows and associated decision variables

mxloc maximum number of locations that should be opened

HUF (Hungarian Forint) is the currency of Hungary. 250 Hungarian Forints equal 1 Euro.

The objective function can be described as follows:

Minimize {milk collection costs + milk transportation costs from region to plant + milk reception costs + production costs + transportation costs of inter-deliveries from plant to plant + transportation costs of finished products from plant to sales region}

In Figure 4 the network of the model is visualized. The model starts at the milk region i. In this region there are costs incurred to collect the milk and transport it to the gravity point of the milk region. The milk collection costs per liter differ per region i and are indicated by mcc_i. After collecting the milk, it is transported from the gravity point of region i to the plant. The transported volume from region i to location j is called X_{ij}. The transportation costs of raw milk are defined as tcm_{ij}. If the binary variable $YMILK_j$ is equal to 1, there is a milk reception unit at location j. This results in set-up costs $smrc$ and variable costs mrc. If the binary variable $YPROD_{jp}$ is equal to 1, there is a production line of product p at location j. This results in set-up costs spc_p and variable costs pc_p for product p. The production volume of product p at location j is called $PROD_{jp}$.

For the production of product p, a certain amount of cream, whey, permeate or buttermilk could be required in addition to raw milk or a certain amount of them could be produced during production of product p. The amount required per location is called $CRREQ_j$, $WHREQ_j$, $PEREQ_j$ or $BUREQ_j$ respectively. When less cream, whey, permeate or buttermilk is required than produced at a location, there is a surplus called $CRSURP_j$, $WHPROD_j$, $PEPROD_j$ and $BUPROD_j$ respectively. The lack of cream, whey, permeate and buttermilk is fulfilled by transporting the required amount from a surplus location. The inter-transportation costs of these 'semi-finished products' from location h to location j

are called tci_{hj}. The amount of semi-finished products transported is indicated by $CREAM_{hj}$, $WHEY_{hj}$, BUT_{hj} and PER_{hj}. In this case study the net amount of cream, whey, permeate and buttermilk in total is a surplus. It is left at the location where it is produced. The surpluses are indicated by $SCREAM_j$, $SWHEY_j$, $SBUT_j$ and $SPER_j$ respectively. In reality the surpluses will be converted into powder or butter in case of cream and are stored or sold to other companies, or the semi-finished products are sold as cattle feed. In the model no cost effects of these surpluses are taken into account, since they are negligible. Finally finished products p are transported to sales regions k, resulting in transportation costs tcf_{jk} and material flows Z_{jkp}.

The model contains 126 binary variables, 2862 continuous variables and another 108 variables, which do not play a role in the objective function but in the restrictions. The optimisation has been carried out with the Xpress – MP optimisation system from Dash Optimization Ltd. Computing times varied from 15–60 minutes depending on the scenario (for scenarios see Table 1).

In mathematical form the objective function can be written as:

$$\min \left\{ \begin{array}{l} smrc \cdot \sum\limits_j YMILK_j + \sum\limits_i \sum\limits_j (mcc_i + mrc + tcm_{ij}) \cdot X_{ij} \\ + \sum\limits_p spc_p \cdot \sum\limits_j YPROD_{jp} \\ + \sum\limits_p pc_p \cdot \sum\limits_j PROD_{jp} \\ + \sum\limits_h \sum\limits_j tci_{hj} \cdot (CREAM_{hj} + WHEY_{hj} + PER_{hj} + BUT_{hj}) \\ + \sum\limits_j \sum\limits_k tcf_{jk} \cdot \sum\limits_p Z_{jkp} \end{array} \right\} \tag{1}$$

3.2 Restrictions

The model should take into account certain restrictions. For example, demand should be satisfied by production and milk supply from a region must not exceed the amount of milk available in that region. There are restrictions for milk transport, milk reception, production, inter-transport, transportation of finished products and possible plant locations. The restrictions are explained below.

Milk transportation and milk reception

The transported amount of milk from region i to location j cannot exceed the amount of milk available in region i:

$$\sum_j X_{ij} \leq mav_i, \qquad \text{for all } i \tag{2}$$

Location j can only process raw milk if it has a milk reception unit:

$$\sum_i X_{ij} \leq bigpos \cdot YMILK_j, \qquad \text{for all } j \tag{3}$$

Production

The supplied amount of milk must equal the amount of milk required for production at location j:

$$skm \cdot \sum_i X_{ij} = \sum_p a_p \cdot PROD_{jp}, \quad \text{for all } j \tag{4}$$

Location j can only produce product group p if it has a production line for p:

$$PROD_{jp} \leq bigpos \cdot YPROD_{jp}, \quad \text{for all } j, p \tag{5}$$

The amount of net cream required or surplus at location j is defined by the following formula:

$$CRREQ_j - CRSURP_j = \sum_p cr_p \cdot PROD_{jp} - \sum_i crm \cdot X_{ij}, \quad \text{for all } j \tag{6}$$

The amount of net whey required or produced at location j is defined by the following formula:

$$WHREQ_j - WHPROD_j = \sum_p wh_p \cdot PROD_{jp}, \quad \text{for all } j \tag{7}$$

The amount of net buttermilk required or produced at location j is defined by the following formula:

$$BUREQ_j - BUPROD_j = \sum_p bu_p \cdot PROD_{jp}, \quad \text{for all } j \tag{8}$$

The amount of net permeate required or produced at location j is defined by the following formula:

$$PEREQ_j - PEPROD_j = \sum_p pe_p \cdot PROD_{jp}, \quad \text{for all } j \tag{9}$$

Since all variables must be positive or zero, each location j has a requirement for semi-finished products or a surplus.

Inter-transport

The transported amount of cream from j to h should not exceed the surplus:

$$\sum_h CREAM_{jh} \leq CRSURP_j, \quad \text{for all } j \tag{10}$$

The transported amount of cream from h to j must satisfy the requirement of cream at location j:

$$\sum_h CREAM_{hj} = CRREQ_j, \quad \text{for all } j \tag{11}$$

The net surplus of cream at location j after transportation to location h is defined by the following formula:

$$SCREAM_j = CRSURP_j - \sum_h CREAM_{jh}, \quad \text{for all } j \tag{12}$$

The transported amount of whey from j to h must not exceed the surplus:

$$\sum_h WHEY_{jh} \leq WHPROD_j, \quad \text{for all } j \tag{13}$$

The transported amount of whey from h to j must satisfy the requirement of whey in location j:

$$\sum_h WHEY_{hj} = WHREQ_j, \quad \text{for all } j \tag{14}$$

The net surplus of whey at location j after transportation to location h is defined by the following formula:

$$SWHEY_j = WHPROD_j - \sum_h WHEY_{jh}, \quad \text{for all } j \tag{15}$$

The transported amount of buttermilk from j to h must not exceed the surplus:

$$\sum_h BUT_{jh} \leq BUPROD_j, \quad \text{for all } j \tag{16}$$

The transported amount of buttermilk from h to j must satisfy the requirement of buttermilk in location j:

$$\sum_h BUT_{hj} = BUREQ_j, \quad \text{for all } j \tag{17}$$

The net surplus of buttermilk at location j after transportation to location h is defined by the following formula:

$$SBUT_j = BUPROD_j - \sum_h BUT_{jh}, \quad \text{for all } j \tag{18}$$

The transported amount of permeate from j to h must not exceed the surplus:

$$\sum_h PER_{jh} \leq PEPROD_j, \quad \text{for all } j \tag{19}$$

The transported amount of permeate from h to j must satisfy the requirement of permeate in location j:

$$\sum_h PER_{hj} = PEREQ_j, \quad \text{for all } j \tag{20}$$

The net surplus of permeate at location j after transportation to location h is defined by the following formula:

$$SPER_j = PEPROD_j - \sum_h PER_{jh}, \quad \text{for all } j \tag{21}$$

Transportation of finished products

The total amount of finished products transported to all gravity points k must equal the production:

$$\sum_k Z_{jkp} = PROD_{jp}, \qquad \text{for all } j, p \tag{22}$$

The production of product p must be equal to the demand of product p (with d_{kp} the demand of product p in gravity point k):

$$\sum_j Z_{jkp} = d_{kp}, \qquad \text{for all } k, p \tag{23}$$

Restricted number of possible plant locations

The maximum number of locations (mxloc) can be established by the following formula:

$$\sum_j YMILK_j \leq mxloc \tag{24}$$

4 Scenarios and results

The model is used to evaluate a number of different scenarios:

- Scenario 0: current situation using Green field production costs.
- Scenario 1: running the model without any additional restrictions to calculate the optimal solution using the current 9 locations.
- Scenario 2: running the model with 4 other potential locations. In the long term, concerning Green field production sites, the dairy company is not bounded to use its existing locations only. If a new plant is build this could take place at another location.
- Scenario 3: running the model with one of the optimal locations left out of the possible locations. One of the locations chosen at the optimal solution is taken out of the possible locations to show the second best option and its cost effects.
- Scenario 4: running the model with production of a certain product group forced at a specific location.
- Scenario 5: running the model with a fixed number of possible plants, namely 1 or 2. With restriction (24) the model is forced to open not more than a certain number of locations. In this case the effect is calculated if $mxloc = 1$ and if $mxloc = 2$.
- Scenario 6: running the model closing one plant in the current situation. This way, the cost effects of closing a certain plant in the current situation are calculated.

Because of the confidentiality of this research, the results cannot be discussed in detail. The results are summarised in Table 1. In Figure 5 an overview of the costs per scenario is presented.

Table 1. Model results of scenario evaluations

Scenario	Number of plants and location	Product portfolio
0	locations A, B, C, D, E, F, G, H and I	Each product group is produced at one or more locations.
1	locations A, B and C	Each product group is produced at one location.
2	locations A, B and C	Each product group is produced at one location.
3	locations A, B and D	Each product group is produced at one location.
4	locations A, B, C and E	Each product group is produced at one location.
5, mxloc = 1	location C	Each product group is produced at one location.
5, mxloc = 2	locations A and C	Each product group is produced at one location.
6	locations A, B, D, E, F, G, H and I	Each product group is produced at one or more locations.

Result scenario 0 and 6:
The model only determines the flow of raw milk, semi-finished products and finished products. The costs are calculated with the nine current locations and for the situation that location C is closed.

Result scenario 1:
The optimal solution in this case-study is three locations. The product groups are allocated in such way that each product group is produced only at one location. A certain level of consolidation and product specialisation is taking place. The locations are specialized in a number of product groups. The product groups requiring a large amount of milk are located near to the largest volume in milk. Product groups requiring less milk are located closer to the market. There is still a certain volume of inter-transport, but less than in the current situation.

Result scenario 2:
The model does not choose to open one of the alternative locations. The results are the same as in scenario 1. The alternative locations considered are not better locations than the already existing locations.

Result scenario 3:
If one of the locations chosen by the model in the optimal solution is taken out of the possible locations (location C), the model determines the second best option. The model still opens three locations, A, B and D. Some product groups of the location taken out are deployed at one of the two other optimal locations. The other product groups are deployed at the new location.

Result scenario 4:
If a certain product group is forced at a certain location, E, the model decides to

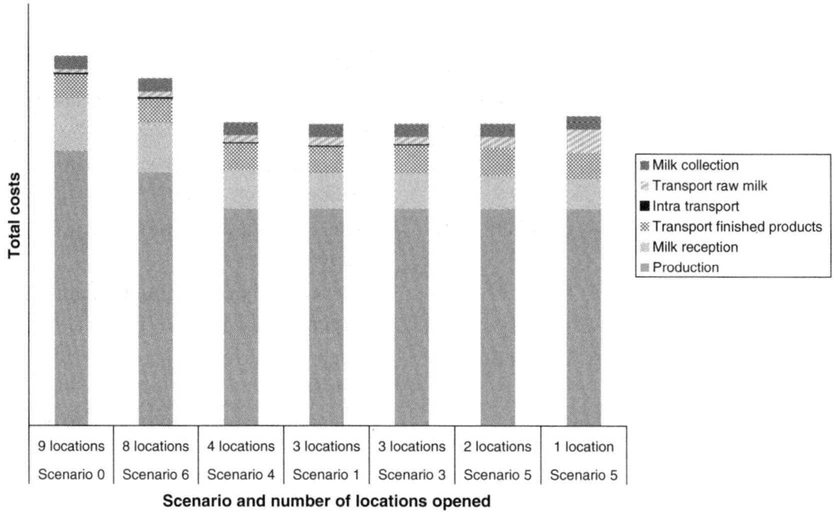

Fig. 5. Production and transportation costs per scenario

open 4 locations, A, B, C and E. The product group forced is the only product group produced at this fourth location. The other product groups are produced at the same location as in the optimal solution.

Result scenario 5:
In this scenario the model is forced to open one location and a second time to open two locations. In case of two locations, the model chooses two of the locations of the optimal solution, A and C. The product groups of the third (closed) location (B) are deployed at the nearest location C. In case of one location, all product groups are allocated also at one of the optimal locations, the one in between the largest milk volume and the largest market, location C.

Reducing the current number of production locations to a lower number of production locations decreases total costs. This applies especially to the production costs due to more product specialisation. If a product group is produced only at one location, the maximum attainable economy of scale is reached. Additional synergy can be reached by more economy of scale in milk reception. The milk reception costs are reduced if the number of locations is reduced, since every location has its own milk reception unit. The larger this milk reception unit, the more economy of scale can be gained. If less than three locations are opened, the milk reception costs decrease but the transportation costs increase. Especially the transportation costs for raw milk. The transportation costs of finished products slightly increase. The inter-transportation costs decrease, but their contribution to the total costs is not substantial. The milk collection costs are approximately the same in every scenario since almost all the milk is used for production.

Sensitivity analysis

Assumptions are made to calculate the transportation and the production costs. Does the result of the model change when these costs are higher or lower than calculated? A sensitivity analysis is carried out to check the consequences of these assumptions.

1. Production and milk reception costs: production is responsible for the major part of total cost. A 10% change in production costs has no impact on the outcome of the model. If the milk reception costs are increased by 30%, only two locations are opened instead of three. The same solution as in scenario 5 with locations A and C is optimal in this case. If the milk reception costs are reduced by 30%, a fourth location is opened (A, B, C and D) and certain product groups are deployed at this fourth location. A change of 10% in milk reception costs showed no effect.
2. Milk transportation costs. If the transportation costs of raw milk are increased by 5% again three locations are opened. However, one of the plant locations is changed as well as the allocation of product groups (locations A, B and D). If the transportation costs of raw milk are increased by 30% a fourth location is opened (locations A, B, C and D). A reduction in raw milk transportation costs showed no effect.
3. If the transportation costs of finished products are reduced by 10%, the same change takes place as in the increase of the transportation costs of raw milk by 5%. An increase up to 30% showed no effect.
4. A change in inter-transportation costs showed no effect.

The model makes a trade-off between transportation costs and economies of scale in production and milk reception. The production costs decrease when product specialisation takes place. The milk reception costs decrease if the number of production locations is reduced. However, the total transportation costs increase when fewer locations are opened. At a certain point the additional transportation costs have reached the same level as the economies of scale advantage in the production costs. If fewer locations are opened for production, the total transportation costs increase more than the milk reception costs decrease, so the total costs increase. At the point of three locations opened, there is an optimum in total costs. This optimum is easily influenced by a small change in transportation costs.

The number of opened locations calculated by the model is 2, 3 or 4 of the locations A, B, C and D. Locations E, F, G, H and I are not chosen by the model, only if the model is forced to. A more refined and detailed study towards transportation costs is necessary before the optimal number and location of production locations can be concluded more precisely. Also important are future expectations on milk supply and market demand, will there be a shift towards other regions of milk suppliers or demand, new product groups, rejection of product groups etc. In this study the current milk volume and sales numbers are taken as granted. If this changes it can certainly influence the optimal solution.

5 Conclusions

This paper discusses a mixed integer linear programming model that was developed in cooperation with the Nutricia Dairy and Drinks Group. The model was used in a real life case to identify the optimal number of plants in Hungary, its locations and the allocation of the product portfolio to these plants, when minimizing the sum of production and transportation costs. The model showed its applicability in the decision making process, although a more refined and detailed study towards transportation costs is necessary as was suggested. The model determined which strategy to follow and at which location. Every investment proposal is compared to see if it fits into this strategy. Step by step a plan will be developed to come closer to the optimal solution. In case of new acquisitions by the NDDG the model can serve again to find the optimal solution including the new milk, product portfolio and market volume acquired together with new possible production locations. The same study could also be carried out for a wider area, e.g. which production strategy to follow in central Europe, western Europe or even whole Europe. In this case more cost factors should be taken into account, e.g. differences per country in milk prices, wages, taxes and in some cases custom duties. Since central European countries will join the EU in the future, the custom duties will disappear.

The model itself can be further extended for future purposes. A first extension could be to decouple the specific preparation and specific packaging process steps in the Green field production costs. Some product groups are using the same preparation but other packaging equipment and other product groups are using other preparation but the same packaging equipment. By combining certain product groups in one factory, more economies of scale in production costs could be realized. Although, there is a certain limit to the economies of scale in production. When a production location produces more and more product groups the managerial complexity increases. This increase in complexity could cause an increase of production costs per unit of product. In the model this could be solved by adding a restriction to the number of product groups produced at one location. A second advantage of decoupling the preparation and packaging process steps is the possibility to switch towards a strategy of process specialisation instead of product specialisation. In the presented model the raw milk is directly linked to the production of product p by restriction (4). There is no inter-transportation of standardised milk built into the model to make process specialisation possible. The model should be adjusted for this and the split between preparation and packaging. This way the inter-transportation will play a more important role.

This model is meant as and is used as a decision-supporting tool. It quantifies, evaluates, verifies and supports ideas and 'gut feeling'. Of course, a lot of other quantifiable (e.g. social costs, disinvestments costs) as well as non-quantifiable factors (e.g. political, social, environmental, quality factors) play a role in decision-making. This model is meant to point out a direction, to get closer to the optimal situation and thus decrease total costs.

References

Cohen MA, Lee HL (1989) Resource deployment analysis of global manufacturing and distribution networks. Journal of Manufacturing and Operations Management 2: 81–104

Eppen GD, Gould FJ, Schmidt CP Moore JH Weatherford LR (1998) Introductory management science, 5th edn. Prentice Hall, Upper Saddle River, NJ

Hax AC, Candea D (1984) Production and inventory management. Prentice Hall, Upper Saddle River, NJ

Krajewski LJ, Ritzman LP (1999) Operations management – strategy and analysis, 5th edn. Addison Wesley Longman, Reading, MA

Slack N, Chambers S, Harland C, Harrison A, Johnston R (1998) Operations management, 2nd edn. Financial Times, Pitman, London

Vidal CJ, Goetschalckx M (2001) A global supply chain model with transfer pricing and transportation cost allocation. European Journal of Operational Research 129: 134–158

Walstra P, Geurts TJ, Noomen A, Jellema A, Van Boekel MAJS (1998) Principles of dairy technology. Dekker, New York

Combined strategic and operational planning – an MILP success story in chemical industry

Josef Kallrath[1,2]

[1] BASF-AG, GVC/S (Scientific Computing) - C13, 67056 Ludwigshafen, Germany
(e-mail: josef.kallrath@basf-ag.de)

[2] Astronomy Department, University of Florida, Gainesville, FL 32661, USA
(e-mail: kallrath@astro.ufl.edu)

Abstract. We describe and solve a real world problem in chemical industry which combines operational planning with strategic aspects. In our simultaneous strategic & operational planning (SSDOP) approach we develop a model based on mixed-integer linear (MILP) optimization and apply it to a real-world problem; the approach seems to be applicable in many other situations provided that people in production planning, process development, strategic and financial planning departments cooperate.

The problem is related to the supply chain management of a multi-site production network in which production units are subject to purchase, opening or shutdown decisions leading to an MILP model based on a time-indexed formulation. Besides the framework of the SSDOP approach and consistent net present value calculations, this model includes two additional special and original features: a detailed nonlinear price structure for the raw material purchase model, and a detailed discussion of transport times with respect to the time discretization scheme involving a probability concept. In a maximizing net profit scenario the client reports cost saving of several millions US$.

The strategic feature present in the model is analyzed in a consistent framework based on the operational planning model, and vice versa. The demand driven operational planning part links consistently to and influences the strategic. Since the results (strategic desicions or designs) have consequences for many years, and depend on demand forecast, raw material availability, and expected costs or sales prices, resp., a careful sensitivity analysis is necessary showing how stable the decisions might be with respect to these input data.

Key words: MILP modelling – Strategic and operational planning – Supply chain – Chemical industry – Transportation

1 Introduction

This contribution evolved from a project in chemical industry in which we developed and successfully applied a mixed-integer model to a real world problem which combines strategic and operational planning aspects in one model (a second example which encourages this approach is described in [9, Sect. 9.2]); operational planning aspects involve decisions on short or mid-term time-scale which are transformed into operational activities, e.g., producing, shipping or selling something. Let us first give some motivation why this simultaneous strategic & operational planning (SSDOP) approach based on mixed-integer optimization may greatly improve a company's situation and let us focus on some problems we might expect.

1.1 Solving design and operational planning problems simultaneously

It is a frequent experience that clients ask for support on a production planning or scheduling problem for a plant or reactor which just went into operation. Often, especially in scheduling problems, it turns out that there exist certain bottlenecks. It would greatly improve the situation if the design of a plant or reactor would be analyzed simultaneously with the planning or scheduling problem. Certainly, this problem is mathematically demanding because scheduling problems alone are already very difficult to solve [10], e.g., because of resources (raw material, machine availability, or personnel) too strongly limited. Thus, if the design and planning/scheduling problem are part of one embracing model this bottleneck situation might be avoided. This simultaneous approach requires the availability of realistic and detailed demand forecast, and expected cost or sales prices, resp., and that the departments being responsible for the design and the planning/scheduling cooperate. The latter problem is by far the more difficult one, especially in large companies. A concrete example of this type, a process design network problem, is discussed in [9, Sect. 9.2].

1.2 Solving strategic and operational planning problems simultaneously

A company running a complex production network consisting of several plants, (see, for instance, [11, Sect. 10.4]) wishes to buy additional plants, open new reactors based on improved technology, or to shut down some older reactors. In their multi-stage production system there might exist logical implications between the status of certain reactors. The data governing the investment or deinvestment decisions are the costs to buy a plant, or the costs to open or shut down a reactor. The investments or deinvestments should be sound over a time horizon of, say, up to 15 years. The best approach to analyze such situations is to develop a quantitative planning model and to enhance it by additional plants or reactors (leading to so-called design variables and constraints) and let the model provide suggestions on optimal design decisions. Regarding the database, it is necessary to provide the full data set (recipes, production rates and capacity, etc.) for all design plants or design reactors. All cost data should be discounted over the time horizon in order to support a net present value analysis. Such a case for a real production division of a chemical company is discussed in Section 2. To model this problem we partly use the model formulation published in [7], [11, Sect. 10.4], or [13]; the current project required major extensions related to design reactors, transport arriving over several time slices, nonlinear pricing structures to purchase raw materials and additional objective functions such as *maximize net profit*, *multi-criteria objectives* (i.e., maximize profit & minimize the quantity of transport in tons), *maximize sales volume* or *maximize turnover*.

Another problem of this type ([8] and [9, Sect. 9.1]), linking strategic and operational aspects is the optimization of a network of processing units at a large production site connected by a system of pipes. The purpose of this model was to design an integrated production network minimizing the costs for raw material, investment and variable costs for re-processing units, and a cost penalty term for low product quality. The investment decisions are considered on a 10-year linear depreciation rate.

1.3 Mathematical problems of combined models: complexity

Although company wide supply chain production models exist, see, for instance, [1] in most cases, because of their high complexity, even pure production planning or supply chain optimization problems are often seen as academic and not suitable for practical application. Instead, either simplifications (like linear relaxations) or simulation approaches are used, especially if there are several plants, tank storages, and the production planning problem is just a part of the comprehensive problem of optimizing the whole supply chain with regard to a detailed description of the commercial environment, such as demands with different prices for different customers, availability of raw material or quality commitments.

Therefore, if one suggests a simultaneous strategic/design & operational planning (SSDOP) approach it is not a surprise that the sceptics might argue *embedding a complex mixed-integer programming model in an even bigger problem including design features "is by far too difficult"*. Indeed, the problem might be large and complex, but as the case discussed in Section 2 shows, it is worthwhile to try and it can be done.

1.4 Mathematical problems of combined models: structure of the objective function

It is important that the problem is – within the limits and the assumptions of the model – approached as an exact optimization problem with a well-defined objective function representing the economic structure of the business process. We need to be able either to prove optimality, or to derive safe bounds enabling us to compare different scenarios reliably and to perform sensitivity analyses. A simulation approach is no substitute because it does not strictly (in the mathematical sense) support these items as it could happen that the scenarios A and B have the optimal solutions 110 and 105 but a simulation based approach produces best evaluations 103 and 104 which would indicate that B is better. In the SSDOP approach there are a few complicating factors related to the structure of the objective function.

Since the objective function may contain terms related to operational planning (variable costs for production and processing, transport, raw material, utilities, inventories, mode-changes, etc.) and the design decision (event costs to close or open reactors, to purchase plants, etc.) the scaling in the objective function terms might be poor. This problem might be overcome by appropriate branching strategies and prioritizing the branching variables. Since the planning horizon considered may cover up to 15 years, nonlinear (usually concave) terms describing a price structure might enter in addition. If these terms are not too complicated they can be described sufficiently accurate as shown in Section 4.2, for instance.

1.5 Mathematical problems of combined models: reliability of data

A point of practical concern is the availability of demand forecast data, costs or sales prices over a long period. People not favoring the SSDOP approach might use this as a strong argument against it. There are two arguments to meet these concerns: a) on what grounds would they base their investment decisions otherwise? (the problem related to accurate data concerns both mathematical planning and non-mathematical planning) and b) the mathematical planning approach supports sensitivity analyses and allows to estimate the stability of the decisions with respect to variability of the forecast data. In addition, when building the model and collecting the data, it is necessary to try to balance the degree of details entering the model. Finally, depending on the specific purpose, the overall model might be adapted to its use on different application levels (pure strategic, pure operational planning with fixed design decisions, etc.) requiring different accurateness of the data.

2 Strategic decisions in a worldwide production-network

The core production planning problem covers large parts of the supply chain including several plants, multi-stage production using multi-purpose reactors with some logical rules in the production scheme, tank storages, transport with a detailed representation of the commercial environment, such as demands with different prices for different customers and availability of raw material. Products are subject to aggregate demand requirements at certain demand points. Variable costs for production, inventory, transport are given. The raw material purchase follows a nonlinear price structure. The client asked to consider the production planning problem as a part of larger problem. This larger problem comprises a production network in which gaseous raw materials along a line of up to six intermediate products are converted into finished products, and in which a production unit from a given set of design units is subject to shutdown or opening, or in which even some whole plants can be purchased. To avoid duplication of material most details of production and other features are specified in Subsection 4.1. The most important objective is to maximize the total net profit (contribution margin minus fixed costs minus investment costs) of the entire production network. In addition the following objective functions are maximized: contribution margin, total sales neglecting cost, turnover and total production. Costs can be minimized and multi-criteria objectives, for instance, maximize profit & minimize transport, are supported. The most relevant decision variables indicate how much time per time-slice a reactor spends in a certain mode, how much of a product is produced, stored or shipped to another location. Binary variables trace the status and mode changes of a reactor. Structurally, most of the constraints are balance equations tracing inventories, connecting production and production recipes over several production levels and tracing mode changes. Other constraints relate production quantities, production rates and available time to each other, or guarantee that capacity limits are observed.

3 The mathematical model: preliminaries

3.1 The structure of the model and its basic objects

The problem sketched above has some common features with modeling *multi-purpose plants* which are frequently used in the food or chemical process industry. In each mode such a reactor can produce several products according to free or fixed recipes (joint production, coproduction) leading to a general mode-product relation described by a set of yield coefficients: in a certain mode several products are produced (with different maximum daily production rates), and vice-versa, a product can be produced in different (but not all) modes. *Mode changes* correspond physically, for instance, to a change of the temperature or pressure of a reactor, put the reactor in a new feasible mode and result in a considerable loss of production time, which in our case is sequence-dependent, and are modeled as a proportional lotsizing and scheduling problem (PLSP, [5, p.150]), i.e., based on time-indexed formulations with at most one setup- or mode-change per period; see also [12] for a survey on lot sizing problems.

We thus transform the problem into a model describing a multi-site production network including multi-stage production in plants (a plant hosting sets of reactors) depending on the mode chosen for the reactors. Parts of the model have already been published in [7], [11, Sect. 10.4], or [13]; the current project required major extensions related to design reactors, transport arriving over several time slices, nonlinear pricing structures to purchase raw materials and additional objective functions such as *maximize net profit*, *multi-criteria objectives* (i.e., maximize profit & minimize the quantity of transport in tons), *maximize sales volume* or *maximize turnover*. We thus extend and develop an elaborated version of the model [11, Sect. 10.4]. In our model we use the following set of objects and indices:

$b \in \mathcal{B}$	$:= \{1,\ldots,N^B\}$	: break points (nonlinear price function)
$c \in \mathcal{C}$	$:= \{1,\ldots,N^C\}$	: sales categories
$d \in \mathcal{D}$	$:= \{1,\ldots,N^D\}$	: demand points
$k \in \mathcal{K}_s$	$:= \{1,\ldots,N_s^K\}$	: production periods at site s
$m \in \mathcal{M}_{sr}$	$:= \{1,\ldots,N_{sr}^M\}$	: modes at site s for reactor r
$p \in \mathcal{P}$	$:= \{1,\ldots,N^P\}$	: products
$r \in \mathcal{R}$	$:= \{1,\ldots,N^R\}$	: reactors
$s \in \mathcal{S}$	$:= \{1,\ldots,N^S\}$	: production sites / plants
$t \in \mathcal{T}$	$:= \{1,\ldots,N^T\}$	: commercial periods

Break points are points at which the unit price as a function of volume changes. *Sales categories* allow, for instance, to model that the first 80% of an order can be purchased at a price of 100 US\$, the next 20% at 90 US\$. *Demand Points* may represent customers, regional warehouse locations or distributors who specify the quantity of a product they request, and are sinks in the supply network, i.e., points where a product leaves the system and is not further traced. Demand may be subject to certain constraints, e.g., satisfying a minimum quantity of demand, observing origins of production or supplying a customer from the same origin.

3.2 Discretization of time

In order to investigate a planning horizon of up to 15 years and to cover the production at a level which is sufficiently detailed we use the time discretization scheme described by [13] using non-equidistant commercial and production time slices (periods); in most cases, the production schedule has a finer resolution than the commercial plans for sales and shipping. Different time scales allow to have smallest time slices relevant to production which may be of the order of just a few days while the commercial periods may cover even a few years. The entire planning horizon is therefore divided into N_s^K production slices of size D_t^P/U_{st} days, where D_t^P is the length of the tth commercial period in days and U_{st} is the number of production slices embedded in that commercial period. Regarding delivery or sale, in typical operational planning, usually a commercial time scale of 12 periods (months) is chosen. Another possible scenario is, for instance, to cover a two year production plan with $N^T = 16$ periods: the first 12 with a length of about 30 days, and four additional ones with a length of 120 days. Commercial and production time slices are linked by the function

$$k_s(s,t) := \begin{cases} 0 & \text{, if } t = 1 \\ k_s(s,t-1) + U_{s,t-1} & \text{, if } t > 1 \end{cases} \quad \forall\{st\} \tag{1}$$

which gives the number (minus one) of the production slice starting at the beginning of commercial period t at site s, and, with u referring to a production slice embedded in the commercial time interval t. The function

$$k(s,t,u) := k_s(s,t) + u \quad \forall\{st\} \tag{2}$$

gives the absolute number $k(s,t,u)$ of the uth production slice in the commercial period t within the production time scale referenced by t and u at plant s, and connects both time scales. For shortness, if sums cover the whole planning horizon, we use k rather than $k(s,t,u)$. Finally, we need the inverse function, $t_k(s,k)$,

$$t_k(s,k) := \min\{t \mid k_s(s,t) < k \leq k_s(s,t) + U_{st}\} \quad \forall\{st\} \tag{3}$$

returning the commercial period that covers the production slice k at plant s. It can be expressed in terms of the functions defined above: For further details on this topic we refer to [13]. A 5-year planning horizon in which the commercial data are available on an annual basis and production should be considered with a fineness of one month, leads to

$$N^T = 5 \quad ; \quad D_t^P = 360 \quad \forall t \quad ; \quad U_{st} = 12 \quad \forall \{st\} \tag{4}$$

while for asset evaluation over a time horizon covering 10 years with two production time slices per year is described by

$$N^T = 10; \quad D_t^P = 360 \quad \forall t; \quad U_{st} = 2 \quad \forall \{st\} \tag{5}$$

3.3 Limits and underlying assumptions

The limits of the model follow from its underlying assumptions which are further discussed in Section 7.1:

1. Only one mode change per production time slice is allowed as the model is formulated as a proportional lotsizing and scheduling problem (PLSP, [5, p.150] or [6]), i.e., based on time-indexed formulations with at most one setup- or mode-change per period. This assumption seems not to be a very serious restriction for operational planning since a production time slice has a length of a week or a month and typically only one setup-change per month occurs. For the long term analysis the description of the mode changing reactors may not be accurate enough. But one should keep in mind that the demand forecast and other input data are of limited accuracy as well; the client regarded the approach as sufficiently accurate and realistic.
2. Transport times are of the order of a few days. In the 12-month scenarios transport times are only considered using a probability assumption. In even longer term scenarios transport times are neglected.
3. Only variable inventory costs are considered. They are based on the capital tied up in inventories and interest rates.
4. Design reactors cannot be subject to intrinsic mode changes. In the current case this limitation did not cause any problems because the approximately 30 reactors subject to design decisions were single product reactors.
5. Production utilization rates and associated constraints refer to utilization per production slice. At present, the model does not include constraints enforcing global utilization aspects of the supply network.
6. In order to support net present value considerations all cost related data are discounted over time using a discount rate of $p\%$.

4 The mathematical model: the operational planning aspects

4.1 Plants, reactors and production

Each plant consists of one or two sets of reactors. While all reactors are subject to multi-stage production requirements and possible coproduction, some are multi-purpose reactors subject to mode changes. Topologically, at each plant, the reactors are arranged in chains, in which each reactor needs only one pre-product; however, the current model formulation does not exploit this features and can also be applied to convergent or divergent material flows and more general topologies. The reactors within a chain operate simultaneously and at different levels of the multi-stage production process. There exist capacity limits for each reactor and task, as well as capacities for the production of each product, and minimum production requirements. Some reactors in the chain are subject to mode changes lasting usually one or two days. Reactors obtain products from preceding reactors or from tanks, and charge the products through pipelines to tanks or to subsequent reactors.

Modeling production involves the concepts of multi-stage production, joint production (coproduction) and mode changes of thc multi-purpose reactors. The description of the mode changes is based on [11, p. 321]; the adoption of this approach to the current problem is as follows. The basic binary variables are the state variables

$$\delta_{srmk} = \begin{cases} 1 \text{ , if reactor } r \text{ is in mode } m \text{ at the end of period } k \\ 0 \text{ , otherwise} \end{cases} \tag{6}$$

It is one of the most fundamental assumptions in this model that there is at most one mode change per period. If the state variables take the values $\delta_{srm_1k-1} = \delta_{srm_2k} = 1$ we have a mode change from mode m_1 to m_2 in time interval k.

Both, the continuity of modes and mode changes are tracked by the binary variables

$$\xi_{srkm_1m_2} = \begin{cases} 1 \text{ , if } \delta_{srm_1k-1} = \delta_{srm_2k} = 1 \\ 0 \text{ , otherwise} \end{cases} \qquad \forall\{srm_1m_2k\} \tag{7}$$

This variable is unity if at the end of period $k-1$ reactor $r \in \mathcal{R}_s \; \left| I_{sr}^{MC} = 1 \right.$ of plant $s \in \mathcal{S}$ is in mode $m_1 \in \mathcal{M}_{sr}$ and at the end of period $k \in \mathcal{K}$ it is in mode $m_2 \in \mathcal{M}_{sr}$. $\xi_{srkm_1m_2}$ is a variable not only describing whether a change-over occurs, but it also indicates whether production continues. If the reactor is in mode m both at the end of period $k-1$ and k then we have $\xi_{srkmm} = 1$.

The state variables and the mode-change variables will now be coupled by some additional binary variables: α_{srmk}, if reactor r at plant s spends some time in period k in mode m; β_{srmk}, if mode m is started on reactor r at site i in period k; and finally, γ_{srmk}, if mode m is terminated on reactor r at site i in period k. These binary variables are related to the mode changing variables by the constraints

$$\beta_{srmk} = \sum_{m_1 \neq m} \xi_{srkm_1m} \quad \forall(srmk) \tag{8}$$

and

$$\gamma_{srmk} = \sum_{m_1 \neq m} \xi_{skmm_1} \quad \forall(srmk) \tag{9}$$

The multi-stage production scheme looks as follows: at a site s a certain reactor r is connected to one (or possibly more) preceding reactors r'; it is assumed that transfer times between reactors are zero. The reactor topology is completely described by the indicator table $I_{sr'r}^{Topo}$ which takes the value 1 if reactor r' can charge to reactor r. Reactor r converts the product p' produced by preceding reactors r' into the product p, or possibly into several coproducts. Actually, this product p' can also be taken from the preceding reactor r' or from an intermediate tank. The total quantity $p_{srp'k}^{U}$ of product p' reactor r uses in period k is therefore

$$p_{srp'k}^{U} = u_{srp'k}^{S} + \sum_{r|I_{srr'}^{Topo}>0 \wedge I_{sr'p'}^{SRP}=1} p_{sr'rp'k}^{D} \tag{10}$$

$$\forall \{srp' \,| I_{srp'}^{PP} = 1 \wedge I_{sr}^{Pipi} = 1\} \quad \forall k \tag{11}$$

where $I_{srp'}^{PP}$ indicates whether reactor r uses p' as a pre-product at all. The indicator table $I_{sr'p'}^{SRP}$ controls which reactor r' at site s is able to produce product p', and $u_{srp'k}^{S}$ is the quantity of product p' taken by reactor r from the tank. Note that $I_{sr}^{Pipi} = 1$ means that reactor r is – via an input pipeline – connected to a tank and is able to extract material from that storage device; in the current case, each reactor needs exactly one pre-product, therefore I_{sr}^{Pipi} does not depend on the product as this is known a priori. $p_{sr'rp'k}^{D}$ is the quantity of product p' charged from reactor r' directly to reactor r and allows also to model convergent material flows if the topology is chosen appropriately. Note that $u_{srp'k}^{S}$ appears as a loss term in the stock balance equation (44). If p' is a raw material, which can only be purchased, (10) reduces to

$$p_{srp'k}^{U} = u_{srp'k}^{S} \quad \forall \{srk \,| I_{srp'}^{PP} = 1 \wedge I_{sr}^{Pipi} = 1\} \tag{12}$$

because p' is not produced on any reactor, and therefore for all reactors r' we have $I_{sr'p'}^{SRP} = 0$. To model raw material availability it has to be ensured in the input data that $u_{srp'k}^{S}$ has no upper bound, or is consistent with the raw material availability, respectively. Finally, if the reactor is not connected to an input tank, (10) reduces to

$$p_{srp'k}^{U} = \sum_{r|I_{srr'}^{Topo}>0 \wedge I_{sr'p'}^{SRP}=1} p_{sr'rp'k}^{D} \quad \forall \{srp'k \,| I_{srp'}^{PP} = 1 \wedge I_{sr}^{Pipi} \neq 1\} \tag{13}$$

The multi-stage production, i.e., the quantity p_{srpk}^{P} of product p produced on reactor r at site s is described by recipe equations of the form

$$\sum_{p\in\mathcal{P}|I_{srp}^{SRP}>0} R_{srp'p} p_{srpk}^{P} = p_{srp'k}^{U} \quad \forall \{srp'k \,| I_{srp'}^{PP} > 0\} \tag{14}$$

It is a matter of taste whether to apply the recipe coefficient to p' or p. Note that reactor r produces several products p using p' simultaneously.

In addition to the multi-stage concept we also have to consider coproduction. If the conversion of product p' produces two products p_1 and p_2 in a fixed ratio $R_{srp_1p_2}^{FR}$, we have

$$p_{srp_1k}^{P} = R_{srp_1p_2}^{FR} p_{srp_2k}^{P} \quad \forall \{srp_1p_2k \,| \exists R_{srp_1p_2}^{FR} > 0\} \tag{15}$$

So (15), means that for each mass unit that is produced of product p_2 one also gets $R^{FR}_{srp_1p_2}$ mass units of product p_1. The quantity p^P_{srpk} of product p which can be produced on reactor r at site s in period k is limited by the capacity R^P_{srp} specified in tons/day. With $P^{\min}_{srpk} := H_{srk}R^{P\min}_{srp} = H_{srk}R^{U\min}_{srp}R^P_{srp}$, where H_{srk} is the number of days available for production in period k and $R^{U\min}_{srp}$ specifies production utilization for reactor r at site s and product p in %, and the production capacity, $P^{\max}_{srpk} := H_{srk}R^P_{srp}$, of reactor r in tons in period k, we get the bounds

$$p^P_{srpk} \le P^{\max}_{srpk} \quad \forall \left\{ srpk \,\middle|\, I^{SRP}_{srp} = 1 \wedge I^{MC}_{sr} \ne 1 \right\} \tag{16}$$

Some additional sets of constraints consider minimum production of a product at a plant over a specific time period. This may be quantified by $R^{U\min}_{srp}$, the production utilization for reactor r at site s and product p in %. If the minimum $P^{\min}_{srpk} := R^{U\min}_{srp}P^{\max}_{srpk}$ is not met, then no production is possible. Therefore, p^P_{srpk} is a semi-continuous variable or subject to the disjunctive constraints, resp.

$$p^P_{srpk} = 0 \quad \vee \quad P^{\min}_{srp} \le p^P_{srpk} \le P^{\max}_{srp} \quad \forall \left\{ srpk \,\middle|\, \exists R^P_{srp} \right\} \tag{17}$$

For those reactors subject to opening or shut-down decisions or belonging to plants which might be purchased the capacity balance reads

$$\begin{gathered} p^P_{srpk} = 0 \quad \vee \quad P^{\min}_{srpk}\theta_{sr} \le p^P_{srpk} \le P^{\max}_{srpk}\theta_{sr} \\ \forall \left\{ srpk \,\middle|\, \exists R^P_{srp} \wedge \exists I^D_{sr} \right\} \end{gathered} \tag{18}$$

where the indicator table I^D_{sr} indicates whether reactor r is a design reactor and the binary variable θ_{sr} specifies whether the reactor is available or not. It is also possible to apply global constraints enforcing that, if a design reactor falls below a certain minimum usage rate, it has to be shut down; however, due to the lack of space we do not present this feature here.

The quantity p^P_{srpk} of output product p is charged to a local tank (site inventory) or charged to subsequent reactors. This is expressed by the distribution equation

$$p^P_{srpk} = p^T_{srpk} + \sum_{r \in \mathcal{R} | I^{Topo}_{srr'} > 0} p^D_{srr'pk} \quad \forall \left\{ srpk \,\middle|\, I^{SRP}_{srp} = 1 \right\} \tag{19}$$

where p^T_{srpk} is the quantity of product p charged to the tank (site inventory) by pipeline, or by

$$p^P_{srpk} = \sum_{r \in \mathcal{R} | I^{Topo}_{srr'} > 0} p^D_{srr'pk} \quad \forall \left\{ srpk \,\middle|\, I^{SRP}_{srp} = 1 \right\} \tag{20}$$

if reactor r has no pipeline connection to an output tank.

4.2 Raw material modeling

Raw materials are in most aspects treated similarly to all other products, for instance, they fulfill the recipe equation (14). However, for raw materials we have to consider purchase cost C_{srp}^{RM} and the availability A_{srpk}^{RM} which describes how much raw material p is available in period k. Raw materials are fed to the reactors by a tank. Thus we model the availability exploiting the tank feature $I_{sr}^{Pipi} = 1$ and the variable u_{srpk}^{s} expressing how much of product p (in this case raw material) reactor r at site s takes from the tank. The variable u_{srpk}^{S} is subject to the availability constraint

$$u_{srpk}^{S} \leq A_{srpk}^{RM} \quad \forall \left\{ srpk \,\middle|\, \exists A_{srpk}^{RM} \wedge I_{srp}^{PP} = 1 \right\} \tag{21}$$

The notation $\exists A_{srpk}^{RM}$ reflects that an inequality is generated only if a table entry for A_{srpk}^{RM} exists. A more elaborated view of the raw materials, controlled by a site and raw material dependent flag R_{sp}^{M}, takes the following features into account:

R_{sp}^{M}	*description*
0	material requirement planning output; no constraints, no costs
1	take as much raw material as needed at a fixed price
2	take as much raw material as needed at a nonlinear price
3	take all available raw material (use or burn) at a fixed price

and generates all constraints only if a raw material tank has been declared, i.e., if $\exists S_{sp}^{SC}$.

If $\mathrm{R}_{sp}^{M} = 0$ no raw material constraints or costs are considered. In the report we just print the raw material which would be required. For modeling the raw material features we introduce the following variables:

- u_{spk}^{R} : usage (in tons) of raw material (product) p in period k
- u_{spbk}^{RB} : usage (in tons) of raw material (product) p in segment b
- b_{spk}^{R} : burned quantity (in tons) of raw material (product) p
- μ_{spbk} : binary var. indicating $R_{spb-1k}^{BPV} \leq u_{spbk}^{RB} \leq R_{spbk}^{BPV}$ in period k,
- ω_{spt} : binary var. indicating the use of raw material p in period t

If Y_{spk}^{RM} specifies the price per ton which can be obtained if the raw material is burnt, it can not be used otherwise, and thus converted into energy. For $\forall \left\{ spk \,\middle|\, \exists A_{spk}^{RM} \wedge \left(\exists r \text{ with } I_{srp}^{PP} = 1 \right) \wedge \exists Y_{spk}^{RM} \right\}$ we apply the following bounds or constraints, resp.:

$$u_{spk}^{R} \leq A_{spk}^{RM} \quad \text{if } \mathrm{R}_{sp}^{M} = 1 \tag{22}$$

$$u_{spk}^{R} + b_{spk}^{R} = A_{spk}^{RM} \quad \text{if } \mathrm{R}_{sp}^{M} = 3 \tag{23}$$

The case $\mathrm{R}_{sp}^{M} = 2$ is more difficult to model and requires the following data

R_{spbk}^{BPC}	specific raw material cost in segment b
R_{spbk}^{BPV}	raw material volume at break point b
R_{spbk}^{ACC}	accumulated raw material cost at break point b
R_{spk}^{FIX}	fixed cost (setup cost) if raw material p is chosen
R_{spk}^{PEN}	penalty cost if raw material p is not used at all

where segment b refers to the range between break point $b-1$ and b, and thus to the interval $[R_{spb-1k}^{BPV}, R_{spbk}^{BPV}]$ the penalty cost R_{spk}^{PEN} apply to the situation that an annual contract with the raw material supplier has been established but the raw material is not used at all, $R_{sp1k}^{ACC} = 0$ and for $b = 2, \ldots, N^B$

$$R_{spbk}^{ACC} = R_{spb-1k}^{ACC} + R_{spbk}^{BPC} \left(R_{spbk}^{BPV} - R_{spb-1k}^{BPV} \right) \tag{24}$$

To select the appropriate segment or interval $[R_{spb-1k}^{BPV}, R_{spbk}^{BPV}]$ for the quantity of raw material purchased we exploit the binary variable μ_{spbk} which indicates that $R_{spb-1k}^{BPV} \leq u_{spbk}^{RB} \leq R_{spbk}^{BPV}$. The constraints

$$\sum_{b=1}^{N^B+1} \mu_{spbk} = 1 \quad \forall \left\{ spk \left| \exists A_{spk}^{RM} \wedge \exists S_{sp}^{SC} \right. \right\} \tag{25}$$

and

$$u_{spbk}^{RB} \geq R_{spb-1k}^{BPV} \mu_{spbk} \quad \forall \left\{ spk \left| \exists A_{spk}^{RM} \wedge \exists S_{sp}^{SC} \right. \right\} \tag{26}$$

$$u_{spbk}^{RB} \leq R_{spbk}^{BPV} \mu_{spbk} \quad \forall \left\{ spk \left| \exists A_{spk}^{RM} \wedge \exists S_{sp}^{SC} \right. \right\} \tag{27}$$

guarantee that u_{spbk}^{RB} falls exactly into one segment. Note that $\mu_{spN^B+1k} = 1$ is used to indicate that no raw material is used at all.

The real quantity u_{spk}^{R} of used raw material is coupled to the segment quantity by

$$u_{spk}^{R} = \sum_{b=1}^{N^B} u_{spbk}^{RB} \quad \forall \left\{ spk \left| \exists A_{spk}^{RM} \wedge \exists S_{sp}^{SC} \right. \right\} \tag{28}$$

As raw material is purchased on the basis of annual contracts, which might force that a certain quantity has to be purchased by the provider, we introduce the binary variable ρ_{spt} indicating whether any quantity of raw material p is used in the commercial period at all

$$\rho_{spt} \geq 1 - \mu_{spN^B+1k} \quad \forall \left\{ sptk \left| \exists A_{spk}^{RM} \wedge \exists S_{sp}^{SC} \right. \right\} \tag{29}$$

If no raw material is used at all, penalty cost may be applied by the provider.

An alternative, and in most cases superior formulation of these raw material aspects replaces (26) to (28) by

$$u_{spk}^{R} = \sum_{b=2}^{N^B} R_{spb-1k}^{BPV} \mu_{spbk} + \sum_{b=1}^{N^B} u_{spbk}^{RB} \quad \forall \{spk \,|\, \exists A_{spk}^{RM} \wedge \exists S_{sp}^{SC}\} \tag{30}$$

$$u_{spbk}^{RB} \leq R_{sp1k}^{BPV} \mu_{spbk} \quad \forall \{spk \,|\, \exists A_{spk}^{RM} \wedge \exists S_{sp}^{SC}\} \tag{31}$$

where u_{spk}^{R} has a slightly different meaning now, and

$$u_{spbk}^{RB} \leq \left(R_{spbk}^{BPV} - R_{spb-11k}^{BPV}\right) \mu_{spbk} \quad \forall \{spk \,|\, \exists A_{spk}^{RM} \wedge \exists S_{sp}^{SC}\} \tag{32}$$

The variable raw material cost are then given by

$$\sum_{s \in \mathcal{S}} \sum_{p \in \mathcal{P}} \sum_{b=2}^{N^B} \sum_{k=1}^{N_s^K} R_{spb-1k}^{ACC} \mu_{spbk} + \sum_{s \in \mathcal{S}} \sum_{p \in \mathcal{P}} \sum_{b=1}^{N^B} \sum_{k=1}^{N_s^K} R_{spbk}^{BPV} u_{spbk}^{RB} \tag{33}$$
$$+ \sum_{s \in \mathcal{S}} \sum_{p \in \mathcal{P}} \sum_{t=1}^{N^T} R_{spt}^{PEN} + \sum_{s \in \mathcal{S}} \sum_{p \in \mathcal{P}} \sum_{t=1}^{N^T} \left(R_{spt}^{FIX} - R_{spt}^{PEN}\right) \rho_{spt}$$

A special ordered set approach involving the μ_{spbk} variables has been tested as well, but did not turn out to be superior compared to the formulation in which the μ_{spbk} variables are just binary variables.

4.3 Transport

The model considers three types of transport: transport of products between production sites and transport of products from production sites to demand points; in some rare cases there is also transport between demand points. Transport costs depend on the source, destination and product, minimal quantities to be observed and transport times (typically a few days). Transportation quantities are expressed by non-negative, dimensionless semi-continuous *transport variables* specifying, when multiplied by the minimum transport quantity, T^M, the quantity of product p shipped:

$$\begin{aligned} &\sigma_{dd'pt}^{DD} && : \ldots \text{ between demand points } d \text{ and } d' \\ &\sigma_{sdpk}^{SD} && : \ldots \text{ shipped from site } s \text{ to demand point } d \\ &\sigma_{ss'pk}^{SS} && : \ldots \text{ shipped from site } s \text{ to site } s' \end{aligned} \tag{34}$$

The semi-continuous variables, for instance, σ_{sdpk}^{SD} are defined

$$\sigma_{sdpk}^{SD} = 0 \ \vee \ 1 \leq \sigma_{sdpk}^{SD} \leq S_{sdpk}^{+} \quad \forall \{sdpk\} \tag{35}$$

with some upper bound S_{sdpk}^{+}, and they enter the inventory balance equation in the form $T_{sdpk}^{MSD} \sigma_{sdpk}^{SD}$ where T_{sdpk}^{MSD} is the minimum transport quantity.

The various time scales in our model SSDOP force us to model transport time very carefully. The actual transport times are a few days, usually between 1 and 6 days. If the transport times are consistent with the time discretization, i.e., the transport time is an integer multiple of the smallest time slice, the transport times can be considered in the index counting the time slice (see [11, Sect. 10.4]). Regarding the length of the planning horizon and the discretization of time we consider two model approaches to transport. In the 15-year planning horizon with half-year time slices and in the short term (operational planning) scenario covering three months the time resolution is one week. That fits transport times which will be reasonably approximated to be 0 or 1 week. In the one-year (operational) planning scenario with time slices of one month this is not accurate enough. We might use smaller time slices (leading to significantly increased CPU times) but in order to avoid this, we suggest to use the following approach based on a probability assumption (for clarity, we neglect some of the indices in this paragraph). The essential idea of this approach is to conserve the flow of materials but to distribute the shipment arrival to two adjacent time slices.

Let D_{sk}^{PK} be the length of the time period k at site s, Δ_{sd} be the time needed for transportation from site s to destination d and let us assume that $\Delta_{sd} \leq \min\left(D_k^{PK}, D_{k+1}^{PK}\right)$ for all k. If we assume that shipments during production period k leave site s with uniform probability in period k, the product will arrive at the destination d with probability

$$\omega_{sdk} := \frac{\Delta_{sd}}{D_{sk}^{PK}} = \frac{\Delta_{sd}}{D_t^P} U_t \tag{36}$$

in period $k+1$ and with probability $1-\omega_{sdk}$ in period k. Therefore, in the one-year planning scenario, we use the following heuristic approach to consider transport in the inventory balance equations. If p_{sd}^T is the quantity to be shipped (appears as a loss term at the site where transport origins) we consider (at the destination of shipment) $(1-\omega_{sdk})p_{sd}^T$ in the balance equation of period k and $\omega_{sdk}p_{sd}^T$ in period $k+1$ as source terms. This approach is not recommended if the planning tool is used operationally. But for long term, e.g., annual planning, it should be perfectly suitable, especially, if there exists some minimum inventory level.

Let us now focus in more detail on products originating from site s in production period k arriving at demand point d in commercial period t. The time, Δ_{sd}, associated with this transport and the length, D_{sk}^{PK}, of the period allow us to compute ω_{sdk} from (36).

At first consider the short term scenario in which Δ_{sd} is an integer multiple of D_{sk}^{PK}, i.e., $T_{sd} = \Delta_{sd}/D_{sk}^{PK}$. Then a shipment originating in period k arrives in period $k+T_{sd}$. If the commercial period t consists of U_{st} production time periods then all shipments originating in periods

$$\left\{k = 1, \ldots, N_s^K \,|t_k(s, k+T_{sd}) = t\right\} \tag{37}$$

arrive in that period.

In the long-term scenario transport is modeled using a probability assumption, i.e., a fraction $1 - \omega_{sdk}$ of the quantity $T_{sdpk}^{MSD}\sigma_{sdpk}$ arrives in period k at the destination d, i.e., the arriving quantity of product p in period k is given by

$$(1-\omega_{sdk})T^{Msd}_{sdpk}\sigma_{sdpk} \quad \{k=1,\ldots,N^K_s\,|t_k(s,k)=t\} \tag{38}$$

and the complementary fraction ω_{sdk} arrives in period $k+1$, i.e.,

$$\omega_{sdk'}T^{Msd}_{sdpk}\sigma_{sdpk'} \quad \{k'=1,\ldots,N^K_s\,|t_k(s,k'+1)=t\} \tag{39}$$

Note that there might occur two terms containing the variable σ_{sdpk} (one associated with the $1-\omega_{sdk}$ term, and another one connected to $\omega_{sdk'}$ for the adjacent period), which in some modeling language may lead to complications (*column appears twice in a row*) if $U_{st}>1$. This problem can be overcome by collecting all coefficients related to the same variable a priori.

4.4 Modeling inventories and stock balances

Inventories and related costs are considered for each product at both plants (inventories at sites) and regional warehouses (inventories at demand points). At demand points it is possible to lease some additional storage capacity.

4.4.1 Inventories at sites At present it is assumed that for each product p there is a dedicated tank available at a site. The stock, s^S_{spk}, for all $\{spk\}$, is described by the balance equation

$$s^S_{spk} = s^S_{spk-1} + p^{ES}_{spk} + \hat{S}^S_{spk} + u^R_{spk} - \sum_{r|I^{Pipi}_{sr}=1} u^S_{srpk} \tag{40}$$

$$+ \sum_{r\in\mathcal{R}|(I^{SRP}_{srp}=1\wedge I^{Pipi}_{sr}=1)} p^T_{srpk} \tag{41}$$

$$- \sum_{s_d\in\mathcal{S}_{sdk}} T^{Mss}_{ss_d}\sigma^{SS}_{ss_dpk} - \sum_{d\in\mathcal{D}_{sdk}} T^{Msd}_{sd}\sigma^{SD}_{sdpk} \tag{42}$$

$$+ \sum_{s_s\in\mathcal{S}} (1-\omega_{s_s sk})T^{Mss}_{s_s s}\sigma^{SS}_{s_s sp,k-(1-\pi)T^{ss}_{s_s s}} \tag{43}$$

$$+ \sum_{s_s\in\mathcal{S}} \omega_{s_s sk}T^{Mss}_{s_s s}\sigma^{SS}_{s_s sp,k-\pi-(1-\pi)T^{ss}_{s_s s}} \tag{44}$$

in which the terms have the following meaning: s^S_{spk-1} is the stock level at the end of the previous period, p^{ES}_{spk} denotes external purchase [see Section 4.6], $\hat{S}^S_{spk}$

$$\hat{S}^S_{spk} := \sum_{s_s\in\mathcal{S}|\exists S^{SS}_{spk}} S^{SS}_{s_s sp k} \tag{45}$$

denotes transport arriving in period k from shipment originating before the first period. The supply, u^R_{spk}, of raw material available in period k appears as a source term, the usage of the product, $\sum u^S_{srpk}$, as a loss term. The sum $\sum p^T_{srpk}$ represents the quantity of product charged from other reactors to this tank. The terms based on the σ-variables have been explained in Section 4.3 describing shipments to other sites and demand points as well as products received from other sites. The sets

$$\mathcal{S}_{sdk} := \left\{ s_d \in \mathcal{S} \left| s_d \neq s \in \mathcal{S} \wedge k + T^{ss}_{ss_d} < N^K \right. \right\} \tag{46}$$

$$\mathcal{S}_{sk} := \left\{ s_d \in \mathcal{S} \left| s_d \neq s \in \mathcal{S} \wedge k > T^{ss}_{ss_d} \right. \right\} \tag{47}$$

$$\mathcal{D}_{sdk} := \left\{ d \in \mathcal{D} \left| k + T^{sd}_{sd} < N^K \right. \right\} \tag{48}$$

and

$$\mathcal{D}_{dk} := \left\{ d_d \in \mathcal{D} \left| d_d \neq d \in \mathcal{D} \wedge t + T^{dd}_{dd_d} < N^T \right. \right\} \tag{49}$$

define the sets of sites or demand points which can be linked by transport to the current location s within the planning horizon.

For $k = 1$ the term s^S_{spk-1} is replaced by the initial product stock S^{SS}_{sp}. The site-balance equation (44) for site inventories (also called local inventories) couples multi-stage production, intermediate storage space and transport (between sites, and between sites and demand points). p^{ES}_{spk} describes external purchase opportunity at site inventories. If no storage capacity is available the variables s^S_{spk} and s^S_{spk-1} drop out in (44).

The maximum stock levels for products read

$$s^S_{spk} \leq S^{CAP}_{sp} \quad \forall \left\{ spk \left| \exists S^{SC}_{sp} \right. \right\} \tag{50}$$

The product stock should never fall below the safety stock, S^{SS}_{sp},

$$s^S_{spk} \geq S^{SS}_{sp} \quad \forall \{spk\} \tag{51}$$

and a possible restriction, S^{LP}_{sp}, in the last period, i.e., as forced by $\mathtt{I}^{RL}_{sp}$=4, is considered as

$$s^S_{spk} \geq S^{LP}_{sp} \quad \forall \{sp\} \quad , \quad k = N^K \tag{52}$$

The user selects different inventory constraints (here just verbally summarized) at the final period by the flag I^{RL}_{sp}

I^{RL}_{sp}	*description*
0	no condition at all
1	total inventory in final period equals total initial inventory
2	specific finished product stock equals the initial inventory
3	specific finished product stock equals pre-given target stock
4	specific finished product stock should be $\geq$ than target stock

4.4.2 Inventories at demand points At demand points there may exist a tank for an end-product p. If so, the stock balance equation reads

$$s_{dpt}^{D} = s_{dpt-1}^{D} + p_{dpt}^{ED} + \hat{S}_{dpt}^{DT} - \sum_{c=1}^{N^C} s_{dpct}^{L} - \sum_{d_d \in \mathcal{D}} T_{dd_d}^{Mdd} \sigma_{dd_d}^{DD} \tag{53}$$

$$+ \sum_{s \in \mathcal{S}} \sum_{\substack{k=1|k+t_1(s,d) \le K_s \\ t(k+t_1(s,d))=t}}^{K_s} (1 - \omega_{sdk}) T_{sd}^{Msd} \sigma_{sdpk}^{SD} \tag{54}$$

$$+ \sum_{s \in \mathcal{S}} \sum_{\substack{k=1|k+t_2(s,d) \le K_s \\ \wedge t(k+t_2(s,d))=t}}^{K_s} \omega_{sdk} T_{sd}^{Msd} \sigma_{sdpk}^{SD} \tag{55}$$

$$+ \sum_{d \in \mathcal{D}|t > t_1(d,d_d)} (1 - \omega_{dd_d t}) T_{dd}^{Mdd} \sigma_{ddp,t-t_1(d,d_d)}^{DD} \tag{56}$$

$$+ \sum_{d \in \mathcal{D}|t > t_2(d,d_d)} \omega_{dd_d t} T_{dd}^{MDD} \sigma_{ddp,t-t_2(d,d_d)}^{DD} \tag{57}$$

for all $\{sr|\exists S_{dp}^{SC}\}$and $t = 2, \ldots, N^T$, in which the terms have the following meaning: s_{dpt-1}^{D} is the stock level at the end of the previous period, p_{dpt}^{ED} denotes external purchase [see Section 4.6], $\hat{S}_{dpt}^{DT}$

$$\hat{S}_{dpt}^{DT} := \sum_{s=1|\exists S_{sdpt}^{PT}}^{N^S} S_{sdpt}^{DT} \tag{58}$$

denotes transport arriving in period k from shipment originating before the first period, and the loss term, $\sum s_{dpct}^{L}$, due to the sales of finished products. The terms based on the σ-variables have been explained in Section 4.3 describing shipments to demand points as well as products received from other sites and demand points.

The functions

$$t_2(x,y) := t_1(x,y) + \pi \quad , \quad t_1(x,y) = (1-\pi) T_{xy}^{xy} \tag{59}$$

are used to select the appropriate indices in time; $\pi = 1$ indicates short term planning (in that case transportation times must correspond to the time discretization) and $\pi = 0$ indicates long term planning (in this case transport arriving is treated using a probability assumption). T_{sd}^{SD} denotes the minimum transport quantity and the time T_{sd}^{P} needed to ship products from site s to demand point d. Note that because T_{sd}^{P} is used in the index it is required that T_{sd}^{P} is measured in units of the period and that it is integral.

For the first period, $t = 1$, s_{dp0}^{D} is replaced by S_{dp}^{DS}, where S_{dp}^{DS} denotes the initial stock of product p at demand point d and

$$\hat{S}_{dpt}^{DT} := \sum_{s=1|\exists S_{sdpt}^{PT}}^{N^S} S_{sdpt}^{DT} \tag{60}$$

denotes transport arriving in period t from shipment not originating within the current planning horizon.

If no inventory is available and there exists a demand D_{dpct}, then the sales variables s^L_{dpct} are immediately coupled to the transportation variables and the terms s^D_{dpt} and s^D_{dpt-1} drop out in (57).

Stock must never exceed the storage capacity, i.e.,

$$s^D_{dpt} \leq S^{CS}_{dp} + s^R_{dpt} \quad \forall\{dpt\} \tag{61}$$

where s^R_{dpt} gives the current additional tank capacity; sometimes it is possible to rent or lease a tank for a short while, e.g., for a few months. The additional stock is usually bounded by

$$s^R_{dpt} \leq S^{CR}_{dp} \quad \forall\{dpt \,|\, \exists S^{CR}_{dp}\} \tag{62}$$

The safety stock bounds are

$$s^D_{dpt} \geq S^{DM}_{dp} \quad \forall\left\{spt \,|\, \exists S^{DM}_{dp}\right\} \tag{63}$$

and the bounds for the last period are

$$s^D_{dpt} \geq S^{DE}_{dp} \quad \forall\left\{spt \,|\, \exists S^{DE}_{dp}\right\} \tag{64}$$

Note that these bounds are only considered if the corresponding stock capacities or safety stocks exist.

4.5 Demand

There are 10 to 20 demand points for the products produced at the production sites located in various parts of the world. The demand points are thought of as regional warehouses. The demand at these points reflects mostly aggregate demand by several customers but also a few large individual customers. The demand for product p is characterized by volume and sales price. This input is generated from an independent sales forecasting model. Sales must not exceed the demand D_{dpct}, i.e.,

$$s^L_{dpct} \leq D_{dpct} \quad \forall\left\{dpt \,|\, \exists D_{dpct}\right\} \tag{65}$$

where D_{dpct} indicates how much of product p at time t is required at demand point d.

In the "maximize contribution margin" scenario there is no need to satisfy demand completely. Therefore, this scenario provides the additional option that a lower limit, D^{LL}_{dpct}, on the demand is considered:

$$s^L_{dpct} \geq \min\left\{D^{LL}_{dpct}, D_{dpct}\right\} \quad \forall\left\{dpt \,|\, \exists D^{LL}_{dpct} \wedge \exists D_{dpct}\right\} \tag{66}$$

In "satisfy demand" scenarios the demand has to be satisfied exactly, i.e.,

$$s^L_{dpct} = D_{dpct} \quad \forall\left\{dpct \,|\, \exists D_{dpct}\right\} \tag{67}$$

Due to lack of production capacity it may happen that not all of the demand can be covered by own production. Therefore, it is very important to provide the option to consider external purchase of products [see Sect. 4.6].

4.6 External purchase

External purchase of products helps to avoid running into situations of not being able to fulfil the demands. External purchase appears as source terms in the inventory balance equations for both site and demand points, and is characterized by the maximum quantity P_{spk}^{ES} $\left(P_{dpt}^{ED}\right)$ available for purchase and the costs/ton C_{spk}^{ES} $\left(C_{dpt}^{ED}\right)$. It is very important that these costs reflect the real business process, and that they are not considered as some artificial penalty costs. In addition to the upper bounds

$$p_{spk}^{ES} \leq P_{spk}^{ES} \quad \forall \{spk \,|\exists\, P_{spk}^{ES}\} \tag{68}$$

$$p_{dpt}^{ED} \leq P_{dpt}^{ED} \quad \forall \{dpt \,|\exists\, P_{dpt}^{ED}\} \tag{69}$$

there might be lower limit P_{spk}^{ES-} or p_{dpt}^{ED-}, resp. The external purchase variables might be interpreted as semi-continuous variables, *i.e.*,

$$p_{spk}^{ES} = 0 \vee p_{spk}^{ES-} \leq p_{spk}^{ES} \leq P_{spk}^{ES} \tag{70}$$

$$p_{dpt}^{ED} = 0 \vee p_{dpt}^{ED-} \leq p_{dpt}^{ED} \leq P_{dpt}^{ED} \tag{71}$$

if the supplier is willing to provide support but asks for a minimum quantity to deliver if support is requested.

5 The mathematical model: the design decisions

Since the network consists of several plants with different product streams it seems adequate to investigate whether it makes sense

- to open certain reactors at a site and to produce additional products,
- to shut-down certain reactors and thus not to produce some products at this site,
- or to buy whole sites.

One way to analyze such questions is to run different scenarios and to pick the best one. This simulation approach, however, might fail to identify the real good candidates and the optimal solution. The alternative approach is to model these questions or corresponding features appropriately and let the optimizer come up with optimal suggestions regarding the design of the production network.

5.1 A simple approach to include design decisions

A simple inclusion of the design features realized in the first phase of the project was to let the optimizer choose whether to open or shutdown a reactor once and for ever at the beginning of the planning horizon. That approach involves the following data: the total cost for buying a plant, C_s^B, the annual cost for buying a plant, C_s^{BA}, the fixed cost to operate a reactor, C_{sr}^{FIX}, the event value cost to open a reactor, C_{sr}^{O}, and the event value cost to shut down a reactor, C_{sr}^{SD}. The design costs to be included in a *net profit objective function* (contribution margin minus fixed costs minus design costs) has the form

$$-z^D := \sum_{s\in\mathcal{S}|\exists C_s^{BA}} C_s^{BA}\eta_s + \sum_{s\in\mathcal{S}} \sum_{r\in\mathcal{R}|\exists C_{sr}^{SDA}} \left(C_{sr}^{SDA} - C_{sr}^{FIX}\right)\varphi_{sr} + \sum_{s\in\mathcal{S}} \sum_{r\in\mathcal{R}|\exists C_{sr}^{SO}} \left(C_{sr}^{O} + C_{sr}^{FIX}\right)\mu_{sr} \quad (72)$$

where the binary variables η_s, φ_{sr} and μ_{sr} indicate whether a plant is bought, a reactor is shut down or opened.

An additional binary variable θ_{sr} is introduced if $I_{sr}^D = 1$;

$$I_{sr}^D = 1 \Leftrightarrow \left\{I_{sr}^{sR} = 1 \wedge \left(\exists C_s^{BA} \vee C_{sr}^{SD} \vee C_{sr}^{O}\right)\right\} \quad (73)$$

where I_{sr}^D indicates whether reactor r at site s is subject to design decisions; most cases we studied involved about 30 *design reactors*. θ_{sr} controls the available capacity of a reactor and is related to the other binary variables as follows:

$$\theta_{sr} = \eta_s \quad \forall\left\{sr\,|\exists\, I_{sr}^{sR} = 1 \wedge \exists C_s^{BA}\right\} \quad (74)$$

$$\theta_{sr} = 1 - \varphi_{sr} \quad \forall\left\{sr\,|\exists\, I_{sr}^{sR} = 1 \wedge \exists C_{sr}^{SD}\right\} \quad (75)$$

$$\theta_{sr} = \mu_{sr} \quad \forall\left\{sr\,|\exists\, I_{sr}^{sR} = 1 \wedge \exists C_{sr}^{O}\right\} \quad (76)$$

The relations (74) to (76) give us a hint on possible priorities (η_s,φ_{sr},μ_{sr},θ_{sr}) for the variables in the branching process. It is assumed that C_{sr}^{SD} and C_{sr}^{O} do not exist simultaneously.

The disadvantage of that approach, in a 10- or 15-year planning horizon, is that a reactor has to be opened or shut down in the first period of the planning horizon and stays in that status for the rest of the planning horizon. A more detailed view described in the next section allows for time dependent and temporary shutdowns and openings.

5.2 Time dependent and temporary shutdowns

In the course of the project it became obvious that more details regarding time resolution are necessary. Therefore, reactors, for which opening or shutdown cost are specified, will be treated as reactors subject to time dependent opening or shutdown decisions. These reactors are modeled similar as the reactors subject to mode changes. For each design reactor we assign two modes ($m = 1$ and 2 corresponding to *on* and *off*), and use the variables α_{sr1k} , β_{sr1k}, γ_{sr1k}, and δ_{sr1k} explained in ([11, Sect. 10.4]). Note that this model approach implies that the design reactors can not be subject to intrinsic mode changes which was not a problem in the current application. However, it is now possible that both C_{srk}^{SD} and C_{srk}^{O} may be different from zero. A further detail to be considered is that the opening decisions need some time to be put into reality. Therefore, a delay time K_{sr}^{FO} might specify which production time slice is the first one in which the reactor could be opened. This conditions, for $k = 1, \ldots, \max(1, K_{sr}^{FO} - 1)$, is realized by the bounds

$$\alpha_{sr1k} = \beta_{sr1k} = \delta_{sr1k} = 0 \quad \forall (sr) \in \mathcal{R}^D := \{(sr) \,|\, I_{sr}^{MC} \neq 1 \wedge \exists I_{sr}^{IBR}\} \tag{77}$$

for all design reactors which are not yet opened.

Although, the economical parameters may already prevent that reactors are opened and shut down wildly and very frequently it might be necessary for managerial reasons to introduce two parameters, K^O and K^C, which specify that if a reactor is opened in production time slice k, it has to be open for the next K^O time slices, and vice versa, if a reactor is shutdown in period k it has to stay closed for the next K^C time slices. These conditions tighten the model constraints by putting the variables α_{sr1k}, β_{sr1k}, γ_{sr1k} and δ_{sr1k} to zero for certain time slices in which a plant cannot operate and are enforced by

$$\sum_{\kappa=k+1}^{k+K^O} \gamma_{sr1k} \leq K^O - K^O \beta_{sr1k} \quad \forall (sr) \in \mathcal{R}^D, \quad \forall k \in \mathcal{K}_s^O \tag{78}$$

$$\sum_{\kappa=k+1}^{k+K^C} \alpha_{sr1k} \leq K^C - K^C \gamma_{sr1k} \quad \forall (sr) \in \mathcal{R}^D, \quad \forall k \in \mathcal{K}_s^C \tag{79}$$

$$\sum_{\kappa=k+1}^{k+K^C} \beta_{sr1k} \leq K^C - K^C \gamma_{sr1k} \quad \forall (sr) \in \mathcal{R}^D, \quad \forall k \in \mathcal{K}_s^C \tag{80}$$

$$\sum_{\kappa=k+1}^{k+K^C} \delta_{sr1k} \leq K^C - K^C \gamma_{sr1k} \quad \forall (sr) \in \mathcal{R}^D, \quad \forall k \in \mathcal{K}_s^C \tag{81}$$

$$\sum_{\kappa=k+1}^{k+K^O} \delta_{sr1k} \geq (K^O - 1) \beta_{sr1k} \quad \forall (sr) \in \mathcal{R}^D, \quad \forall k \in \mathcal{K}_s^O \tag{82}$$

$$\sum_{\kappa=k+1}^{k+K^C} \gamma_{sr1k} \leq K^C - K^C \gamma_{sr1k} \quad \forall (sr) \in \mathcal{R}^D, \quad \forall k \in \mathcal{K}_s^C \tag{83}$$

$$\sum_{\kappa=k+1}^{k+K^C} \beta_{sr1k} \leq K^O - K^O \beta_{sr1k} \quad \forall (sr) \in \mathcal{R}^D, \quad \forall k \in \mathcal{K}_s^O \tag{84}$$

with $\mathcal{K}_s^C := \{1, \ldots, \max(1, N^K(s) - K^C)\}$ and $\mathcal{K}_s^O := \{1, \ldots, \max(1, N^K(s) - K^O)\}$. Finally, when a reactor is open, i.e., $\alpha_{sr1k} = 1$, it is ensured that a certain minimum quantity is really produced, i.e.,

$$p_{srpk}^P \geq P_{srpk}^{MIN} \alpha_{sr1k} \quad \forall \left\{ srpk \,|\, I_{sr}^{MC} \neq 1 \wedge \exists I_{sr}^{IBR} \wedge \exists I_{srp}^{SRP} \right\} \tag{85}$$

Note that (85) cannot be applied to reactors which are design reactors and are already open at the beginning of the planning horizon. The reason is that for such reactors we necessarily have $\alpha_{sr1k} = 1$. However, if such a reactor is subject to a shutdown decision it may be not even able to produce and thus cannot meet the minimum production requirement (85).

The basic cost terms are the costs for shutdown and opening adjusted for discounted cash flows and depreciation (see below), i.e., terms such as

$$c^{D_1} := \sum_{(sr)\in\mathcal{R}^D} \sum_{k=1}^{N_s^K} C_{srk}^{SD}\gamma_{sr1k} \quad , \quad c^{D_2} := \sum_{(sr)\in\mathcal{R}^D} \sum_{k=1}^{N_s^K} C_{srk}^{O}\beta_{sr1k} \tag{86}$$

The improved design model also considers the residual book value V_{srk} [see Equation (90)] of an opened reactor, i.e., a reactor which caused opening cost) at the end of the planning horizon. This adds the term

$$z^{D_2} := - \sum_{\{sr\}\in\mathcal{R}^D} \sum_{k=1}^{N_s^K} V_{srk}\beta_{sr1k} + \sum_{\{sr\}\in\mathcal{R}^D} \sum_{k=1}^{N_s^K} V_{srk}\gamma_{sr1k} \tag{87}$$

to the objective function. V_{srk} is based on the *net present value* and on the depreciation. At present we apply a linear depreciation rate. The discounted opening cost in production time slice k are D_t^P/U_{st}

$$C_{srk}^{O} = C_{sr}^{O}/D^{k-1} \quad , \quad D := 1 + \frac{p}{100U_{st}} \tag{88}$$

where p is the discount rate in percent. The calculation of the depreciation depends on the opening time and thus we obtain the depreciation factor

$$F_s^D := 1 - \frac{N_s^K - k}{U_{st}T^D} \tag{89}$$

where T^D denotes the depreciation time in years; a typical value is $T^D = 15$ years. Formula (89) for the computation of the depreciation factor assumes that the depreciation time is measured in units of the commercial time scale; typically the discretization of the commercial time periods is one year and so the depreciation time is also given in years. Based on these assumptions, if the opening occurs in period k the residual book value is

$$V_{srk} = C_{sr}^{O}/D^{k-1}\left(1 - \frac{N_s^K - k}{U_{st}T^D}\right) \tag{90}$$

If we assume that a reactor which causes opening cost is not shut down, i.e., it is opened at most once, the term added to the objective function is

$$\sum_{\{sr\}\in\mathcal{R}^D} \sum_{k=1}^{N_s^K} V_{srk}\alpha_{sr1N_s^K}\beta_{sr1k} = \sum_{\{sr\}\in\mathcal{R}^D} \sum_{k=1}^{N_s^K} V_{srk}\beta_{sr1k} \tag{91}$$

Therefore, the total contribution of β_{sr1k} to the design term in the objective function representing the opening cost and the residual book value is

$$z^{D_2} := \sum_{\{sr\} \in \mathcal{R}^D} \sum_{k=1}^{N_s^K} \left(V_{srk} - C_{srk}^O \right) \beta_{sr1k} \tag{92}$$

An additional cost term is included to describe reactors that are subject to a shutdown decision in the first period but require continuation cost C_{sr}^{CNT}, e.g., for maintaining or upgrading existing facilities, if the reactor is not shut down in the first period. This feature is considered by the binary variable

$$\sigma_{sr} \quad \forall \left\{ sr \,\middle|\, (sr) \in \mathcal{R}^D \wedge \exists C_{sr}^{CNT} \right\} \tag{93}$$

the constraint

$$\sigma_{sr} \geq 1 - \gamma_{sr11} \quad \forall \left\{ sr \,\middle|\, (sr) \in \mathcal{R}^D \wedge \exists C_{sr}^{CNT} \right\} \tag{94}$$

and the costs term

$$c^{D_3} := \sum_{(sr) \in \mathcal{R}^D | \exists C_{sr}^{CNT}} C_{sr}^{CNT} \sigma_{sr} \tag{95}$$

in the *net profit* objective function.

The total contribution of the design reactors to the objective function is

$$z^D := -c^{D_1} + z^{D_2} - c^{D_3} \tag{96}$$

6 The mathematical model: the objective functions

The model covers eight different objective functions invoked by `OBJTYPE=`n, where n is a number between 1 and 8:

1. "max. contribution margin" ([11, Sect. 10.4])
2. "max. contribution margin while guaranteeing minimum demand"
3. "min. cost while satisfying full demand" ([13])
4. "max. total sales neglecting cost" ([13])
5. "max. net profit" (the detailed and full design problem)
6. "multi-criteria objectives" (maximize profit & minimize transport)
7. "max. total production"
8. "max. total production of products for which demand exists"

Only the fifth one exploits all design features; all other assume a fixed design. The first one maximizes the contribution margin, $y - c$, and includes the yield, y, calculated on the basis of production and the associated sales prices and the sum of all variable cost c. The second one minimizes the variable cost, c, while satisfying demand. In the cases 1 to 6 the following variable cost are involved: variable production cost, change-over cost, transport between sites, between sites and demand points, and between demand points, inventory cost for products, and cost for external purchase of products; in case 5 these cost terms are discounted according to Section 5.2 which leads to the *net profit* objective function

$$\max z \quad , \quad z := y - c + z^D \tag{97}$$

The objective functions 7 and 8 are used in an initial phase to test the data and to derive the theoretical capacity of the production network. The total net profit objective function contain the design cost terms which have quite different scaling characteristics compared to the standard production planning terms. Using variable directives in the B&B scheme, i.e., prioritizing the design decisions, it was possible to cope with this problem. The multi-criteria objectives scenario is solved by a goal programming approach as described in [9].

7 Computational issues, implementation and results

The model has been coded and the MILP problem has been solved using Dash's modeling language and MILP-solver `XPRESS-MP` 10.60 ([2], [3] and [4]). During some first numerical experiments it was observed that the objective function in the scenarios 4, 7 and 8 was dually degenerated. Thus, a significant speed-up was achieved using the primal Simplex algorithm to solve these scenarios. Some DOS-based procedures have been programmed to automate the process of accessing the data from an `EXCEL` spreadsheet, generating the matrix, solving the problem and returning the results into the `EXCEL` spreadsheet. Especially, some batch files developed enabled us to pass appropriate command streams to the solver depending on the objective function scenario.

7.1 Model assumptions and validation

The first step of model validation was to review and summarize the assumptions and limitations of our mathematical model. They are already given in Section 3.3 but in this practical case, only the implementation of the model showed that they have been reasonable and acceptable. The model was carefully validated by the client, i.e., by an experienced production planner and a financial expert running the model and the software under different circumstances. Especially, the opening and shut-down of reactors were carefully traced and subject to plausibility checks by the financial planners, for instance, taking into account the contribution margin related to a newly opened reactor compared to its investment cost. It was interesting to inspect solutions in which a reactor which was no longer profitable was shut-down immediately, after a reactor based on a new technology was opened. Transport between sites and demand points was an issue which sometimes lead to solutions non-intuitive to the people responsible for the production planning in single plants, and usually needed clarification.

7.2 Computational issues

To give an example of the problem size and some solution characteristics we quote a typical scenario (S_5) with about 30 design reactors covering 10 years with 10 commercial and 20 production time slices, for which we derived production and design plans maximizing total net profit. Using Dash's MILP-solver XPRESS-MP 10.60 ([2], [3] and [4]) for a problem with $n_c = 26941$ continuous, $n_c = 5100$ binary and $n_{sc} = 1100$ semi-continuous variables and $c = 28547$ constraints, we got the following results (including the integer solution number IP, number of nodes n_n, run time τ on a 266 MHz Pentium II PC in minutes, best upper bound z^U, best lower bound z^L and integrality gap $\Delta := 100\frac{z^U - z^L}{z^L}$):

	IP	n_n	τ	z^U	z^L	Δ
S_5	1	218	4	140.8	137.5	2.4
S_5	2	1794	34	138.4	138.4	–

The first feasible integer solution is usually found within 10 minutes after exploring about 300 nodes in the B&B tree. Usually, for pure operational planning, this solution is accepted and the tree search is terminated. This heuristic is justified if Δ is of the order of a few percent (well within the error associated with the input data) because it eliminates the need for the time consuming complete search for the absolute optimal solution via the B&B algorithm. In the SSDOP approach this is only valid, if Δ is less or equal, say, 1%.

It is remarkable that this model even when all design features are exploited is able to find the first feasible integer solution, usually after 200 or 300 nodes, within a few minutes (integrality gap between 1 and 3 percent) and is able to prove optimality in most cases within 30 minutes (3000 nodes typically). If all design features are fixed, the problem is solved to optimality within a few minutes (1000 to 2000 nodes).

In order to carry out sensitivity analyses of the solution with respect to the input data, especially to the demand data and costs or sales prices, it is important that we are able to conduct the complete search via the B&B algorithm or to get an integrality gap Δ less than one percent. The sensitivity analysis was carried out by the client by varying the input data by up to 20 percent and inspecting the objective function value and the design decisions.

7.3 Commercial results, benefits and experience

The client reports cost savings of several millions of US$. These cost savings were achieved via a reduction in transportation cost compared to the previous year when the model was not in use. The solution for a one year planning horizon allowed the company to better understand and forecast the flow of products between North America, Europe and Asia. This knowledge was then used to reduce the need and cost of urgent shipments.

The results obtained with the use of the design feature allowed the business team to clearly demonstrate the value of its investment plan to senior management.

The comprehensive results obtained from the model allowed the team to focus its recommendation on facts and quickly address questions concerning product flows, production sourcing, capacity constraints, and working capital needs in addition to investment capital requirements. Moreover, it was beneficial to the client to see that the design solutions (which reactors to be opened or to be closed) were stable against up to 20% changes in the demand forecast.

It was vital to the project that on the client's side people responsible for the operational planning and colleagues from the financial planning department cooperated with each other. It was very interesting during the modeling phase to see how know-how from quite separate areas merged and lead to a complex model providing just thc right dcgree of detail to satisfy all parties involved.

8 Conclusions

We have developed and applied a model that combines operational planning with strategic aspects having consequences for years. The decisions determine the infrastructure and aggregate production plans for a horizon up to 15 years. It is important that realistic and detailed demand forecast, and, possibly, cost and sales prices are available, and also that a sensitivity analysis is performed showing how stable the optimal solution is with respect to changes in these input data. Especially, for problems with such a long planning horizon and great financial impact it is vital that the optimality of a solution can be proven, or that at least some safe bounds can be specified.

As far as good modeling practice is concerned, we learned from the discussion and communication with the client during the model building phase that great attention has to be paid to achieving a good balance of details entering the model. Some focus had also to be given to the structure of the objective function which might contain terms of very different size, and thus may lead to bad scaling. A useful extension of the model might be to include tax and depreciation related features considering special rules for specific countries and their tax rates.

Provided that optimality is proven or safe bounds are derived, combining strategic or design aspects with operational planning is an elegant approach taken by some chemical companies; it can save huge quantities of money and also supports an analysis related to the stability of solutions.

Last but not least we want to stress one crucial fact learned from two projects (one described in this paper, for the other one see [9, Sect. 9.2]): the simultaneous strategic/design & operational planning approach requires that the departments being responsible for the strategic or design decisions and the planning/scheduling cooperate; our experience is that this problem is by far more difficult to solve than the mathematical or technical ones, especially in large companies and their cultural and organizational structures. But especially, in large organizations, for instance, if new sites are established in otherwise nonindustrial areas (e.g., the construction of new plants in South East Asia) or new structures have to be embedded into existing sites, appropriate models combining operational planning with strategic or design planning almost certainly lead to great financial savings, and therefore, the simultaneous approach should be attractive to many other, especially large companies.

References

1. Ahmadi J, Benson R, Supernaw-Issen D (1999) Mixed-integer nonlinear programming applications. In: Ciriani TA, Gliozzi S, Johnson EL, Tadei R (eds) Operational research in industry, pp 199–231. Macmillan, Houndmills, Basingstoke, UK
2. Ashford RW, Daniel RC (1987) LP-MODEL XPRESS-LP's model builder. Institute of Mathematics and its Application Journal of Mathematics in Management 1: 163–176
3. Ashford RW, Daniel RC (1991) Practical aspects of mathematical programming. In: Munford AG, Bailey TC (eds) Operational research tutorial papers, pp 105–122. Operational Research Society, Birmingham
4. Ashford RW, Daniel RC (1995) XPRESS-MP Reference Manual. Dash Associates, Blisworth House, Northants NN73BX, http://www.dash.co.uk
5. Domschke W, Scholl A, Voß S (1997) Produktionsplanung, 2nd edn. Springer, Berlin Heidelberg New York
6. Drexl A, Haase K (1995) Proportional lotsizing and scheduling. Int J of Prod Econ 40: 73–87
7. Kallrath J (1995) Diskrete Optimierung in der chemischen Industrie. In: Bachem A, Jünger M, Schrader R (eds) Mathematik in der Praxis – Fallstudien aus Industrie, Wirtschaft, Naturwissenschaften und Medizin, pp 173–195. Springer, Berlin Heidelberg New York
8. Kallrath J (1999) Mixed-integer nonlinear programming applications. In: Ciriani TA, Gliozzi S, Johnson EL, Tadei R (eds) Operational research in industry, pp 42–76. Macmillan, Houndmills, Basingstoke, UK
9. Kallrath J (2002) Gemischt-ganzzahlige Optimierung: Modellierung und Anwendungen. Vieweg, Wiesbaden
10. Kallrath J (2002) Planning and scheduling in the process industry. OR Spectrum 24: 219–250
11. Kallrath J, Wilson JM (1997) Business optimisation using mathematical programming. Macmillan, Houndmills, Basingstoke, UK
12. Kuik R, Solomon M, van Wassenhove, LN (1994) Batching decisions: structure and models. European Journal of Operational Research 75: 243–263
13. Timpe C, Kallrath J (2000) Optimal planning in large multi-site production networks. European Journal of Operational Research 126(2): 422–435

An integrated system solution for supply chain optimization in the chemical process industry*

Guido Berning, Marcus Brandenburg, Korhan Gürsoy, Vipul Mehta, and Franz-Josef Tölle

Bayer AG, Bayer Technology Services, Supply Chain Optimization,
51368 Leverkusen, Germany (e-mail: {guido.berning.gb, marcus.brandenburg.mb, korhan.guersoy.kg, vipul.mehta.vm, franz-josef.toelle.ft}@bayer-ag.de)

Abstract. This paper considers a complex scheduling problem in the chemical process industry involving batch production. The application described comprises a network of production plants with interdependent production schedules, multi-stage production at multi-purpose facilities, and chain production. The paper addresses three distinct aspects: (*i*) a scheduling solution obtained from a genetic algorithm based optimizer, (*ii*) a mechanism for collaborative planning among the involved plants, and (*iii*) a tool for manual updates and schedule changes. The tailor made optimization algorithm simultaneously considers alternative production paths and facility selection as well as product and resource specific parameters such as batch sizes, and setup and cleanup times. The collaborative planning concept allows all the plants to work simultaneously as partners in a supply chain resulting in higher transparency, greater flexibility, and reduced response time as a whole. The user interface supports monitoring production schedules graphically and provides custom-built utilities for manual changes to the production schedule, investigation of various what-if scenarios, and marketing queries.

Key words: Supply chain management – APS-system – Collaborative planning – Optimization – Genetic algorithm

1 Introduction

In recent years, there is a renewed interest in Advanced Planning and Scheduling (APS) applications (cf. [21]). This is evident from a range of new generation APS tools available in the market which combine the increased computational power

* The authors would like to thank Hans-Otto Günther and Roland Heilmann for helpful comments on draft versions of this paper.

with state-of-the-art optimization and database management techniques. Having implemented an Enterprise Resource Planning (ERP) system (see e.g. [20] for a more detailed description), many companies are interested in integrating their supply chain internally and externally with the aid of the internet-based methodologies (see e.g. [12, 18, 22, 25]).

Although the scope of applications varies, the primary needs for implementing APS tools remain the same: To plan future activities while anticipating the market but at the same time being flexible enough to handle changes in the short term and minimizing the effects of these changes on the long term perspective. The added emphasis on meeting conflicting business objectives, such as reducing response time, meeting due dates, increasing customer satisfaction, and reducing inventory, is an additional reason for using an APS tool. Sometimes, with all the hype surrounding the modern APS tools, these basic goals are forgotten.

Long term rough-cut planning essentially looks ahead to plan roughly and prepare the business well in advance for the upcoming market trends while meeting the primary goals of the business in terms of improving overall business efficiency. *Short term* scheduling, on the other hand, considers decisions on production sequencing, minimizing idle time and due date violations, and keeping the inventory level low, at the same time respecting all production constraints.

In practice, however, it is often not possible to represent all the constraints in a mathematical model or to provide a realistic estimate of the specific cost effects. Due to uncertainty in cost and production parameters the methods that strive for the true optimum are condemned to fail in practice and hence are rendered less effective in handling industrial applications. Many production constraints, such as sequence dependencies, lot size restrictions or resource allocations, lead to $\mathcal{NP}$-hard optimization problems (cf. [16, 19]), for which cost optimum solutions cannot be computed within a reasonable CPU time. Moreover, automated system solutions without user interface are of little use in many practical applications. Instead, flexibility based on qualitative factors such as market and production dynamics and multiple conflicting business objectives is important. A holistic approach is needed, which combines the computational power and the flexibility of the modern APS tools with the possibility of user-driven interactions and manual interventions.

There are many articles in the scientific literature which deal with mathematical optimization problems of planning and scheduling in the chemical industry (see e.g. [3, 5, 10, 24]). Other papers present case study oriented research contributions (e.g. [1, 7, 15]) or focus on specific methodological aspects (e.g. [2, 4, 11, 17, 23]). According to our practical experience at Bayer AG, a successful planning application, however, involves three major components: (i) A comprehensive mathematical backbone to compute an optimized solution which can then be improved either through further optimization on specific sub-problems or through manual interaction, (ii) a mechanism for transparency, collaboration, information sharing and conflict management, and (iii) a powerful interface for manual interaction and problem visualization.

This paper emphasizes various aspects of an integrated planning and scheduling problem occuring in the process industry. Problems involving short and long term scheduling and planning with product chains passing through multiple batch production plants and consisting of many intermediate production steps are highlighted (see Sect. 2). All the three aspects mentioned above were considered in the industrial application discussed in this paper: (i) A *genetic algorithm* (*GA*) was adopted instead of relying on other optimization techniques such as MILP or MINLP (see e.g. [13, 14] for an overview). The mathematical model and the implemented algorithm are described in Section 3. (ii) Increased transparency and reduced response time for changes in the production schedule were achieved by implementing a *collaborative planning model* which allows distributed decision making and access to the production schedule at the various plants involved. Additional system support for conflict management was provided by introducing the concept of a so-called "chain planner" who is authorized to look at production schedules of multiple plants simultaneously. The concept and architecture are presented in Section 4. (iii) Using a commercial APS tool, an application specific scheduling model was implemented to allow *manual updates and changes*, as described in Sections 5 and 6.

2 Problem statement

2.1 Application environment

We consider a network of multi-purpose production plants that produce a variety of finished products through multi-stage production processes. The plants are linked by supplier-customer relations, i.e. one plant may produce intermediate goods that are processed further by other plants. The customer-supplier relations depend on the material flow given by the recipes of the final products. Furthermore, each plant interacts with external suppliers and customers. To fulfill customer orders, a variety of processing steps and a number of plants have to be considered depending on the recipe of the final products. Figure 1 depicts the plant network and material flow arising from customer orders for two plants, x_{10} and x_{11}.

Each plant comprises a large number of multi-purpose production *facilities* (or *equipment units*) at which production takes place in *batch* mode. An ordered set of batches produced consecutively (without any delay in between) at the same facility is called a *campaign*. In general, the size of a campaign is limited from below.

A *product* is the outcome of a unique *process*. Each process involves various types of facilities (e.g. mixers, reactors, dryers) which have to be passed in a given order. The processing of a campaign at a facility is called an *activity*. Hence, a process can be defined as an ordered set of activities. Since a campaign consists of a set of batches which have to be processed one after another, an activity can be described as an ordered set of *operations*. An operation is defined as the processing of a batch.

Due to the peculiarity of the process industry the number of feasible combinations of process-specific equipment units (*routings*) is limited. Also the type of equipment feasible for a specific product is limited. The alternative product-equipment assignments are defined by a routing structure which consists of a source and a sink node and paths indicating the alternative production flows (see Fig. 2a).

Before starting a process, the necessary equipment units have to be assembled, i.e. the various facilities have to be connected (*routed*) to each other (as required

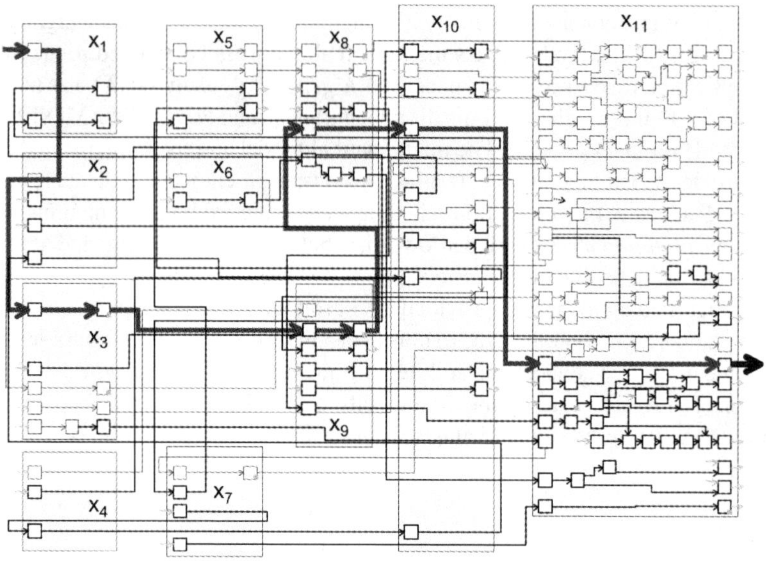

Fig. 1. Network consisting of plants x_1 to x_{11} and material flows between plants

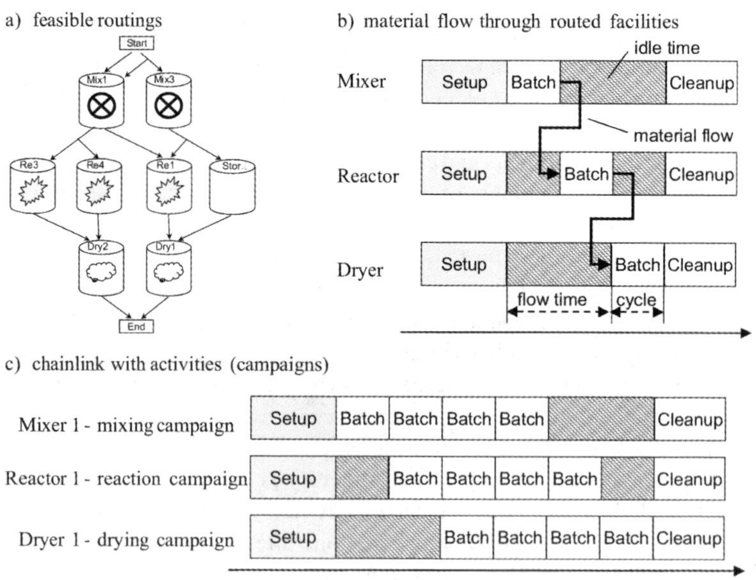

Fig. 2. Routing superstructure, material flow and formation of production campaigns

by the process) and to be cleaned. After termination of the process, the equipment is cleaned again. Therefore, a *setup* time before a process and a *cleanup* time after a process is required.

The time between the beginning of the first activity and the beginning of the last activity of a process is defined as the *flow time*. The duration of an operation is characterized by a given *cycle time* which indicates the time span required to complete a batch at a facility (see Fig. 2b). The cycle time depends on the process and the selected routing, but is identical for every operation carried out at the routed facilities.

The *material flow* between production facilities can be linear, convergent or divergent (due to coupled production). In the practical application considered here two types of demand can be distinguished, forecasts and customer orders. Each *demand* (or *requirement*) *element* is given by a required quantity of a specific product and a specific due date.

2.2 Formal description of the decision problem

2.2.1 Basics. The problem addressed here can be roughly stated as follows:

Given

(a) a set of production plants, their production equipment and corresponding capacities,
(b) a set of products, that can be processed in each plant, associated constraints, production parameters, routings, and bills of material (BOM),
(c) a set of demand elements (both forecasts and customer orders), and
(d) a set of penalties and cost functions.

Provide

1. a feasible schedule which simultaneously ensures that
   (a) enough material at each BOM level is produced to satisfy the demands,
   (b) early warnings for shortages of intermediates and raw materials are provided,
   (c) all production constraints, resource availabilities, and business requirements are respected, and
   (d) good solutions in a practical sense with regard to the multiple conflicting objectives are obtained;
2. a platform and a mechanism allowing
   (a) every plant to operate independently, however to be able to collaborate with other plants for increased transparency and information sharing,
   (b) reduction in response time to any changes in the production schedule of a plant acting as the internal supplier or customer, and
   (c) conflict management and monitoring;
3. a system support and a user friendly tool for
   (a) easy visualization of the production schedule,
   (b) manual updates and interactive changes of production schedules,
   (c) answering marketing queries,
   (d) maintaining latest inventory positions, and
   (e) maintaining master and transactional data.

Next, with the above explanations in mind, we give the notation used for the formulation of the optimization problem.

2.2.2 Problem parameters (master data). Let $\mathcal{X}$ denote the set of multi-purpose, multi-product plants and $\mathcal{F}$ the set of facilities. Each plant $x \in \mathcal{X}$ involves a set $\mathcal{F}_x \subseteq \mathcal{F}$ of facilities, where $\mathcal{F} = \cup_{x \in \mathcal{X}} \mathcal{F}_x$ and for $x, x' \in \mathcal{X}$ with $x \neq x'$ we have $\mathcal{F}_x \cap \mathcal{F}_{x'} = \emptyset$. $\mathcal{P}$ denotes the set of products (raw material, intermediates or finished goods) that may be produced or consumed by processes.

The set of all processes is denoted by $\mathcal{A}$. Associated with each process $a \in \mathcal{A}$ are a set $\mathcal{P}_a^-$ of input and a set $\mathcal{P}_a^+$ of output products with $\mathcal{P}_a^- \cap \mathcal{P}_a^+ = \emptyset$. For any pair $\mathcal{P}_a^+, \mathcal{P}_{a'}^+$ with $\mathcal{P}_a^+ \neq \mathcal{P}_{a'}^+$ we have $\mathcal{P}_a^+ \cap \mathcal{P}_{a'}^+ = \emptyset$, i.e., each product $p \in \mathcal{P}$ is manufactured in a unique way. For product $p \in \mathcal{P}_a^-$ ($\mathcal{P}_a^+$), $\delta_{a,p}^-$ ($\delta_{a,p}^+$) denotes the input (output) fraction of product p for process a.

In order to manufacture a final product $p \in \mathcal{P}$ one has to produce a set $\mathcal{P}_p \subseteq \mathcal{P}$ of products consisting of p and the intermediates according to the recipe of p. Let $\mathcal{A}_p \subseteq \mathcal{A}$ denote the corresponding set of processes.

A routing $r \subseteq \mathcal{F}$ is an ordered set of facilities that have to be passed when manufacturing product p by process a_p. r_i is the i-th routed facility. The set of feasible routings associated with $a \in \mathcal{A}$ is given by $\mathcal{R}_a$.

The size b of a batch depends on process a as well as on the chosen routing $r \in \mathcal{R}_a$: $b = b(a, r)$. The same holds for the minimum campaign size $m = m(a, r)$, the cycle time $z = z(a, r)$, the flow time $f = f(a, r)$, the setup time $s = s(a, r)$, and the cleanup time $c = c(a, r)$. For the sake of simplicity, the dependence on a and r is omitted in the following if no misunderstanding can occur.

2.2.3 Instance parameters (transactional data). The customer orders and forecast demands are reflected by a set $\mathcal{E}$ of requirement elements $e = (\pi, q, t_d)$. $\pi \in \mathcal{P}$ is the product required, q the corresponding quantity, and t_d denotes the due date. In general, to satisfy demands, a process $a \in \mathcal{A}$ has to be performed several times. In order to distinguish between these occurrences of a we use the term *chainlink* C_a, which denotes a demand-induced set of batches of process a, all performed in a single campaign using the same routing. To manufacture π, each process $a \in \mathcal{A}_\pi$, i.e. a corresponding chainlink C_a, has to be performed. According to the recipe of π, several processes $a', a'', \ldots$, i.e. chainlinks $C_{a'}, C_{a''}, \ldots$, have to be performed in a certain order. The resulting (partially) ordered set $\{C_{a'}, C_{a''}, \ldots\}$ of chainlinks is named *(product) chain* $\mathcal{C}_e$.

In the following, we assume that for all points in time t we have $t \geq 0$ where t is not allowed to exceed a given *time horizon*. The *inventory level* of product $p \in \mathcal{P}$ is denoted by $i(p, t)$. $i(p, 0)$ is the *initial inventory level.*

The occupation of a facility $f \in \mathcal{F}$ is reflected by a *resource (occupation) object* $o_f = (t_s, t_c, n, v, w, a, r)$. t_s (t_c) denotes the start (completion) time of chainlink C_a at each facility $f \in r$ or, in other words, the start (completion) time of the activity at facility $f = r_1$ ($r_{|r|}$). n denotes the campaign size. v and w are binary variables indicating whether a setup or cleanup takes place ($v = 1$ and $w = 1$, respectively) or not ($v = 0$ and $w = 0$, respectively). (Later on, we show, how setup and cleanup can be saved.) Hence, all $f \in r$ are occupied during the time interval $[t_s - v \cdot s, t_c + w \cdot c[$. Besides the occupation of f by a chainlink, o_f might also reflect the occupation of f by a *shutdown* (e.g. for maintenance). In this case,

Table 1. Size of problem instances – master data

	Average value	Maximum value
$\lvert\mathcal{F}\rvert$	1,000 – 1,300	1,300
$\lvert\mathcal{A}\rvert$	600 – 650	700
$\lvert\mathcal{R}_a\rvert$	2 – 4	150
$\lvert r\rvert$	3 – 8	10
$\lvert\mathcal{A}_p\rvert$	3 – 6	10
$\lvert\mathcal{C}_e\rvert$	6 – 25	9,000

Table 2. Size of problem instances – transactional data

	Maximum value
Time horizon	600
Objects in $\hat{O}$	1,200
$\lvert\mathcal{E}\rvert$	950
Inventory changes in $\hat{O}$	45,000

we set $n = v = w = a = r = 0$. $\mathcal{O}_f$ denotes the set of all occupations of f and the vector $O = (\mathcal{O}_f)_{f\in\mathcal{F}}$ is called *(production) schedule*.

Upon generating a new schedule one has to consider that there are some chain-links and shutdowns, reflected by resource objects $\hat{o}_f$, which have already been scheduled in the past and may not be changed. These objects form the *initial schedule* $\hat{O} = (\hat{\mathcal{O}}_f)_{f\in\mathcal{F}}$. Moreover, for some products $p \in \mathcal{P}$ a so-called *frozen zone* $[0, f_p[$, i.e. a kind of quarantine time where no scheduling of C_{a_p} is allowed, might be given.

To give a short impression of the size of real-world instances of the optimization problem, some key figures are listed in Tables 1 and 2.

2.3 Properties of a feasible solution

A schedule O is called *feasible*, if the following conditions hold (with $o_f = (t_s, t_c, n, v, w, a, r)$ and $o'_f = (t'_s, t'_c, n', v', w', a', r')$):

1. O contains $\hat{O}$, i.e.:
$$\mathcal{O}_f \supseteq \hat{\mathcal{O}}_f \ \ (f \in \mathcal{F}). \tag{1}$$
2. At each facility $f \in \mathcal{F}$, at most one resource object is scheduled at a time, i.e. for any pair o_f, o'_f with $t_s \leq t'_s$:
$$t_c + w \cdot c \leq t'_s - v' \cdot s' \ (o_f, o'_f \in \mathcal{O}_f, f \in \mathcal{F}). \tag{2}$$
3. The frozen zones $[0, f_p[$ are obeyed, i.e.:
$$t_s - v \cdot s \geq f_p \ (p \in \mathcal{P}; o_f \in \mathcal{O}_f \backslash \hat{\mathcal{O}}_f, f \in \mathcal{F}). \tag{3}$$
4. The minimum campaign sizes m are respected, i.e.:
$$n \geq m \ \ (o_f \in \mathcal{O}_f, f \in \mathcal{F}). \tag{4}$$

5. The operations of an activity are performed one after another without preemption, i.e.:

$$t_c = t_s + f + n \cdot z \quad (o_f \in \mathcal{O}_f, f \in \mathcal{F}). \tag{5}$$

6. Each activity uses a feasible routing, i.e.:

$$r \in \mathcal{R}_a \quad (o_f \in \mathcal{O}_f, f \in \mathcal{F}). \tag{6}$$

7. The activities of a chainlink run in parallel and the setup and cleanup variables take the same values, i.e. for each pair $o_f, o_{\tilde{f}}$ with $o_{\tilde{f}} = (\tilde{t}_s, \tilde{t}_c, \tilde{n}, \tilde{v}, \tilde{w}, \tilde{a}, \tilde{r})$ and $\tilde{f} \in r$:

$$t_s = \tilde{t}_s, t_c = \tilde{t}_c, n = \tilde{n}, v = \tilde{v}, w = \tilde{w}, a = \tilde{a}, r = \tilde{r} \\ (o_f \in \mathcal{O}_f, f \in \mathcal{F}; o_{\tilde{f}} \in \mathcal{O}_{\tilde{f}}, \tilde{f} \in r). \tag{7}$$

8. If two resource objects o_f, o'_f are scheduled consecutively at the same facility f w.r.t. the same process a and routing r with $t_s \leq t'_s$, the cleanup after the activity corresponding to o_f and the setup before the activity corresponding to o'_f can be saved, i.e.:

$$r = r' \wedge a = a' \Rightarrow v = 0 \wedge w' = 0 \quad (o_f, o'_f \in \mathcal{O}_f, f \in \mathcal{F}). \tag{8}$$

9. No negative inventory level may occur, i.e.:

$$i(p,t) \geq 0 \quad (p \in \mathcal{P}; t \geq 0) \tag{9}$$

with

$$i(p,t) = i(p,0) + \sum_{f \in \mathcal{F}} \sum_{\substack{o_f \in \mathcal{O}_f: \\ f = r_{|r|}}} \sum_{\nu \in \{1,\dots,n\}} H(t_s + f + \nu \cdot z, t) \cdot \delta^+_{a,p} \cdot b \\ - \sum_{f \in \mathcal{F}} \sum_{\substack{o_f \in \mathcal{O}_f: \\ f = r_1}} \sum_{\nu \in \{1,\dots,n\}} H(t_s + (\nu - 1) \cdot z, t) \cdot \delta^-_{a,p} \cdot b \tag{10}$$

where

$$H : \mathrm{IR} \times \mathrm{IR} \rightarrow \{0,1\}, H(t,t') = 1 \Leftrightarrow t \leq t'. \tag{11}$$

2.4 Solution methodology

As mentioned before, the application requirements are considered in three distinct aspects: (i) *optimization*, (ii) *collaborative planning*, and (iii) *manual interaction*.

2.4.1 Optimization. Due to the size and complexity of realistic problem instances, exact solution methods are not applicable. Moreover, the problem structure does not allow to sub-divide the problem by plants, facilities or manufacturing levels, or to represent capacities, resources or the planning horizon in an aggregated way. An optimization method has to be chosen which quickly computes initial solutions and shows a strong convergence behaviour. As *genetic algorithms* (*GA*) tend to have this property, it has been decided to use them as a basis for the implemented solution strategy. (For an overview of genetic algorithms and their applications see [6].)

2.4.2 Collaborative planning. Using AspenMIMI© from AspenTech as a commercially available APS tool and a common server-client network structure, a platform for collaborative planning was established where local models are able to retrieve information from a central database. The data transfer and data consistency are realized with the help of remote function calls and macros.

2.4.3 Manual interaction. A user friendly tool was developed for easy and fast manual interaction and monitoring the production schedule. Again, AspenMIMI© from AspenTech was used, which – due to its toolbox structure with graphical user interface oriented programming – allows a high degree of freedom in customizing and development. A high level of user feedback has been considered in developing application specific utilities.

3 Solution approach

Let $e = (\pi, q, t_d) \in \mathcal{E}$ be a requirement element and $\mathcal{R}_e = \{R^1, \ldots, R^{n_e}\}$ a set of *routing combinations* R determining for each process $a \in \mathcal{A}_\pi$ which routing $r \in \mathcal{R}_a$ to choose: $R^i \in \times_{a \in \mathcal{A}_\pi} \mathcal{R}_a$ $(i = 1, \ldots, n_e)$ where $n_e = \prod_{a \in \mathcal{A}_\pi} |\mathcal{R}_a|$. R_a denotes the routing $r \in \mathcal{R}_a$ to be chosen for all processes a in chainlink C_a. Hence, associated with each routing combination $R \in \mathcal{R}_e$ is a chain $\mathcal{C}_e$ (cf. Fig. 3). The set $\{\mathcal{C}_e^1, \ldots, \mathcal{C}_e^{n_e}\}$ of all chains associated with $\mathcal{R}_e$ is denoted by Γ_e. $\mathcal{C}_e$ has the following properties:

- Each chainlink $C \in \mathcal{C}_e$ uses the routing given by R.
- The output of each chainlink $C \in \mathcal{C}_e$ is sufficient to satisfy the primary demand q and the secondary demands induced by q.

In the following, we show how the chains $\mathcal{C}_e \in \Gamma_e$ are generated based on the determination of $\mathcal{R}_e$ (cf. Sect. 3.1), computed (cf. Sect. 3.2), and scheduled (cf. Sect. 3.3), and how the obtained schedule can be improved (cf. Sect. 3.4). In Section 3.5, we show how all the components of GA work together. Section 3.6 is devoted to the parameter setup and in Section 3.7, the implementation issues are presented.

3.1 Chain generation

Prior to the computation of $\mathcal{R}_e$ and Γ_e the sets $\mathcal{P}_\pi$ and $\mathcal{A}_\pi$ as well as the manufacturing levels l_p $(p \in \mathcal{P}_\pi)$ of product p are determined by means of Algorithm 1: Here, $\mathcal{P}'$ denotes an auxiliary set and l_e is the maximum manufacturing level associated with e.

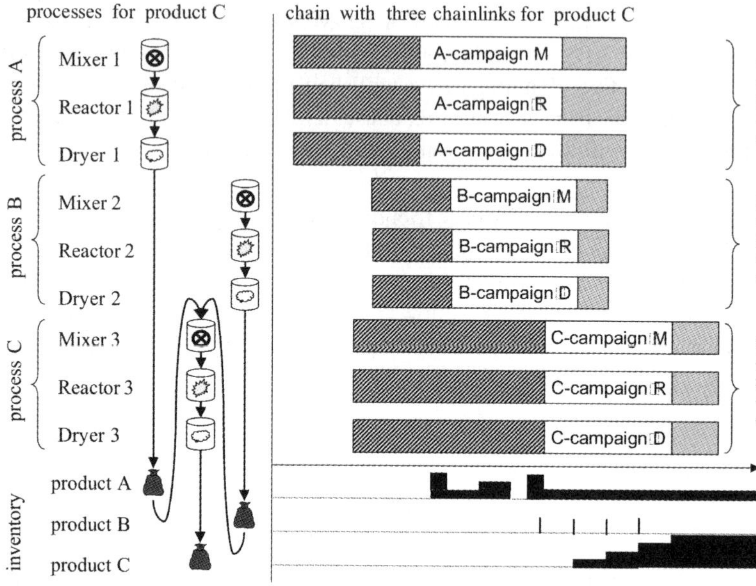

Fig. 3. Example of a chain consisting of three chainlinks

Algorithm 1 (Determination of $\mathcal{P}_\pi$, $\mathcal{A}_\pi$, and manufacturing levels)

$\mathcal{P}_\pi := \{\pi\}$; $\mathcal{A}_\pi := \{a_\pi\}$; $\mathcal{P}' := \{\pi\}$; $l_\pi := 0$

WHILE $\mathcal{P}' \neq \emptyset$ **DO**

$\mathcal{P}' := \mathcal{P}' \backslash \{p\}$; $\mathcal{A}_\pi := \mathcal{A}_\pi \cup \{a_p\}$

FOR $p' \in \mathcal{P}^-_{a_p}$ **DO**

$\mathcal{P}' := \mathcal{P}' \cup \{p'\}$; $l_{p'} := l_p + 1$

IF $p' \notin \mathcal{P}_\pi$ **THEN** $\mathcal{P}_\pi := \mathcal{P}_\pi \cup \{p'\}$,

$l_e := \max_{p \in \mathcal{P}_\pi} l_p$

Now, the set $\mathcal{R}_e$ can be easily obtained by setting $\mathcal{R}_e := \emptyset$ and adding all possible routing combinations R. A routing combination R is yielded by selecting exactly one routing $r \in \mathcal{R}_a$ for each process $a \in \mathcal{A}_\pi$.

3.2 Chain computation

To compute a chain $\mathcal{C}_e$ w.r.t. routing combination R the following computations for each chainlink $C_a \in \mathcal{C}_e$ have to be performed:

First, the size n_{C_a} of the campaigns associated with C_a are determined such that the total (primary and secondary) demand q_p of each product $p \in \mathcal{P}_\pi$ is satisfied (cf. Algorithm 2, where $r := R_a$). Let $\mathcal{P}_l = \{p \in \mathcal{P}_\pi | l_p = l\} \subseteq \mathcal{P}_\pi$ be the set of all products in $\mathcal{P}_\pi$ with manufacturing level l.

Algorithm 2 (Determination of campaign sizes for chainlinks)

$q_\pi := q$; **FOR** $p \in \mathcal{P}_\pi \backslash \{\pi\}$ **DO** $q_p := 0$

FOR $l := 0$ **TO** l_e **DO**

  FOR $p \in \mathcal{P}_l$ **DO**

    $n_{C_a} := \lceil \max\{q_p, \delta^+_{a_p,p} \cdot m \cdot b\}/(\delta^+_{a_p,p} \cdot b) \rceil$

    FOR $p' \in \mathcal{P}^-_{a_p}$ **DO** $q_{p'} := q_{p'} + \delta^-_{a_p,p'} \cdot b \cdot n_{C_a}$

Second, the set $Pred_{C_a}$ ($Succ_{C_a}$) of *predecessor* (*successor*) *chainlinks*, i.e. chainlinks $C_{a'}$ ($C_{a''}$) with $\mathcal{P}^+_{a'} \cap \mathcal{P}^-_a \neq \emptyset$ ($\mathcal{P}^+_a \cap \mathcal{P}^-_{a''} \neq \emptyset$) and the corresponding minimum time lags or *delays* $t_{C_{a'}C_a}$ ($t_{C_aC_{a''}}$) between the start time of $C_{a'}$ (C_a) and the start time of C_a ($C_{a''}$) have to be determined by means of Algorithm 3 (with $r' := R_{a'}$, $b' = b'(a', r')$, and $z' = z'(a', r')$). Here, $n_{p,i}$ denotes the minimum number of batches to be completed by C_a (with $a = a_p$) before the i-th batch of $C_{a'} \in Succ_{C_a}$ can be started w.r.t. the demand for (input) product $p \in \mathcal{P}^-_{a'}$.

Algorithm 3 (Determination of predecessor and successor chainlinks and delays)

FOR $C_a \in \mathcal{C}_e$ **DO**

  $Pred_{C_a} := \emptyset$; $Succ_{C_a} := \emptyset$

  FOR $C_{a'} \in \mathcal{C}_e$ **DO**

    IF $\mathcal{P}^+_{a'} \cap \mathcal{P}^-_a \neq \emptyset$ **THEN**

      $Pred_{C_a} := Pred_{C_a} \cup \{C_{a'}\}$; $Succ_{C_{a'}} := Succ_{C_{a'}} \cup \{C_a\}$

  FOR $C_{a'} \in Succ_{C_a}$ **DO**

    $t_{C_aC_{a'}} := 0$

    FOR $p \in \mathcal{P}^+_a \cap \mathcal{P}^-_{a'}$ **DO**

      $n_{p,i} := \min\{j \in \mathbb{N} | \delta^+_{a,p} \cdot b \cdot j \geq \delta^-_{a',p} \cdot b' \cdot i\}$

      $t_{C_aC_{a'}} := \max\{t_{C_aC_{a'}}, \max\{t_i | i \in \{1, \ldots, n_{C_a}\}\} + f\}$

      with $t_i := n_{p,i} \cdot z - (i-1) \cdot z'$

Note that each chain $\mathcal{C}_e \in \Gamma_e$ is computed independently from all other chains $\mathcal{C}_{e'}$ which have already been scheduled, i.e. without consideration of inventory levels at the start times of the chainlinks $C \in \mathcal{C}_e$. Moreover, the minimum campaign sizes might lead to overproduction, i.e. producing more than required to cover total demands. (In Sect. 3.4 it will be shown how overproduction can be decreased.)

3.3 Chain scheduling

3.3.1 Strategies. Having explained how chains are generated and computed, it will now be described how they are scheduled. Let a requirement element $e = (\pi, q, t_d)$ and the corresponding sets $\mathcal{R}_e$ and Γ_e be given. A chain $\mathcal{C}_e \in \Gamma_e$ is scheduled into an existing schedule O by iteration on the chainlinks $C_a \in \mathcal{C}_e$. For each chainlink C_a, a start time t_s is determined in such a way that

- each facility $f \in r := R_a$ is available within the time interval $[t_s - s, t_s + f + n \cdot z + c[$ with $n := n_{C_a}$,
- possible savings of setup and cleanup are realized,
- the delays $t_{C_{a'}C_a}$ with $C_{a'} \in Pred_{C_a}$ and $t_{C_a C_{a''}}$ with $C_{a''} \in Succ_{C_a}$ are respected,
- a violation of due date t_d is avoided (if possible),
- activities being performed consecutively at the same facility are synchronized, i.e. the time lags between their start times are kept short, and
- the obtained schedule O' is feasible (cf. Sect. 2.3).

After the determination of t_s a resource object o_f reflecting the occupation of f by C_a is created for each facility $f \in r$ and added to $\mathcal{O}_f$. Due to material flows between a chainlink and preceding and subsequent chainlinks, the chainlinks have to be scheduled in a certain order. We distinguish between two strategies:

- *Backward scheduling strategy*: The chainlinks C_a are ordered by increasing manufacturing levels l_p with $p \in \mathcal{P}_a^+$. Starting at time t_d, they are scheduled iteratively by this order.
- *Forward scheduling strategy*: The chainlinks C_a are ordered by decreasing manufacturing levels l_p with $p \in \mathcal{P}_a^+$. Starting at time 0, they are scheduled iteratively by this order.

As it is possible that $\mathcal{C}_e$ cannot be scheduled without exceeding t_d, the backward scheduling strategy does not lead to a feasible schedule O' in any case. In these cases, the forward scheduling strategy generates schedules with exceeded due dates. Then, both strategies are combined in a backward-and-reverse scheduling strategy that first tries to schedule a chain just-in-time using the backward scheduling strategy and, in case of failure, switches to the forward scheduling strategy.

3.3.2 Scheduling algorithm. A feasible schedule O is obtained from the initial schedule $\hat{O}$ by scheduling the chains $\mathcal{C}_e \in \Gamma_e$ $(e \in \mathcal{E})$ iteratively following an order given by a permutation ε of $\mathcal{E}$. If the surplus amount $i(\pi, t_d) - \sum_{e' \in \mathcal{E}: \pi' = \pi} H(t'_d, t_d) \cdot q'$ of product π being available at time t_d is not sufficient to satisfy q, the corresponding chain $\mathcal{C}_e$ is scheduled w.r.t. a routing combination $R \in \mathcal{R}_e$. R is taken from a vector ϱ of routing combinations. (A detailed description follows in Sect. 3.5.)

As mentioned above, we first try to schedule $\mathcal{C}_e$ according to the backward scheduling strategy. Before, the *current schedule* $\tilde{O} := O$ is saved and may be called again if the backward strategy fails and one has to schedule $\mathcal{C}_e$ according to the forward scheduling strategy.

When a chainlink C_a has been scheduled with start time t_s, the latest start time of each preceding chainlink $C_{a'} \in Pred_{C_a}$ with start time t'_s is updated as follows: First of all, the earliest precedence-feasible start time is determined:

$$t'_s := \min\{t'_s, t_s - t_{C_{a'}C_a}\}. \tag{12}$$

Afterwards, the resource constraints are taken into account by setting

$$t'_s := \max\{t \le t'_s \,|\, [t - s', t + n' \cdot z' + f' + c'[\,\cap\, [t_s - v \cdot s, t_c + w \cdot c[= \emptyset \; (o_f \in \mathcal{O}_f, f \in r)\}, \tag{13}$$

with $n' := n_{C_{a'}}$ and $r' := R_{a'}$. Provided that

$$t'_s - s' < 0 \text{ or } t'_s - s' < \max_{p' \in \mathcal{P}^+_{a'}} f_{p'} \tag{14}$$

(i.e. a frozen zone of an output product of a' is violated), the backward scheduling strategy is stopped and the forward scheduling strategy is restarted with $\tilde{O}$. Else, possible savings of setup and cleanup are realized: For each $f \in r'$, let o_f reflect the first occupation of f after t'_s. If $a = a'$ and $r = r'$ hold for each facility $f \in r'$, then we set $w := 0$, $v' := 0$, and increase the start time $t'_s := t'_s + c' + s$. (If the forward scheduling strategy has been applied, the first occupation before t'_s is examined and the time spans c and s' are reduced, if possible.)

Next, a resource object o'_f with $t'_c := t'_s + n' \cdot z' + f'$ is added to $\mathcal{O}_f$ ($f \in r'$). Moreover, the inventory levels of all input products $p \in \mathcal{P}^-_{a'}$ and output products $p \in \mathcal{P}^+_{a'}$ are updated according to (10).

3.4 Schedule improvement

3.4.1 Order aggregation. As mentioned before, the constraints on minimum campaign sizes may cause overproduction. As the position of each chain is determined in the scheduling step, inventory levels cannot be considered in the previous chain computation step which computes each chain $\mathcal{C}_e$ independently from all others. Furthermore, scheduling chainlinks with overproduction (i.e. where the size n of the corresponding campaigns is too long) may result in a poor synchronization of the chainlinks $C_a \in \mathcal{C}_e$ because – with increasing n – it becomes more difficult to find a feasible start time t_s for C_a.

To consider these effects, chains are not generated and computed for single requirement elements $e \in \mathcal{E}$. Requirement elements for the same product are cumulated until minimum campaign sizes for all the required intermediates are reached. Basically, order aggregation applies a well-known procedure of material requirements planning (see e.g. [8]), where the total demand for a product is computed based on the primary demands and the matrix of input factors.

At first, all requirement elements w.r.t. π that can be fulfilled using the initial inventory $i(\pi, 0)$ are removed from $\mathcal{E}$ and the inventory level for π is updated. Let $\mathcal{E}_\pi \subseteq \mathcal{E}$ be the set of all requirement elements for π ordered by increasing due dates. An *aggregate order* $\tilde{e} = (\tilde{\pi}, \tilde{q}, \tilde{t}_d)$ w.r.t. an (ordered) subset $\tilde{\mathcal{E}}_\pi \subseteq \mathcal{E}_\pi$ is given by

$$\tilde{\pi} := \pi, \quad \tilde{q} := \sum_{e' \in \tilde{\mathcal{E}}_\pi} q', \text{ and } \tilde{t}_d := \min_{e' \in \tilde{\mathcal{E}}_\pi} t'_d. \tag{15}$$

Define the matrix A of (modified) input factors by $A := (a_{pp'})_{p,p' \in \mathcal{P}_\pi}$ with $a_{pp'} := \delta^-_{a,p} / \delta^+_{a,p'}$. Furthermore, let $\hat{q} = (\hat{q}_p)_{p \in \mathcal{P}_\pi}$ be the vector of *primary demands* with $\hat{q}_p = \tilde{q}$ for $p = \tilde{\pi}$ and 0, otherwise. Then, following [8], the vector $\bar{q} = (\bar{q}_p)_{p \in \mathcal{P}_\pi}$ of *total demands* can be obtained as follows:

$$\bar{q} := (I - A)^{-1} \cdot \hat{q}, \tag{16}$$

where I denotes the *identity matrix*.

To obtain aggregate orders for π, the first elements of $\mathcal{E}_\pi$ are aggregated until either

$$\bar{q}_p \geq \delta^+_{a,p} \cdot b \cdot m(p \in \mathcal{P}_\pi) \tag{17}$$

is fulfilled or all requirement elements of $\mathcal{E}_\pi$ have been aggregated. In the former case, the next aggregated order is generated by starting with the first element of $\mathcal{E}_\pi$ that has not been aggregated yet.

3.4.2 Campaign reduction. Neither the chain computation nor the order aggregation considers available inventory for intermediate products. Furthermore, the order by which requirement elements are aggregated does not affect the order by which the respective chains are scheduled nor the start times of the corresponding chainlinks. Therefore, it is necessary not only to check the campaign size prior to scheduling, but to adjust it after the scheduling of a chain in order to avoid overproduction.

After a chain $\mathcal{C}_e$ is scheduled, it is checked for all chainlinks $C_a \in \mathcal{C}_e$ if the size n of the campaigns associated with C_a can be reduced, starting with C_a having the lowest manufacturing level.

To check a chainlink C_a with start time t_s and completion time t_c, the surplus inventory i^+_p being available at t_c is determined for each output product $p \in \mathcal{P}^+_a$:

$$i^+_p := \min_{t \geq t_c} i(p,t) - \sum_{e \in \mathcal{E}:\pi=p} q. \tag{18}$$

The maximum number n^- by which each campaign corresponding to C_a can be reduced is given by

$$n^- := \min\{n, \min_{p \in \mathcal{P}^+_a} \{\lfloor i^+_p / \delta^+_{a,p} \cdot b \rfloor\}\}. \tag{19}$$

If $n^- = n$, then all campaigns corresponding to C_a can be saved and the resource objects o_f $(f \in r)$ associated with C_a can be removed from the schedule. In addition, for each $f \in r$, we have a look at the first resource object o'_f (o''_f) before (after) t_s and adjust the setup and cleanup indicators accordingly: $w' := 1$ $(v'' := 1)$. Then, we check for campaign reduction w.r .t. the chainlinks associated with o'_f (o''_f).

If $n^- < n$, then all campaigns corresponding to C_a are reduced by $\min\{n^-, n-m\}$ batches. In this case, the completion time t_c of C_a is reduced by $\min\{n^-, n-m\}$ and the resource objects o_f $(f \in r)$ associated with C_a are adjusted accordingly. Moreover, for each facility $f \in r$ the setup indicator of the first resource object o'_f after t_s is updated by setting $v' := 1$.

After C_a has been reduced, the inventory levels of all input and output products $p \in \mathcal{P}^-_a \cup \mathcal{P}^+_a$ are updated and campaign reduction is performed for all $C_{a'} \in Pred_{C_a}$. Having checked the whole chain $\mathcal{C}_e$, the delay times between its chainlinks have to be updated.

3.4.3 Move-algorithm. After a chain $\mathcal{C}_e$ is scheduled and all campaigns are reduced, $\mathcal{C}_e$, i.e. some of its chainlinks, might be moved backward or forward. By moving a chainlink, idle time on the corresponding facilities is reduced and the synchronization of subsequent chainlinks is improved. To move a chainlink, first the direction of the possible move is determined. In case the due date t_d of the requirement element e is exceeded, it is tried to move all chainlinks backward, starting with a chainlink with the highest manufacturing level. In the other case, a forward move starting with a chainlink with the lowest manufacturing level is attempted.

A forward move of a chainlink $C_a \in \mathcal{C}_e$ begins with the determination of the latest possible start time $\bar{t}_s$ considering the existing schedule O, due date t_d, and delay times $t_{C_a C_{a'}}$ for all subsequent chainlinks $C_{a'} \in Succ_{C_a}$. After determining $\bar{t}_s$, it is checked if the inventory levels of all input and output products $p \in \mathcal{P}_a^- \cup \mathcal{P}_a^+$ remain nonnegative, provided that C_a is moved from t_s to $\bar{t}_s$. In that case, the move is performed, i.e. the start and completion time of each resource object o_f $(f \in r)$ associated with C_a, the inventory levels of all input and output products $p \in \mathcal{P}_a^- \cup \mathcal{P}_a^+$, and all affected setups and cleanups are modified accordingly.

3.5 Optimization algorithm

In the following, starting with an initial schedule $\hat{O}$, a feasible schedule O is obtained for an instance of the optimization problem using GA. Algorithm 4 shows how all components of GA work together (with $\mathcal{E}'$ denoting the initial requirement set).

O is determined w.r.t. a pair (*individual*) (ε, ϱ) consisting of a permutation $\varepsilon = (\varepsilon_1, \ldots, \varepsilon_{|\mathcal{E}|}) \in E$ (with E denoting the set of all permutations of $\mathcal{E}$) and a vector $\varrho \in P = \{(\varrho_1, \ldots, \varrho_{|\mathcal{E}|}) | \varrho_e \in \mathcal{R}_e, e = 1, \ldots, |\mathcal{E}|\}$ of routing combinations (with ϱ_e determining the routing combination R to be used for chain $\mathcal{C}_e$ $(e = 1, \ldots, |\mathcal{E}|)$ and P denoting the set of all possible vectors of routing combinations).

For each $e \in \mathcal{E}$, we determine the routing combination $R := \varrho_e$ to be used for chain $\mathcal{C}_e$ and compute $\mathcal{C}_e$ w.r.t. R. Next, the chains $\mathcal{C}_{\varepsilon_1}, \mathcal{C}_{\varepsilon_2}, \ldots, \mathcal{C}_{\varepsilon_{|\mathcal{E}|}}$ are scheduled following the order induced by ε, and a feasible schedule $O = O_{\varepsilon,\varrho}$ is obtained. With an objective function F defined on the set of all feasible schedules, a fitness value $\tilde{F}((\varepsilon, \varrho)) := F(O_{\varepsilon,\varrho})$ can be assigned to each individual $(\varepsilon, \varrho) \in E \times P$.

Using appropriate methods of *replication* and *mutation* of individuals and *selection* to choose individuals from a *population* $\mathcal{I} \subseteq E \times P$, the strategy of genetic algorithms can easily be applied to obtain improved feasible schedules. In a preprocessing step, the sets $\mathcal{R}_e$ and P are generated. The optimization starts with a randomly generated initial population.

A fitness value $\tilde{F}((\varepsilon, \varrho))$ assigned to each individual $(\varepsilon, \varrho) \in \mathcal{I}$ is used to select a set $\mathcal{I}_s \subseteq \mathcal{I}$ of *survivors* from the population. By replication and mutation, a new population $\mathcal{I}'$ for the next iteration of the GA is generated from the survivors in $\mathcal{I}_s$. The iteration stops if a termination criterion is fulfilled, and the schedule $O_{\varepsilon,\varrho}$ corresponding to the individual (ε, ϱ) with the best fitness value $\tilde{F}((\varepsilon, \varrho))$ is returned.

Algorithm 4 (GA)

STEP 0 (*Initialization*)

Determine $\mathcal{E}$ by applying *order aggregation* to $\mathcal{E}'$

Determine $\mathcal{R}_e$ $(e \in \mathcal{E})$ by *chain generation*

Select $\mathcal{I} \subseteq E \times P$; $\mathcal{I}' := \emptyset$; $O := \hat{O}$; $F^\star := \infty$

STEP 1 (*Determination of new schedule*)

IF $\mathcal{I} \neq \mathcal{I}'$ **THEN**

  Select $(\varepsilon, \varrho) \in \mathcal{I} \backslash \mathcal{I}'$; $\mathcal{I}' := \mathcal{I}' \cup \{(\varepsilon, \varrho)\}$; $O := \hat{O}$; $i := 1$

  WHILE $i \leq |\mathcal{E}|$ **DO**

    Compute $\mathcal{C}_e$ w.r.t. $e := \varepsilon_i$ and $R := \varrho_{\varepsilon_i}$; $\tilde{O} := O$

    IF $\mathcal{C}_e$ can be scheduled *backward* into $\tilde{O}$ **THEN** $O := \tilde{O}$

    ELSE Schedule $\mathcal{C}_e$ *forward* into O

    Perform *campaign reduction* to $C_a \in \mathcal{C}_e$

    Perform *move-algorithm* to $C_a \in \mathcal{C}_e$

    $i := i + 1$

  $\tilde{F}((\varepsilon, \varrho)) := F(O_{\varepsilon,\varrho})$

  IF $\tilde{F}((\varepsilon, \varrho)) < F^\star$ **THEN** $F^\star := \tilde{F}((\varepsilon, \varrho))$; $O^\star := O$

  GOTO STEP 1

ELSE

  IF Termination criteria fulfilled **THEN RETURN** $O^\star$

  ELSE

    STEP 2 (*Determination of new population*)

    Determine $\mathcal{I}_s \subseteq \mathcal{I}$ by *selection*

    Determine $\mathcal{I}' \subseteq E \times P$ from $\mathcal{I}_s$ by *replication*

    Perform *mutation* to $\mathcal{I}'$

    $\mathcal{I} := \mathcal{I}'$; $\mathcal{I}' := \emptyset$; **GOTO STEP 1**

3.5.1 Objective function. In the industrial application considered, the planning horizon is significantly longer than the time between two consecutive planning runs. Thus, the distant part of the schedule will be rescheduled in later planning runs. Therefore, most terms of the objective function F are weighted by a time-dependent factor $e^{\omega \cdot t}$ with a degression parameter $\omega < 0$. As a consequence, the impact of a decision on the objective function value (e.g. violation of a due date) depends on the corresponding point in time t.

The first evaluation criterion for a schedule is given by the development of *inventory levels* over time. With $\Delta_t := \min_{t' \in T: t' > t} t' - t$ (where T denotes the set of all points in time t at which $i(p,t)$ changes) and $\Delta_{i(p,t)} = i(p,t) - \sum_{e \in \mathcal{E}:\pi=p} H(t_d,t) \cdot q$, the inventory level of a product $p \in \mathcal{P}$ can be quantified by

$$\sum_{t \in T} \max\{0, \Delta_{i(p,t)}\} \cdot \Delta_t \cdot e^{\omega \cdot t}. \tag{20}$$

For a requirement element $e = (\pi, q, t_d)$, let t_s (t_c) be the start (completion) time of chainlink $C_a \in \mathcal{C}_e$ with $a = a_\pi$. The *earliness* and *tardiness* for e are measured by

$$\max\{0, t_d - t_c\} \cdot q \cdot e^{\omega \cdot t_d} \tag{21}$$

and

$$\max\{0, t_c - t_d\} \cdot q \cdot e^{\omega \cdot t_d}, \tag{22}$$

respectively. The durations of the *setups* and *cleanups* are measured by

$$v \cdot s \cdot e^{\omega \cdot (t_s - s)} \tag{23}$$

and

$$w \cdot c \cdot e^{\omega \cdot (t_c + c)} \tag{24}$$

respectively. The *makespan*

$$\max_{f \in \mathcal{F}} \max_{o_f \in \mathcal{O}_f} (t_c + w \cdot c) \tag{25}$$

of schedule O is also considered. Moreover, without going into detail, a *priority value* for each routing has been taken into account when computing $F(O)$.

With w_e, w_t, w_p, w_s, w_c, and w_m denoting weights corresponding to earliness, tardiness, (product) inventory levels, setup, cleanup, and makespan, the objective function value $F(O)$ is computed as follows:

$$\begin{aligned} F(O) := & \sum_{p \in \mathcal{P}} w_p \cdot (20) + w_e \cdot \sum_{e \in \mathcal{E}} (21) + w_t \cdot \sum_{e \in \mathcal{E}} (22) \\ & + w_s \cdot \sum_{f \in \mathcal{F}} \sum_{o_f \in \mathcal{O}_f} (23) + w_c \cdot \sum_{f \in \mathcal{F}} \sum_{o_f \in \mathcal{O}_f} (24) + w_m \cdot (25). \end{aligned} \tag{26}$$

In the following, it will be described, how a new population $\mathcal{I}'$ can be obtained from a given population $\mathcal{I}$ with $|\mathcal{I}'| = |\mathcal{I}| = \lambda$.

3.5.2 Selection. After determining the fitness value $\tilde{F}((\varepsilon, \varrho))$ of each individual $(\varepsilon, \varrho) \in \mathcal{I}$ as described above, a set $\mathcal{I}_s \subseteq \mathcal{I}$ of μ *survivors* is selected from the population $\mathcal{I}$ or from the populations $\mathcal{I}$ and the survivors $\mathcal{I}_s'$ of the former iteration. This selection can be performed either in a deterministic manner, choosing the first μ individuals with the best fitness value, or in a stochastic manner, choosing the individuals (ε, ϱ) randomly, each with a probability $\tilde{F}((\varepsilon, \varrho)) / \sum_{(\varepsilon', \varrho') \in \mathcal{I}} \tilde{F}((\varepsilon', \varrho'))$.

The set $\mathcal{I}_s$ is the basis for the generation of $\mathcal{I}'$. The individuals of $\mathcal{I}'$ are yielded by replication and mutation.

3.5.3 Replication. The replication method consists of two strategies: *reduplication* and *recombination*. The former creates a new individual (ε, ϱ) by copying one randomly chosen survivor $(\varepsilon', \varrho') \in \mathcal{I}_s$: $(\varepsilon, \varrho) := (\varepsilon', \varrho')$. The latter generates a new individual (ε, ϱ) from two randomly chosen survivors $(\varepsilon', \varrho'), (\varepsilon'', \varrho'') \in \mathcal{I}_s$.

To recombine (ε', ϱ') and $(\varepsilon'', \varrho'')$, a *cross-over position* $i \in \{1, \ldots, |\mathcal{E}|\}$ is determined randomly. The permutation ε of the new individual is obtained by copying the first i elements of ε' to ε and adding all missing elements of $\{1, \ldots, |\mathcal{E}|\}$ to ε in the order given by ε''. The reduplication of ϱ is simply done by copying the first i elements of ϱ' and the last $|\mathcal{E}| - i$ elements of ϱ'' to ϱ. The ratio of the number of elements generated by recombination and the number of elements generated by reduplication is given by $\zeta \in [0, 1]$.

3.5.4 Mutation. The mutation method modifies a replicated individual (ε, ϱ) to become a different individual (ε', ϱ'). Several *mutation parameters* are assigned to (ε, ϱ) to control the mutation: the *mutation rates* $\alpha_\varepsilon = \alpha_\varepsilon(\varepsilon, \varrho)$ and $\alpha_\varrho = \alpha_\varrho(\varepsilon, \varrho)$, which determine the number of mutations that will be performed to ε and ϱ, respectively, the *mutation step length* $\beta = \beta(\varepsilon, \varrho)$, and the *strategy selection probability* $\gamma_j = \gamma_j(\varepsilon, \varrho)$, which equals the probability for mutating ε by strategy j. Mutating an individual, these mutation parameters are mutated, too. Note that, if an individual (ε, ϱ) is obtained by recombination of two individuals (ε', ϱ') and s$(\varepsilon'', \varrho'')$, its mutation parameters equal the arithmetic mean of the parameters of (ε', ϱ') and $(\varepsilon'', \varrho'')$.

To mutate ε, four different mutation strategies are applicable, and in each of the α_ε mutations, exactly one strategy j is chosen randomly with probability γ_j:

- The *exchange strategy* exchanges the elements ε_i and ε_j at two randomly chosen different positions $i, j \in \{1, \ldots, |\mathcal{E}|\}$.
- The *inversion strategy* inverts the order given by ε between two randomly chosen different positions $i \in \{1, \ldots, |\mathcal{E}|\}, j \in \{i-\beta, \ldots, i+\beta\} \cap \{1, \ldots, |\mathcal{E}|\}$.
- The *insert strategy* inserts the element ε_i at a randomly chosen position $i \in \{1, \ldots, |\mathcal{E}|\}$ into a randomly chosen target position $j \in \{i - \beta, \ldots, i + \beta\} \cap \{1, \ldots, |\mathcal{E}|\}$. This can be done either by *left insert* $(i > j)$ or by *right insert* $(i < j)$. The choice between left and right insert is done randomly with equal probability.
- The *move strategy* moves k randomly chosen subsequent elements of ε to other k randomly chosen subsequent positions. The positions $i \in \{1, \ldots, |\mathcal{E}|\}$ and $j \in \{i - \beta, \ldots, i + \beta\} \cap \{1, \ldots, |\mathcal{E}|\}$ as well as the length $k \in \{1, \ldots, |i - j|\}$ are chosen randomly with equal probability. The move strategy can be performed either to the left or to the right.

To mutate ϱ, we first set $\varrho' := \varrho$. Then, α_ϱ times, an element ϱ'_e is chosen randomly from ϱ' and changed to a randomly determined value which is taken from the set $\{1, \ldots, |\mathcal{R}_e|\}$.

The mutation parameters of an individual are mutated by considering two parameters $\phi \in [0, 100]$ and $\chi \in [0, 1]$. Each mutation parameter is either increased or decreased by ϕ per cent, and the probability for an increase or decrease equals χ. Hereafter, the mutated probabilities γ'_j are standardized so that they sum up to 1.

3.6 Calibration of the optimization method

To tune the parameters for optimization properly, two classes of them have been considered: First, the quality of the final schedule and the convergence bahaviour of the algorithm can be influenced by the *GA parameters* and the way of creating the initial population. Second, the *objective function parameters* (weights) have to be set appropriately in order to reflect the overall business objectives.

3.6.1 GA parameters. As there is a trade-off between the computational time and the quality of the schedule obtained, the choice of appropriate termination criteria is highly important. We have experienced that evaluating not more than 200–500 populations and allowing a maximum computational time of 7,200 seconds lead to good results for problem instances of realistic size.

Settings of λ and μ influence the convergence behaviour of the algorithm, possible avoidance of local optima, and computational time for each iteration. A high ratio λ/μ makes it difficult to find good individuals in an early iteration, i.e. the algorithm will converge slowly. A low ratio λ/μ makes it more difficult to leave local optima. Good results have been obtained by choosing $\lambda/\mu \in [10, 20]$. Furthermore, it has been considered that the absolute values of λ and μ influence the computational time spent on each iteration of GA and therefore the time required for obtaining a first feasible solution. Numerical experiments revealed that not more than 40 individuals ought to be evaluated in each iteration.

The generation of an *initial population* can be performed by creating each individual either randomly or deterministically. Tests of both methods show that it is recommendable to create initial individuals (ε, ϱ) with ε being given by ordering the requirement elements in $\mathcal{E}$ by increasing due date t_d and with ϱ being obtained by choosing a chain using routings with highest priority. The mutation parameters for each initial individual are set deterministically, too: The mutation rates α_ε and α_ϱ are set to 1, i.e. ε and ϱ are mutated exactly once. Since the mutation of ε should consider the whole permutation, step length β is set to $|\mathcal{E}|$. Each mutation strategy should be chosen with equal probability.

For *selection*, tests have shown that a stochastic selection strategy has a negative impact on the selection pressure because individuals with a poor fitness value may be chosen to survive with a certain probability. For that reason, a deterministic strategy has been applied.

For *replication*, we have found out that better results are obtained if many individuals contain information given by two surviving individuals. Therefore, the recombination parameter ζ is set to a rather high value, i.e. $\zeta \geq 0.6$.

The *mutation* of an individual (ε, ϱ) can be performed by either mutating only ε or ϱ in one single mutation step or by mutating both. Best results are yielded by deciding randomly with equal probability if either ε or ϱ should be mutated, with the exception that both are mutated in the first iteration. For the mutation of ε, all four mutation strategies should be applied. The parameters ϕ and χ ought to be set in such a way that the probabilities for the insert and move strategy slightly increase while that of the invert and exchange strategy slightly decrease. The setting of the mutation parameter for β should ensure that it is expected to increase β by 25 per cent. Hence, even in later iterations it is probable that the whole permutation ε is considered during mutation.

3.6.2 Objective function parameters. Good results are obtained with a degression parameter $\omega \in [1/60, 1/80]$. The weights w_s and w_c are taken from the interval $[1,000; 3,000]$ and the other weights are chosen from $[1, 3]$.

3.7 Implementation issues

The optimization algorithm has been coded in C++ using a two-processor Intel Pentium machine with 500 MHz clock pulse and 2 GB memory under the operating system Windows NT© 4.0. The duration of setting up an optimization model for a problem instance (with a size as given in Tables 1 and 2) and of the following optimization run is in the range of 45 minutes to 120 minutes, depending on the parameter settings for the termination criteria and the size of the problem instance.

It is appealing to implement a GA on computers with parallel CPUs since each individual of a population can be treated independently from others. One way to do that (in an object-oriented manner) is to use *threads* (cf.[9]). Threads are in some way independent processes which use the same memory in the RAM for the calling process. The key issue in using threads is to avoid *deadlocks*. Deadlocks can occur when more than one thread tries to access the same element in memory. This problem can be solved by encapsulating elements by *semaphores*. A thread can access an element only by taking its semaphore before and relasing it after the access. Threads that try to use the same element at the same time have to wait until the semaphore is free. The threads, which are thrown in a loop over all individuals, are scheduled by the dispatcher of the optimization system.

If there are several CPUs, the threads are scheduled in parallel at the same time on different CPUs. All terminated threads are caught in another loop. The next population can be computed after all threads have been terminated and the new population has been generated. The server used for the optimization algorithm is equipped with two CPUs and allows speed improvements of up to 1.6 times compared to a single-CPU system. The theoretical improvement of factor 2 cannot be reached because of communication and memory overhead.

4 Collaborative planning

4.1 The approach

The optimization model comprises an integrated scheduling model and is run in a nightly batch job. The solution provided by the optimization run, however, is subject to manual changes and updates (cf. Sect. 5). Due to the integrated structure of the production system and interdependencies of plant schedules, changes in one plant affect multiple other plants, and the ripple effects can increase the magnitude of changes in downstream plants. Figure 1 shows some of the representative production chains for the finished products in plants x_{10} and x_{11} involving product chains which run through other plants. In the industrial application, 11 plants are integrated, and the plant schedules are highly interdependent. More than 50 per cent of the end products are produced in chains involving more than 4 plants. For this reason, it is important to consider the latest schedule situation in other plants while making changes in a given plant. Due to performance and flexibility issues these changes are carried out locally by the plant planner. However, a plant planner is allowed to change only part of the schedule. For this purpose, the integrated model containing the results of the optimization run is broken down into plant-specific models. Individual plant planners can use the reduced models for manual changes to the solution proposed by the system. Each sub-model is a copy of the integrated model with only the plant-specific resources.

From a plant scheduler's point of view, the studies on the integrated model brought out the need for a model architecture, where the individual plant scheduler should

- be able to work on a smaller model with the facilities the planner is allowed to schedule but at the same time be able to share and view information with and from others plants,
- be able to see the schedule changes to the relevant production steps in other plants,
- make other plants aware of the schedule changes, and
- with the help of a chain planner, reduce conflicts and find a mutually agreeable solution for product chains running through multiple plants.

The model that was implemented in this context is called *collaborative planning model*. From the business point of view, the collaborative planning model serves the following goals:

- transparency of production schedules and shared information among all the plants,
- reduction in response time for reacting to any changes in the production schedule of upstream or downstream plants,
- minimization of conflicts among involved plants by simultaneously considering the production schedule of a specific subset of plants, and
- effective use of the global inventory maintained in the individual plants.

The objectives of shared information can be served by having one centralized model which can be accessed by the different plant planners, who are allowed to operate only on the authorized part of the schedule. However, this approach was found unacceptable due to the poor system performance arising from the projected size and complexity of the integrated model, selected APS system, and links with the optimizer. Hence, an alternative approach was adopted. The approach considers each plant as a separate entity or partner in a supply chain which needs to be integrated, and all partners work in collaboration. The collaboration is achieved by sharing the relevant part of the production schedule from each plant with every other plant. This architecture allows breaking up of the integrated production schedule model involving all plants into separate plant specific models. As the separate model for each plant is a reduced model, the approach guarantees a good system response time for manual changes.

The adopted approach required an architecture and a mechanism to gather information about the latest production schedule in all other partner plants and to publish production schedule changes to the partner plants. This transfer must maintain the consistency of data and schedule. In addition, taking advantage of the fact that all the plants are part of the same organization, a mechanism was also established to prepare a virtual plant involving multiple plants to consider decisions for chain production.

4.2 System architecture

The system architecture is based on a client-server-concept. It involves plant specific models on the local client communicating with an integrated model on a server (see Fig. 4). The architecture consists of the following modules, which are briefly explained in the following:

- *master production model* (*MM*),
- *plant production models* (*PMx*),
- *master view model* (*MVM*) for monitoring,
- *chain planner models* (*CPM*s),
- *communication mechanism* among these models, and
- *login and authorization concept* for data consistency and updates.

4.2.1 Master production model. The *master production model* (*MM*) contains information related to all the master data and the current integrated schedule of all the plants called *master schedule* (*MS*). This model is updated during the night for the latest data related to (i) products, (ii) bills of material (BOMs), (iii) demands, (iv) material receipts, and (v) inventory positions from the ERP system.

MM is used for the nightly optimization run and for rolling the production schedule forward each night. Then, copies of plant specific models are created from it for manual changes. This ensures that the models are synchronized at the beginning of each day. During the day, *MM* serves as a focal point to receive and send latest changes in the schedule and master data in each of the individual plants. This model is used for communication from and between the plant and chain scheduler(s). Any changes made within the plant models or in the chain planner models are broadcast back to *MM*. By collecting all the changes during the day, it provides the latest information on the production schedule and the master data for all the involved plants. It thus acts as a central data repository which always holds the most up-to-date information. By receiving the information from *MM*, the plant model remains synchronized and the data consistency among the models is maintained. The latest information on the production schedule in the other plants and other changes in the dynamic data are also received from *MM*.

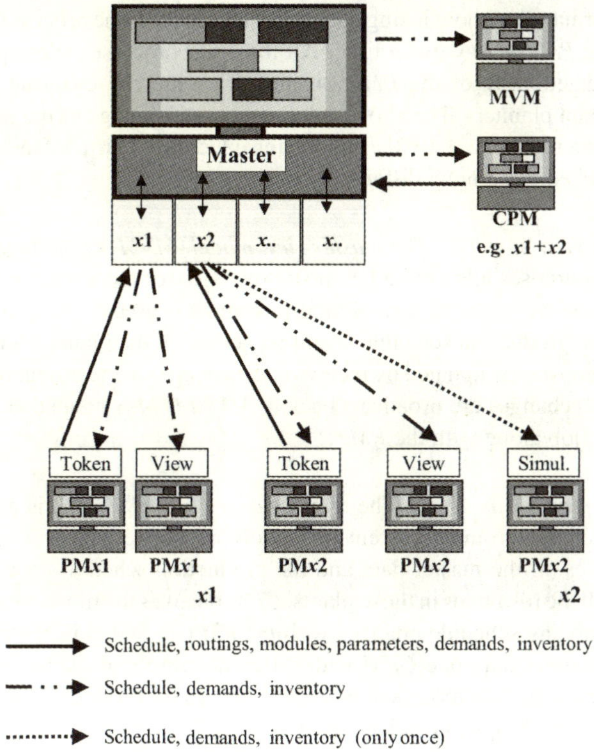

Fig. 4. Collaborative planning system architecture

4.2.2 Plant model. The *plant model* for plant x (*PM*x) contains information related to the schedule of plant x and the relevant master data. Apart from the detailed information on the production schedule for plant x, it also contains production schedule information published by all the other plants $x' \neq x$. The *PM*x are generated daily from the *MM* after re-optimizing *MS*. During the day the plant planner updates the *PM*x on the local client machines.

The authority to change the plant schedule remains with the plant itself. However, the schedule information published outside a plant can be viewed by the partners to consider if the changes are in accordance with the existing schedules in other plants. Any plant schedule considers the resources of the current plant x. The production schedules from other plants are shared using a set $\mathcal{F}_v$ of *virtual facilities*. For each product p produced in plant x, a virtual facility $f_p \in \mathcal{F}_v$ is introduced in plant x. f_p depicts the production schedule information for product p in plant x. Using the virtual facilities, the production schedule is shared without revealing the exact information about the real facilities used within the plant.

Any schedule or master data relevant changes within a *PM*x are broadcast to *MM*. Other plant planners from the same plant work with the updated *PM*x containing the latest master data. This way it is avoided that for every change in the plant master data a completely new *PM*x is created from *MM*. This improves the system performance without losing data consistency. Only one process is authorized to change the *PMx* at any given time. All the other processes subsequently work with the updated copy of the *PMx* on the server thereby ensuring consistency among the plant planners. The above solution takes advantage of the exclusivity of the master data related to the *PMx* and the introduction of virtual facilities to share production schedules among different plants.

4.2.3 Master view model. The *master view model* (*MVM*) is a reduced model for "view only" purposes allowing a simplistic overview of the integrated production schedule. It involves only the virtual facilities, and it is used by the chain scheduler to monitor the production schedule spanning across all the plants. During the day, the model is constantly updated by receiving the changes in the production schedule from *MM*. No changes are broadcast back to *MM*. *MVM* is created at the end of a nightly batch job along with the *PMx*.

4.2.4 Chain planner model. The *chain planner model* (*CPM*) is a model for a virtual plant created from the current *MM* involving a selected subset of plants. The model involves all the master data and the production schedules for the selected plants with all the resources in those plants. *CPM* receives the most current schedule from *MM*, and any schedule changes within *CPM* can be published back to *MM*. There can be more than one *CPM* with different subsets of plants involved. The *CPM* supports schedule modifications involving chain products and the affected plants and thereby helps to resolve conflicts.

4.2.5 Communication mechanism. The communication mechanism among *MM*, the *PMx* and the *CPM*s involves a series of complex macros using remote function calls to send and receive the information. Table 3 gives an overview of communication channels among the models. The type of data that are exchanged between the models can be broadly classified into the following three categories:

- *Master data* (*M*) refer to (*i*) products, (*ii*) facilities, (*iii*) routing information, (*iv*) production parameters, (*v*) BOMs, (*vi*) restricted sets for special planning board views, and (*vii*) plant working hour calendar.
- *Schedule relevant data* or *dynamic data* (*S*) consist of (*i*) production schedule and capacity allocation, (*ii*) information related to each chainlink, i.e. production rates, chainlink specific BOMs, batch numbers, etc., (*iii*) initial inventory, (*iv*) demands (orders, net forecast, and ATP orders), and (*v*) batch numbers and other characteristic information related to campaigns.
- *Authorization and data consistency relevant data* (*A*) for *PMx*, *CPM*, and *MM* consist of time and date of last modification and flags and sets that keep information regarding the authorized and active users.

Only the selected information passes between any pair of models as shown in Table 3. The following points have to be obeyed:

- *MM* can send the information to every other model.
- Every model can communicate with *MM*.

Table 3. Communication mechanism

From/to	*MM*	*PMx*	*MVM*	*CPM*
MM	–	*S/A*	*S/A*	*S/A*
PMx	*M/S/A*	–	–	–
MVM	*A*	*A*	–	*A*
CPM	*S/A*	–	–	–

- *PMx* can broadcast information only to *MM*. There is no direct communication between two *PMx* and/or *CPM*s and information is shared only via *MM*.
- Only *PMx* can broadcast master data back to *MM*.
- *MVM* can change the information related to authorization on *MM* and it can directly communicate with the *PMx* and *CPM*s.

4.2.6 Login and authorization concept. The architecture allows more than one scheduler in a plant to monitor and analyze the production schedule. Even a single user can work with multiple copies of the same model and such a system session is called *process*. To maintain data consistency and to avoid conflicts, the concept of *token* was introduced. A token is assigned to each plant in the integrated *MM*. The process that is currently allowed to broadcast information is the one which has the token. This concept ensures that at any given point in time only one process per plant is authorized to broadcast. Such a working mode is called *token mode*. The plant token can be acquired by a valid and authorized plant planner in a *PMx*, by a chain planner in a *CPM* or in an *MVM*. Though the *MVM* can acquire the token for plants, no master or schedule data can be changed in this case. All the other processes without token can simultaneously receive the data from *MM*. However, these processes cannot broadcast any changes back to *MM*. This working mode is called *view mode* as it is used to monitor and track changes in the production schedule in other plants. To investigate alternative scenarios to production changes within the plant, the planner is allowed to work in isolation and such a working mode is called *simulation mode*. The activities and changes that are allowed in each of the three working modes w.r.t. different models are summarized in Table 4.

From Table 4 one can see that – for the *PMx* and *CPM*s –

- one can broadcast information to *MM* only in the token mode,
- master data and manual updates of the inventories are allowed only in token or simulation mode,
- information can be received from *MM* either in token or view mode, and
- once in simulation mode, the model works as an isolated system where no broadcast to or from *MM* is made.

For the *MVM*, there is only one working mode, as this model is solely used for monitoring purposes. No changes are ever broadcast back to *MM*, and only the latest schedule information is received from *MM*.

Table 4. Working modes and functions for collaborative planning

Model	Mode	Master data and schedule update functions					Broadcast		Receive
		master data	schedul. utilities	manual update current chainlk.	order manag.	plan. board views, reports	schedule, invent., demands, batch no.	master data	schedule, invent., demands, batch no.
PMx	token	X	X	X	X	X	X	X	X
	view	–	X	–	X	X	–	–	X
	simul.	X	X	X	X	X	–	–	–
MVM		–	X	X	X	X	–	–	X
CPM	token	–	X	X	X	X	X	X	X
	view	–	X	–	X	X	–	–	X
	simul.	–	X	X	X	X	–	–	–

5 Manual planning interaction

5.1 Basics

System-generated schedules need frequent manual changes. There are many reasons for such manual changes, such as the breakdown of an equipment, the acceptance of an unexpected customer order, or the necessity to rework some production output. To handle such changes and to analyze their consequences, an APS system has to allow manual interaction in the plant. In the industrial application considered, the commercially available APS tool AspenMIMI© from AspenTech was used to represent the schedule and to enable manual interaction. It allows a high degree of freedom in customizing and development due to its toolbox structure with application specific programming. The tool provides a customizable planning board in form of a Gantt chart and built-in scheduling functions to support the development of an application-specific planning and scheduling system.

The built-in support functions provided by the APS tool are used both to analyze the schedule and to enable manual interaction functions. The most important built-in functions are

- the planning board with a high flexibility for customizing,
- forward and backward tracing of production activities, which allows the user to follow the production sequences and material flow,
- and multi-level BOM explosion, which enables viewing the material consumption even between the plants.

With the support of these scheduling functions, an APS tool for manual user interaction was developed. This tool provides many utilities such as *available to promise* (ATP), *capable to produce* (CTP), inventory updates, batch identification through numbering, manual changes to campaigns with capacity considerations, status management for interaction with the ERP system, maintenance of the master production data and routing settings, and general modifications to a chainlink, including creating, deleting, splitting, combining, adding batches, and specifying a chainlink specific recipe. Some of the important functions are briefly described below.

5.2 Utilities

5.2.1 ATP and CTP. The *ATP* and *CTP utilities* allow the user to inquire about the availability of a product. The user can check the unallocated inventory of a product around a specified date or inquire about the earliest date for the availability of a specified quantity. If the product is available at the requested date, an order can be placed to deliver the product exclusively from inventory (ATP). If the unallocated inventory is not sufficient, the user can inquire for a capacity check to produce the missing quantity (CTP). Provided that there is enough capacity, the order can be placed, and the corresponding chain can be planned. A major restriction of this utility is that a planner is allowed to use it only for products or chains which do not involve other plants. This avoids conflicts arising from inter-plant planning. The ATP/CTP utility brings benefits both as an immediate query tool and as a simulation help.

5.2.2 Manual update of current production. The automatic *roll-forward utility* updates the plan overnight and rolls the date forward by one day. However, this is based on standard production parameters. The planner revises the status of the current chainlinks using the *manual update utility* which allows the planner to enter the actual production situation into the schedule. This update of the production is an essential part of the daily manual planning in the plant and also allows a better inspection of the problems in the integrated schedule and of the conflicts between the plants.

5.2.3 Status management. Though the APS tool has the priority over the ERP system in the process of production planning, the ERP system has to be informed about the actual production schedule. The ERP system requires that the activities have different confirmation states. For example, chainlinks in the short term plan have to be frozen or confirmed so that the purchase orders for raw materials can be generated through the ERP system. The planner may set and change the status of an activity using a *status management function*. The status is considered when modifying a production schedule with other utilities such as drag and drop, as there might be restrictions to such modifications due to the confirmed status. The status management function ensures the parallelism between the ERP system and the APS tool and plays an essential role in giving logistics management the most up to date picture to support them in their internal and external purchasing and delivery decisions.

5.2.4 Drag and drop of activities. Perhaps one of the most important features of a manual planning tool is the so-called *drag and drop* of production activities in the Gantt chart. However, such a simple mouse action has its consequences on the capacity utilization and material availability. Therefore, the developed *post-procedures* of moving an activity on the planning board from one facility to another or from one time slice to another check the availability of capacity at the target facility and target time slice. In addition, available routings, possible parameter changes and requirements for setup and cleanup are considered during a drag and drop action to ensure the feasibility of the resulting schedule. In case of a conflict, the utility offers alternative or complementary actions which might possibly solve the conflict, e.g. by suggesting another target facility or moving the conflicting activities simultaneously to solve the capacity problems. The drag and drop function is a very practical tool for the daily manual planning, e.g. when simulating various scenarios to solve resource conflicts.

5.2.5 Routing data maintenance. The *routing data maintenance utility* provides a set of user friendly matrix-based interfaces which help the planner to create or modify routings, the facilities involved, and the related production parameters. The aim is to keep this vital information as compact and simple as possible for processing and easily accessible for the planner. The processes are implicitly defined in this structure, where each of them must comprise at least one equipment unit. The decision which routing to use for the production depends on the priority and availability of the equipment units. The utility provides checks for cycles and any inconsistency in the routing structure and parameters. All routing data are only available from the plant, and are not part of the shared information between the APS tool and the ERP system.

5.2.6 Batch identification. Each batch of a campaign gets a number which will stay with it from the actual production until it leaves the shop floor. The *batch numbering utility* allows the user to define the first number to be given to the first batch in the schedule. All batches following the first one automatically get ascending numbers. Batch numbers assigned to a campaign are considered when modifying production schedule with other utilities such as drag and drop to ensure that batches still have ascending numbers. Batch identification is essential to tracking batches from production to shop floor.

6 Interface with the ERP system and the optimizer

The APS tool communicates with the ERP system and the optimizer using two separate interfaces. Figure 5 shows the interfaces as part of the overall system landscape. These interfaces were developed using a programming language of the APS-tool, a certified interface between the ERP system and the APS tool, batch scripts, and simple text files in ASCII format. These interfaces do not work in real time but are activated once a day by the system. They operate in an event-driven sequence and are controlled by a batch script.
The *APS-ERP interface* consists of two parts:

- transfer of master product data and of dynamic sales and operations data from the ERP system to the APS system and

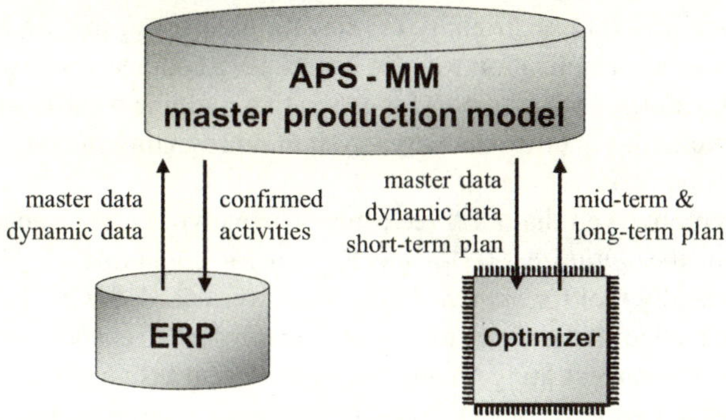

Fig. 5. System landscape

- transfer of the confirmed activities of the production plan from the APS system to the ERP system.

The *APS-optimizer interface* also consists of two parts:

- transfer of product master data, dynamic sales and operations data, capacity availability information, routings and production parameters from the APS tool to the optimizer and
- transfer of the schedule information from the optimizer to the APS tool and further processing to integrate the results of the optimization run with the rest of the plan and data in the model.

7 Conclusions

The paper has demonstrated a practical way of approaching a complex scheduling problem in the chemical process industry involving batch production. The successful implementation has lead to both quantitative and qualitative benefits. It has increased overall customer satisfaction and stability without sacrificing flexibility in production schedule changes. Moreover, it has reduced conflicts, and improved communication among different business functions.

A tailored optimization algorithm has been developed to handle a network of production plants with interdependent production schedules, multi-stage production with multi-purpose facilities and chain production. The solution procedure tends to (i) improve the utilization of available resources, (ii) reduce projected inventories, and (iii) minimize due date violations. The algorithm considers production schedules of all plants involved simultaneously and thus provides a better overall solution than on e obtained by individually optimizing production schedules.

A collaborative planning model allows (i) transparency, (ii) shared information on production schedule, (iii) reduction in response time for reacting to any changes upstream or downstream, (iv) system support for conflict management, (v) monitoring and analyzing the latest production schedule of the integrated model, and (vi) better use of globally available inventory. The custom-built tool for manual interaction provides (i) user friendly utilities for changes in the production schedules, (ii) a useful visualization tool for monitoring and analyzing plant specific production schedules, (iii) a support for generating multiple what-if scenarios, (iv) answers to marketing queries, and (v) a system representing the current situation in the production plant.

It has been observed that all three aspects, namely (i) a high quality solution obtained from the *optimizer*, (ii) a mechanism for *collaborative planning*, and (iii) a user friendly tool for *manual interaction*, were needed within an integrated framework to realize all the above mentioned benefits and to ensure user acceptance of the APS tool. Concentrating on any one of these aspects and achieving the best result in it but ignoring the others would have lead to a failure in realizing the full potential for improvements.

References

1. Blömer F, Günther HO (1998) Scheduling of a multi-product batch process in the chemical industry. Computers in Industry 36: 245–259
2. Blömer F, Günther HO (2000) LP-based heuristics for scheduling chemical batch processes. International Journal of Production Research 38: 1039–1051
3. Burkard RE, Hujter M, Klinz B, Rudolf R, Wenning M (1998) A process scheduling problem arising from chemical production planning. Optimization Methods & Software 10: 175–196
4. Burkard RE, Kocher M, Rudolf R (1998) Rounding strategies for mixed integer programs arising from chemical production planning. Yugoslav Journal of Operations Research 8: 9-23
5. Garcia-Flores R, Wang XZ (2002) A multi-agent system for chemical supply chain simulation and management support. OR Spectrum 24: 343–370
6. Gen M, Cheng R (2000) Genetic algorithm and engineering optimization. Wiley, New York
7. Grunow M, Günther HO, Lehmann M (2002) Campaign planning for multi-stage batch processes in the chemical industry. OR Spectrum 24: 281–314
8. Günther HO, Tempelmeier H (2002) Produktion und Logistik. Springer, Berlin Heidelberg New York
9. Hughes C, Hughes T (1997) Object-oriented multithreading using C++. Wiley, New York
10. Kallrath J (2002a) Planning and scheduling in the process industry. OR Spectrum 24: 219–250
11. Kallrath J (2002b) Combined strategic and operational planning – An MILP success story in chemical industry. OR Spectrum 24: 315–341
12. Knolmayer G, Mertens P, Zeier A (2000) Supply Chain Management auf Basis von SAP-Systemen. Springer, Berlin Heidelberg New York
13. Kolisch R, Hartmann S (1998) Heuristic algorithms for solving the resource-constraint project scheduling problem - Classification and computational analysis. In: Weglarz J (ed) Handbook on recent advances in project scheduling, pp 147–178. Kluwer, Dordrecht
14. Kolisch R, Padman R (2001) An integrated survey of deterministic project scheduling. Omega - The international Journal of Management Science 29: 249–272
15. Mendez CA, Cerda J (2002) An MILP-based approach to the short-term scheduling of make-and-pack continuous production plants. OR Spectrum 24: 403–429

16. Monma C, Potts C (1989) On the complexity of scheduling with batch setup times. Operations Research 37: 798–804
17. Neumann K, Schwindt C, Trautmann N (2002) Advanced production scheduling for batch plants in process industries. OR Spectrum 24: 251–279
18. O'Leary DE (2000) Supply chain processes and relationships for electronic commerce. In: Shaw M, Blanning R, Strader T, Whinston A (eds) Handbook on electronic commerce, pp 431–444. Springer, Berlin Heidelberg New York
19. Pinedo M (1995) Scheduling - theory, algorithms, and systems. Prentice Hall, Englewood Cliffs, NJ
20. Schumann A (1997) SAP-R/3 in process industries: expectations, experiences and outlooks. ISA Transactions 36: 161–166
21. Stadtler H, Kilger C (eds) (2000) Supply chain management and advanced planning. Springer, Berlin Heidelberg New York
22. Tan GW, Shaw MJ, Fulkerson W (2000) Web-based global supply chain management. In: Shaw M, Blanning R, Strader T, Whinston A (eds) Handbook on electronic commerce, pp 457–478. Springer, Berlin Heidelberg New York
23. Timpe C (2002) Solving planning and scheduling problems with combined integer and constraint programming. OR Spectrum 24: 431–448
24. Westenberger H, Kallrath J (1994) Formulation of a job shop problem in process industry. Unpublished working paper, Bayer AG, Leverkusen, and BASF AG, Ludwigshafen
25. Yang R (2000) Supply chain management: developing visible design rules across organizations. In: Shaw M, Blanning R, Strader T, Whinston A (eds) Handbook on electronic commerce, pp 445–456. Springer, Berlin Heidelberg New York

Plant co-ordination in pharmaceutics supply networks

M. Grunow, H.-O. Günther, and G. Yang

Department of Production Management, Technical University of Berlin, Wilmersdorfer Straße 148, 10585 Berlin, Germany
(e-mail: {M.Grunow, HO.Guenther, G.Yang}@ww.tu-berlin.de)

Abstract. The production of active ingredients in the chemical-pharmaceutical industry involves numerous production stages with cumulative lead times of up to two years. Mainly because of rigorous purity requirements and the need of extensive cleaning of the equipment units, production is carried out in campaigns, i.e. multiple batches of the same product type are produced successively before changing to another product type. Each campaign requires a specific configuration of equipment units according to the recipes of the particular chemical process. In the chemical-pharmaceutical industry, production stages are often assigned to different locations, even different countries. Hence the co-ordination of plant operations within the resulting multi-national supply network is of major importance. A key issue is the co-ordination of campaign schedules at different production stages in the various plants. In practice, it is almost impossible to determine exact optimal solutions to the corresponding complex supply network problem with respect to overall logistics costs. In order to reduce the required computational effort, we introduce several aggregation schemes and a novel MILP model formulation which is based on a continuous representation of time. Moreover, we propose an iterative near-optimal solution procedure which can be successfully applied to even exceptionally large real life problem instances. The applicability of the approach suggested is shown using a case study from industry.

Key words: Supply network planning – Pharmaceutical industry – Mixed-integer linear programming model – Aggregation techniques – Hierarchical production planning

Correspondence to: H.-O. Günther

1 Introduction

In recent years, supply network planning emerged as one of the most challenging problems in the chemical-pharmaceutical industry, especially when multi-stage production systems and multi-purpose equipment are used. Such supply networks are particularly complex in the production of active ingredients, where numerous material flows between different production locations occur resulting in cumulative lead-times of up to two years. As a consequence, effective inter-plant co-ordination of production activities within the entire supply network is of utmost importance in order to avoid unnecessary inventories, inefficient capacity utilisation, and uneconomical allocation of resources. In this paper, a practical modelling framework and an efficient solution procedure for supply network planning in the chemical-pharmaceutical industry are presented. The production of active ingredients is used as a complex industrial application for evaluating the approach suggested.

Due to the chemical peculiarity in processing active ingredients, production is carried out in batch mode, which offers the advantage of an increased flexibility with respect to product variety, production volume, and the range of recipes that can be processed by the particular equipment (Grunow et al., 2002; Kallrath, 2002b). However, production planning and scheduling is significantly complicated by the large number of stages and batches involved, the dissimilarity of the production paths, and short-term variations in product demand. Since multi-purpose equipment is employed, clean-out of equipment units is required when changing to another product type. In case of pharmaceutical ingredients, the processing mode (such as temperature, pressure, batch size, feasible equipment etc.) and cleaning operations are subject to strict legal regulations. For the sake of pureness, extensive cleaning operations are prescribed, which may require up to two weeks.

Time and cost-intensive cleaning operations as well as the necessity of detailed quality control after each batch are the major motivations for operating production in so-called campaign mode. The units required by a particular type of product or process are set up according to the corresponding recipe. The resulting configuration remains unchanged for a number of successive batches, i.e. the campaign, and is then reconfigured for another type of chemical process. Typically, only a few campaigns of a certain product type are produced each year. For some product types there may be only one set-up per year. To improve operating efficiency, minimum and maximum campaign lengths are usually defined by management.

To produce high value-added pharmaceuticals, major enterprises tend to build up multi-national supply networks with plants located in different countries. This development is primarily motivated by the legal and taxation systems in different countries and financial incentives for investments in economically low developed regions. Because of the small transportation volume of high-value active ingredients, transportation costs of products are of minor relevance. However, transit times between different plants need to be considered.

In this paper a real world case study of producing active ingredients is investigated. The production network under consideration comprises eight plants, which are located at four different production sites, one in the UK and three in continental Europe. Figure 1 shows the entire production network which is characterised by the large number of production stages involved and the customer-supplier relationships between the plants. For some of the products, the production process involves three or four different plants in several countries. Raw materials are transformed into final products (active ingredients) through a number of production stages, so-called manufacturing pipelines. A pipeline consists of multiple stages in which various intermediates are processed and, finally, one or more types of active ingredients are obtained. A pipeline may have a divergent or convergent material flow structure. Further characteristics of the supply network are the following:

- The equipment consists of 30 different resource groups which carry out 44 individual processing tasks.
- The eight types of active ingredients produced are made from six types of raw material and 23 intermediates.
- A recipe for a final product includes up to 8 production stages.
- Due to the decentralized production system, 14 intermediates must be transferred between plants.
- Production volumes per campaign may differ between the various plants and production stages due the individual equipment capacities and the complex network of material flows.

In the industrial application, production planning currently consists of two stages: material requirements planning carried out by a central logistics unit for the entire supply network and local production scheduling in each plant. A major drawback of this procedure is that capacity constraints are not considered during the material requirements planning phase. As a result, production schedules are often not feasible at the plant level. Therefore, several planning cycles and a tedious co-ordination of schedules through manual communication are necessary. As a consequence, unnecessary inventories are built up and delivery dates are often not met. Thus, an appropriate procedure for integrated supply network planning is desired, which performs material and capacity planning simultaneously. In particular, supply network planning has to fulfil the following requirements.

- The demand of all final products has to be satisfied without backordering. The size (number of batches) and the sequence of the various campaigns within a plant as well as their starting and finish times have to be determined accordingly.
- The production plans of all plants have to be co-ordinated, i.e. the entire consolidated production plan must be feasible with respect to the quantities of material available from the predecessor campaigns at upstream plants.
- Specific resource constraints, e.g. the availability of multi-purpose equipment at each plant, must be considered. Thus, the detailed assignment of equipment units to campaigns must be determined. Usually, resource groups are reconfigured after a campaign has been completed.
- The objective is to minimize total logistics costs, including production costs, inventory costs for final products, transportation costs for intermediates and costs for set-up and clean-out operations.

In the next section, an overview of the relevant literature is given. In Section 3, we outline our basic modelling concept and introduce different aggregation

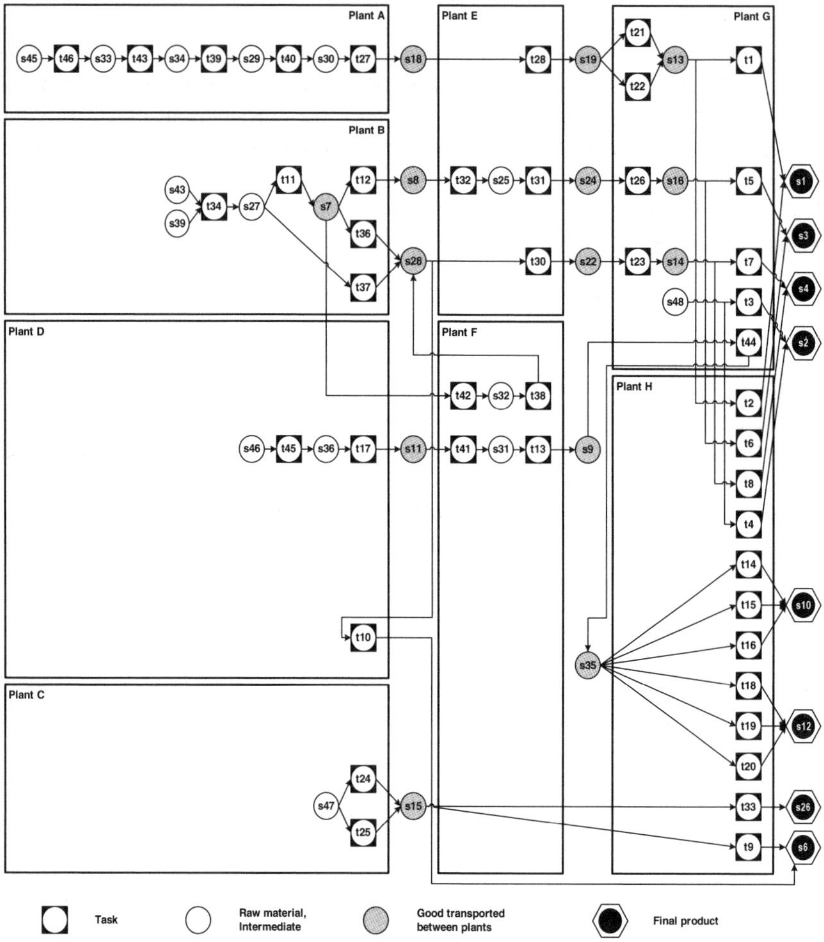

Fig. 1. Entire production network of producing active ingredients

techniques which help to reduce the number of entities to be considered in the supply network planning model. The case study example is used to demonstrate the effectiveness of these aggregation techniques. In Section 4, a novel MILP model formulation based on a continuous time representation is developed. Section 5 summarizes the numerical results obtained for the case study. Finally, some concluding remarks are given and directions for future research indicated.

2 Literature review

A large body of literature on production planning and logistics in the process industry has evolved over the last two decades (see e.g. Kallrath, 2002b, for a recent review). For the problem at hand, the relevant literature can be classified into three groups: *strategic network design*, *supply network planning*, and *batch scheduling*.

This classification reflects the hierarchical structure, which is typical of today's advanced planning systems applied in many enterprises in the chemical and pharmaceutical industry. As Kallrath (2002b) points out, the distinction between planning and scheduling in the process industries is very vague. Therefore, integrated optimisation models which are based on powerful *aggregation techniques* are of particular interest. Such papers constitute a fourth group of literature to be reviewed in the following.

Strategic network design

In the long-term, decisions on the design of the production and logistics network, the capacities and the configuration of the plants within the network have to be made. Among the many publications available, the papers by Sahinidis et al. (1989), Sahinidis and Grossmann (1991), Norton and Grossmann (1994), and Iyer and Grossmann (1998) shall be mentioned which focus on process technology aspects arising in the chemical industry. Two industrial applications are reported by Oxé (1997), who describes a procedure for deciding on the reduction of overcapacities in a chemical production network, and by Kallrath (2002a), who presents a model supporting decisions on opening/closing reactors and entire plants. All these approaches require the solution of MILP models which are based on a discrete representation of time with periods of predefined length. Liu and Sahinidis (1995) discuss the effects of time representation in strategic planning models. Another relevant aspect is the integration of long term strategies for the development of new products and the corresponding investments in production equipment (see e.g. Maravelias and Grossmann 2001; Papageorgiou et al., 2001). These authors also consider the enormous financial risks involved in the development of new drugs in the pharmaceutical industry.

Supply network planning

In the process industries, production stages are often assigned to different plants at geographically distributed locations. Production activities within such multi-site logistics networks have to be well co-ordinated. An industrial case study of supply network planning for the production of chemical commodities is provided by Timpe and Kallrath (2000). In the application considered, production processes for individual products have to be assigned to individual sites in the network. Major modelling assumptions are that at most two different types of products are produced per period and that no transportation of intermediates between plants takes place. An industrial application from the specialty chemicals industry is investigated by Berning et al. (2002). They consider a complex supply network optimization problem at Bayer AG and apply a tailor-made optimization algorithm in order to determine production paths and facilities for each campaign. Their approach is embedded into a collaborative planning concept which allows all plants working together as partners in a supply chain. Grunow et al. (2002) also consider a case study from the specialty chemicals industry. They address the problem of co-ordinating campaigns within a multi-site production network. Aggregation techniques and a hierarchical model formulation based on mathematical and constraint programming are suggested, which allow the user to determine campaign schedules and corresponding equipment configurations with only reasonable computational effort.

Batch scheduling

In the chemical-pharmaceutical industry, batch processing mode prevails. Batch scheduling is often complicated by the diversity of processing tasks, the necessity of cleaning operations and constraints arising from individual process plans. In the academic literature, various approaches based on mixed-integer linear programming (MILP) model formulations are developed, e.g. Grossmann et al. (1996), Pinto and Grossmann (1998) and Blömer (1999). They propose model formulations and related solution techniques for scheduling problems arising in multi-product/multi-purpose batch plants. Kondili et al. (1993) introduced the so-called *state-task-network* (STN) concept, a representation scheme for chemical batch processes, which has since found widespread acceptance. A state-task-network consists of three types of elements: *state nodes*, representing the feeds, intermediates, and final products, *task nodes*, representing the processing steps, and *arcs*, which indicate the material flow (cf. Fig. 1). Pantelides (1994) extended the concept of STN to also include production equipment. Just like feeds, equipment units are regarded as input and output of a processing task. The concept is thus named *resource-task-network* (RTN).

The resulting MILP-based scheduling approaches may be classified according to the type of time domain representation being used. In a *discrete* time domain representation, the time horizon is sub-divided into a number of time slots of equal and fixed duration (e.g. Kondili et al., 1993; Blömer, 1999). From a practical point of view, the major difficulty with a discrete time representation is that a sufficiently large number of time slots must be used in order to achieve a suitable approximation of the time domain. Hence, a large number of binary variables and constraints is required when industrial problems are tackled. To overcome this difficulty, Blömer and Günther (1998, 2000) propose several LP-based heuristics which allow a more compact model formulation and considerable savings in computational time. The work by Papageorgiou and Pantelides (1996a) on campaign planning is also based on the STN concept. Even for a single production site, their approach leads to very large MILP models such that a complex decomposition scheme is required (see Papageorgiou and Pantelides, 1996b).

In a *continuous* time domain representation, time slots have variable length. Hence, events are not limited to take place at the boundaries of pre-defined time intervals. Recently, various model formulations and scheduling methodologies based on a continuous time domain representation have been developed (e.g. Hui et al., 2000; Hui and Gupta, 2000; Kim et al., 2000; Mendez and Cerda, 2000, 2002; Mendez et al., 2000, 2001; Orcun et al., 2001; Lin and Floudas, 2001). Another approach to reduce the size of the MILP model is to use a non-uniform discretization of time that regards the timing of all events as decision variables (e.g. Zentner at al., 1994; Zentner and Reklaitis, 1996; Mockus and Reklaitis, 1997; Pinto et al., 1998). In addition, solution procedures based on project scheduling methods have been developed, which are very efficient for certain types of problems (e.g. Schwindt and Trautmann, 2000; Neumann et al., 2002). Finally, hybrid solution methods have been used, e.g. combining classical branch-and-bound techniques with constraint programming (Timpe, 2002).

Aggregation techniques

Most of the supply network planning and batch scheduling approaches published rely on MILP model formulations based on a discrete representation of time. To overcome the difficulties associated with a large number of time slots, several authors apply temporal aggregation. For instance, Wilkinson et al. (1995) in their approach on batch scheduling join micro-periods used in the scheduling model into macro-periods of greater length in the aggregate plant model. In Wilkinson et al. (1996) their approach has been extended to the case of a three-plant supply network, which is in fact very small compared to the case study example presented in the previous section. Moreover, restrictive assumptions are made, e.g. inter-site transport of intermediates is not allowed and no co-ordination of the campaigns at the various plants is taken into account. Another approach of time aggregation is provided by McDonald and Karimi (1997) who introduce minimum run length constraints to be included in an MILP model for scheduling a network of single-stage continuous production sites. As in Wilkinson et al. (1995, 1996) their approach has been tested only for very small sample problems. Bassett et al. (1996a, 1996b) present concepts for aggregating STNs for large scale problems. Their main focus is on time-based aggregation and disaggregation which is performed similar to Wilkinson et al. (1995). For integrating detailed scheduling with medium term aggregate planning they present the concept of a reverse rolling time window. Only for the time window considered, a detailed schedule of the processing operations based on a finer time discretization is developed. Starting at the end of the planning horizon, the time window is shifted backwards period by period. Bassett et al. (1996a, 1996b) also propose a task-unit aggregation, which generates so-called *super-tasks* and *super-units*. This concept, however, has not been tested for supply networks of major size with a complex network of material flows. A similar scheme of rolling time windows is proposed by Dimitriadis et al. (1997). Their concept is based on fixing decision variables in the MILP model for the time window under consideration.

3 Modelling approach and aggregation techniques

3.1 Overview

The major difficulty with formulating MILP models for supply network planning is associated with the large number of entities involved, mainly processing tasks, equipment units, and transportation activities. Another difficulty arises from the length of the planning horizon. Hence, MILP models which have been developed for batch and campaign scheduling in the chemical industry (e.g. Grunow et al., 2002), are hardly applicable to the case study problem considered here. Compared to supply network planning in chemical batch processing, the production of active ingredients involves a considerably longer throughput time and planning horizon because of the far larger number of stages at different production sites. In addition, due to the large number of processing tasks involved, the definition of a common time grid is extremely difficult. In order to avoid infeasibilities as a result from rounding of processing times (cf. Zentner et al., 1994), an extraordinarily large number of planning periods and corresponding binary decision variables had to be considered. Altogether, these reasons condemn an MILP modelling approach based on a discrete representation of the time domain to failure.

We therefore decided to develop an MILP model formulation based on a continuous representation of time. Apart from avoiding the drawbacks of a discrete representation mentioned above, this approach also simplifies the modelling of time minimum and maximum campaign length constraints, which are a core issue in the highly regulated pharmaceutics industry. The MILP model to be presented in the next section is very efficient in terms of computational tractability. However, an application to an exceptionally large industrial case such as the one investigated here requires the use of adequate aggregation techniques to ensure that computational times are within the limits acceptable for an interactive decision support tool.

In this paper, four major aggregation techniques are employed. First, consecutive processing tasks, which are required to produce a particular product, are merged into so-called *cascades*. In the mathematical model, we then refer to the cascades rather than to individual processing tasks. Second, *sub-plants* comprising several equipment units are defined in order to reduce the number of decision variables for process-equipment assignments. Third, *patterns of material flows* between campaigns are introduced which replace individual material flows in the model formulation. Finally, a practical procedure for *disaggregating demand data* is proposed. In the following, the different aggregation techniques are explained in greater detail.

3.2 Formation of cascades

According to the particular recipe, very often a consecutive number of processing tasks has to be performed using a specific group of equipment units. In general, batches are completed after a given cycle time, which is determined by the material flow in the state-task-network, the batch sizes, and the processing time of the involved tasks. Moreover, fixed start-up and clean-out times have to be considered. Thus, the time required to produce a campaign consists of a fixed and a variable component. The latter depends on the number of batches within the campaign. In order to reduce the number of variables and constraints in the MILP model, a series of tasks can be aggregated into a so-called *cascade*. For sequential processing tasks, the transformation is straight forward. For the example of tasks t46, t43, t39, t40, and t27 to be performed in plant A (see the upper part of Fig. 1), the definition of the resulting cascade is illustrated in Figure 2.

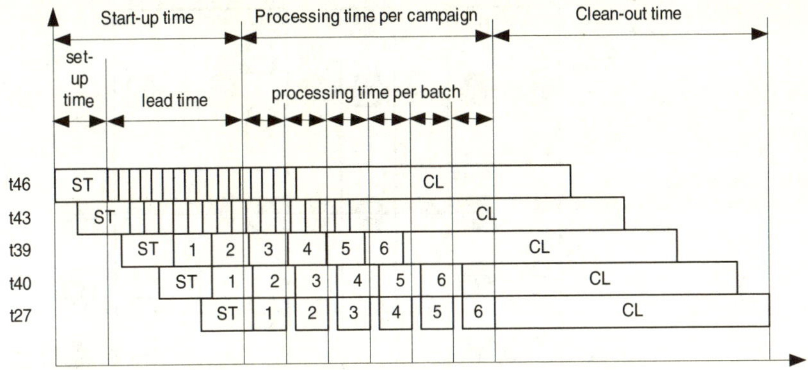

Fig. 2. Example of a cascade

Once two batches have been produced in task t46, enough input material is available to start the first batch of task t43. Likewise, task t39 may begin as soon as three batches have been completed in task t43. Task t39 is then succeeded by tasks t40 and t27. Depending on the individual batch sizes and batch processing times, some tasks may have to wait until the preceding task has been completed (e.g. task t27 succeeding t40). Thus, the processing time per batch of the entire cascade is determined by the batch processing time of task t40. As can be seen from Figure 2, a cascade comprises two fixed time intervals, i.e. the start-up and the clean-out time. The start-up time is determined by the set-up time of the various processing tasks and the lead time between starting batches in the head task t46 and the tail task t27. Similarly, there is a fixed clean-out time after completion of the final batch in the tail task.

In the case study example of Figure 1, cascades have been introduced in order to aggregate processing tasks in plants A, B, D, E, and F. The resulting aggregate state-task-network representation is displayed in Figure 3. As can be seen, the number of states (products) decreases from 29 to 20, a figure that can no further be reduced as it is made up of the six different raw materials and the 14 intermediates which are transferred between plants. Accordingly, the number of individual tasks decreases from 44 to 35.

The primary purpose of introducing cascades as an aggregation technique is to reduce the complexity of the optimisation model. However, each alternative production path, i.e. the sequence of tasks for an intermediate which is shipped to downstream plants for further processing, has to be reflected by a corresponding cascade. Clearly, cascade formation only makes sense, if the number of resulting cascades is considerably smaller than the number of corresponding tasks. In the case of numerous alternative production paths, cascade formation may not be effective. In such cases, the generation of sub-plants should be preferred, which is described in the following.

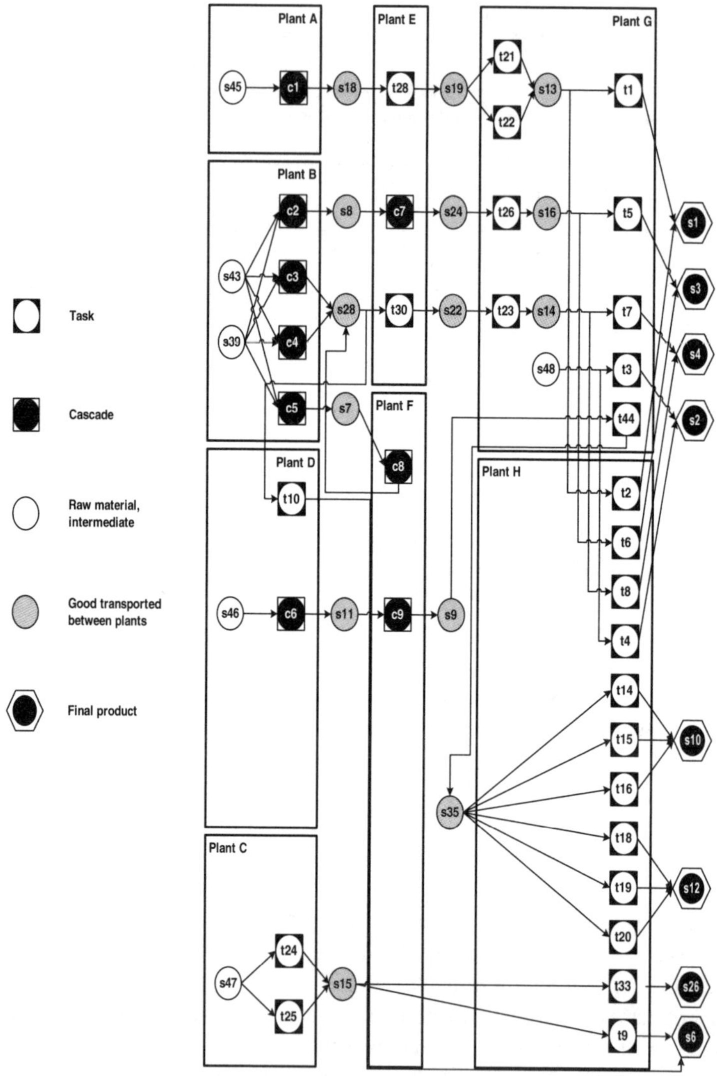

Fig. 3. The aggregate production network after introducing cascades

3.3 Generation of sub-plants

The idea behind the generation of sub-plants is to reduce the number of process-equipment assignments and thus the number of corresponding decision variables in the MILP model formulation. A sub-plant is defined as an equipment configuration which is capable of performing a distinct set of chemical processes. This aggregation technique is explained using an elementary example. Let us assume that a plant carries out four different chemical processes and that the maximum number of campaigns is eight. Each campaign requires a specific configuration of equipment

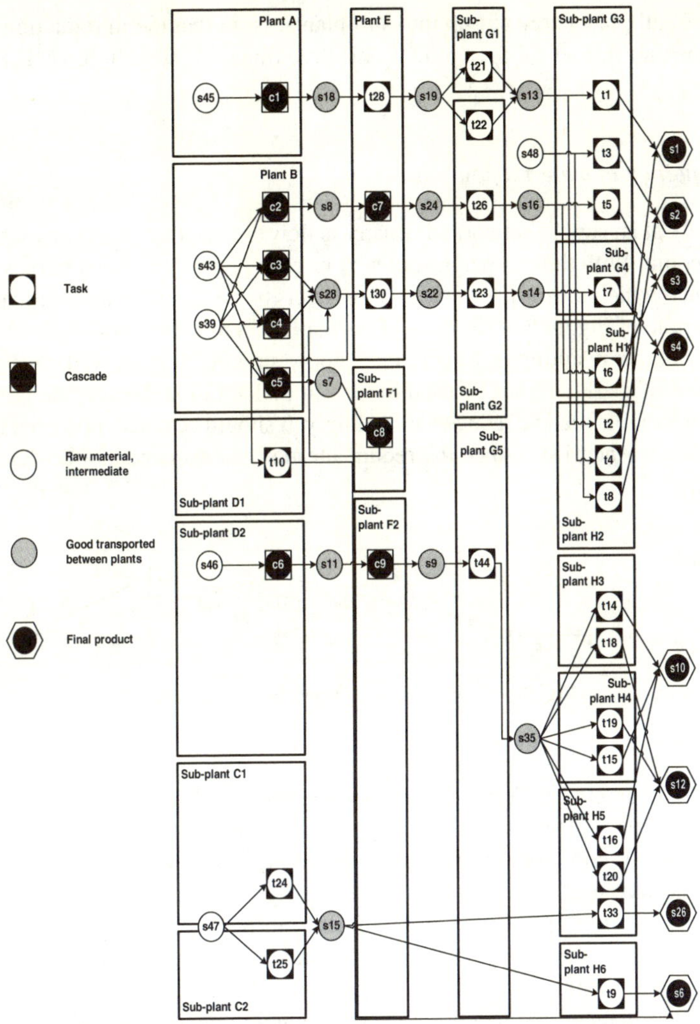

Fig. 4. The aggregate production network after introducing cascades and sub-plants

units. Then 32 feasible process-equipment assignments have to be considered. Let us further assume that the equipment available may be grouped into two distinct configurations such that two campaigns may run concurrently. Then also the set of processes to be carried out in that plant can be sub-divided into two disjunctive sub-sets, one for each equipment configuration. If, for instance, two of the four processes considered may be performed in one configuration and two in the other, then two separate sub-plants can be defined, each of which carries out at most four campaigns. Hence, the sub-plant approach requires only eight process-equipment assignments per sub-plant, i.e. 16 in total.

An analysis of the feasible process-equipment assignments in the case study example yields the sub-plant structure presented in Figure 4. Apart from plants A,

B, and E, all plants are split up into sub-plants. Note that the introduction of sub-plants has no impact on the optimal objective function value in the MILP model solution.

3.4 *Definition of material flow patterns*

In a multi-plant supply network, the material flow between the plants has to be taken into account. With the campaigns defined for each plant, decision variables have to be defined which mirror the transfer of products between individual campaigns. In the example of Figure 5, three campaigns in plant A and five campaigns in plant B are considered, resulting in 15 potential supply streams. It is most likely, however, that some of these will not be used in the optimal solution. Hence, the number of decision variables can be reduced accordingly. It should be noted, however, that this aggregation is heuristic, since the redundant material flows cannot be determined in advance.

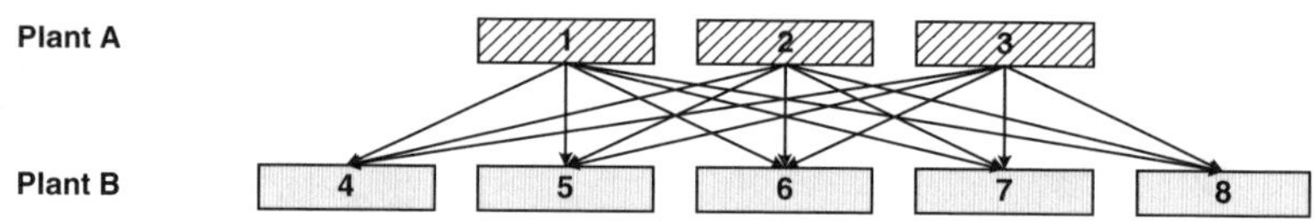

Fig. 5. Potential material flows at campaign level

In the example of Figure 5, for instance, the link between campaigns 1 and 8, i.e. the supply of input material from the first campaign in plant A to the last one in plant B, could be omitted, because the material transferred would have to be stocked for an extremely long time. Similarly, supplying the first campaign in plant B (campaign 4) from the last campaign in plant A (campaign 3) seems to be unrealistic, as all campaigns in plant B would have to wait until the production in plant A terminates. Thus, eliminating a number of links between campaigns in a controlled manner and deleting the corresponding decision variables will most likely enhance the tractability of the MILP model without impairing its solution quality.

In practice, a limited number of reasonable material flow patterns (MFPs) would be defined drawing on the expertise of a human planner. For the case study example considered here, the algorithm indicated in Figure 6 has been developed in order to systematically generate feasible MFPs. They differ in the density of the network of permitted supply links und thus also in the computational effort for solving the corresponding MILP model.

Figure 7 demonstrates the results of the algorithm for the pipeline producing final product s1 (cf. upper part of Fig. 4). Consider the first two processing stages, i.e. plants A and E. The algorithm of Figure 6 starts with the first campaign in plant E ($s(i(E)) = 1$) and links it to the first campaign in plant A. Setting the parameter *MFP* equal to 1, the most basic MFP 1 is generated. In this pattern, for instance, only one additional campaign, i.e. the second campaign in plant A, is linked to the

```
MFP: number of material flow pattern

FOR all plants p
   FOR all campaigns i(p) in plant p
      Determine position s(i(p)) of campaign i(p) in the sequence of
      campaigns defined for plant p
      FOR all upstream plants pu(p) of plant p
         FOR all campaigns i(pu) in upstream plant pu(p) of plant p
         which are in position s(i(p)) to s(i(p)) + MFP in the
         sequence of campaigns defined for upstream plant pu
            {link campaign i(p) to campaign i(pu)}
         IF campaigns without link between p and pu remain
         (the numbers of campaigns in p and pu are not identical)
          THEN
          {link the unconnected campaigns in either p or pu to
          campaigns of the other plant in the last MFP positions}
         END
      END
   END
END
```

Fig. 6. Algorithm for generation of MFPs

first one in plant E. The procedure is repeated for the second campaign in plant E which is linked to the second and third campaign in plant A. The third campaign in plant E, however, can only be linked to the last campaign in plant A. The fourth and fifth campaign in plant E remain idle. They are therefore linked to the last campaign in plant A (if-clause in Figure 6). In the same way, MFP 2 is generated using the parameter value $MFP = 2$. For reasons of comparison, the complete MFP is depicted in the lower part of Figure 7.

3.5 Demand disaggregation

In the case study example, total quantities of all final products to be produced until the end of the planning horizon are given. The satisfaction of these demand figures can be considered as a hard constraint. To facilitate the generation of campaigns and to reduce the computational burden for solving a large-scale MILP model, total demand is broken up on a monthly basis. Meeting monthly demand figures can thus be regarded as a soft constraint. As a result, more freedom is given for co-ordinating the campaigns within the supply network and for reducing network-wide inventories.

4 Mathematical model formulation

As discussed in Section 2, MILP models for supply network planning can be formulated based on a discrete or a continuous representation of time. For the type of planning problem considered, we have presented arguments which favour a continuous time based model formulation. However, an MILP model comprising all relevant entities and their relationships may still become too large to be solved within reasonable computational time. Thus, the MILP model formulation presented in the following exploits the human planner's expertise in order to reduce

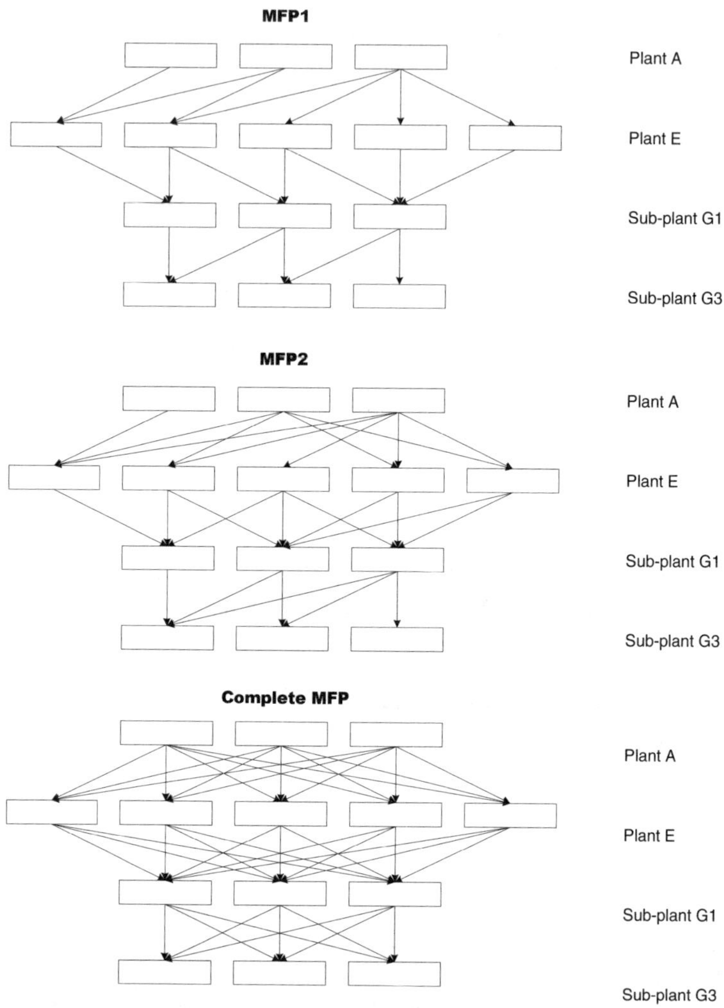

Fig. 7. Material flows for the pipeline producing final product s1

the complexity of the optimisation model in a controlled manner. In particular, we make use of the following ideas.

- Adequate aggregation strategies are employed which reduce the number of decision variables and constraints (see Sect. 3).
- Alternative material flow patterns can be defined by the human planner in order to control the supply links between campaigns in the network (see Sect. 3.4).
- A fixed maximum number of campaigns is allocated to each plant leaving open the particular chemical process assigned to it and the number of batches to be produced. At the same time, the minimum and maximum number of batches per campaign may be pre-defined depending on factors, such as the flexibility of the

production equipment, demand volume, batch sizes, start-up and clean-out time, and holding costs of the products.

- Moreover, a reasonable time window for each campaign can be defined, which also considers the necessary lead and follow-up times at upstream and downstream plants, respectively.

As a result, the structure of the network-wide campaign schedule is pre-determined to some extent. However, the cumbersome decisions on the individual task-equipment assignments, the detailed scheduling of campaigns and batch processing tasks, as well as the determination of the material flows are left to the optimisation model.

In the following, a novel mixed-integer linear programming (MILP) model for supply network planning based on a continuous time representation is presented. The notation used in the model formulation is given as follows:

Indices, sets

$p \in P$	plants
$i \in I$	campaigns
$i \in I_p$	ordered set of campaigns in plant p, with $\lvert I_p \rvert$ = maximum number of campaigns in plant p
$i \in IF$	campaigns producing final products
$m \in IR_i$	campaigns which receive feeds from campaign i
$i \in IS_m$	campaigns which supply campaign m
$l = 1, \ldots, MC_p$	concurrent campaigns in plant p
$j \in J$	chemical processes (tasks and cascades)
$j \in J_i$	processes feasible for campaign i
$j \in J_u$	processes which require equipment unit u
$j \in JO_k$	processes which produce product k
$j \in JI_k$	processes which consume product k
$k \in K$	raw materials, intermediates, products
$k \in KF$	final products
$k \in KI_j$	products consumed in process j
$k \in KO_j$	products produced in process j
$u \in UC_p$	common units available at plant p, i.e. units which can be assigned to different resource groups
$t \in T$	periods ($t = 1 \ldots H$)

Parameters

α_{jk}	proportion of input of product k in process j
β_{jk}	proportion of output of product k in process j
τ_j^f	fixed start-up time per campaign of process j
τ_j^c	clean-out time per campaign of process j
τ_j^v	processing time per batch of process j
a_{im}	transit time between campaigns i and m
b_j	batch size of process j
$\underline{B}_j$, $\overline{B}_j$	minimum and maximum number of batches of process j, respectively
c_{kim}	cost per unit of product k for production and transit from campaign i to m
ccf_j	start-up and clean-out costs for a campaign of process j
d_{kt}	cumulative net requirements of final product k up to period t
h_k	holding costs per unit of final product k
M	sufficiently large number
MC_p	maximum number of concurrent campaigns in plant p
N_p	minimum number of campaigns to be operated in plant p
$\underline{TS}_i$, $\overline{TS}_i$	earliest and latest feasible starting time of campaign i, respectively
H	planning horizon

Decision variables

$x_{ij} \in \{0,1\}$	$= 1$, if process j is assigned to campaign i (0, otherwise)
$y_{ij} \in \{0,1,2,\ldots\}$	number of batches of process j in campaign i
$z_{il} \in \{0,1\}$	$= 1$, if campaign i uses an equipment unit which is also required for the predecessor campaign $i-l$ (0, otherwise)
$\lambda_i \in \{0,1\}$	$= 1$, if campaign i is active, i.e. a process is assigned to it (0, otherwise)
$\pi_{im} \in \{0,1\}$	$= 1$, if campaign i precedes campaign m (0, otherwise)
$f_{it} \in \{0,1\}$	$= 1$, if period t succeeds the completion of campaign i (0, otherwise)
$q_{kim} \in (0,\infty]$	quantity of product k that flows from campaign i to campaign m
$p_{ikt} \in (0,\infty]$	quantity of final product k produced in campaign i in period t
$T_i \in (0,\infty]$	duration of campaign i
$ts_i \in (0,\infty]$	starting time of campaign i
$v_{kt} \in (0,\infty]$	inventory of product k at the end of period t

The optimal campaign schedule for the entire supply network can be determined by solving the following MILP model. The objective function aims at minimising total inventory costs for final products, costs for start-up and clean-out operations of all active campaigns, as well as production and transportation costs including transit costs for intermediates. Note that inventory holding costs for intermediates are not explicitly considered. They are, however, accounted for in the definition of the MFP (cf. Sect. 3.4).

Objective function

Minimise

$$\sum_{k \in KF} \sum_{t \in T} h_k \cdot v_{kt} + \sum_{(i,j) | i \in I \wedge j \in J_i} x_{ij} \cdot ccf_j + \sum_{k \in K} \sum_{i \in I} \sum_{m \in IR_i} c_{kim} \cdot q_{kim} \tag{1}$$

subject to the following constraints

Processes – campaigns

In the model formulation, *active* and *non-active* campaigns are distinguished. A campaign is active, if a process is assigned to it. Non-active campaigns remain idle. For each plant, a minimum number of active campaigns is given. Additional campaigns are termed *optional* campaigns. The following constraints are used to model the process-campaign assignments.

$$\lambda_i \leq \lambda_{i-1} \qquad p \in P,\ i = first(I_p) + N_p, \ldots, last(I_p) \tag{2}$$

Campaign i is only active, if its predecessor campaign $i-1$ is active. In other words, non-active campaigns are assigned to the end of the schedule. As the minimum and maximum number of campaigns to be operated in each plant p (N_p and $|I_p|$, respectively) are assumed to be given, the above constraints only apply to the set of optional campaigns $i = \mathit{first}(I_P) + N_p \ldots \mathit{last}(I_p)$. Also variables λ_i are only defined for optional campaigns.

$$\sum_{j \in J_i} x_{ij} = 1 \qquad p \in P,\ i = first(I_p), \ldots, first(I_p) + N_p - 1 \tag{3a}$$

$$\sum_{j \in J_i} x_{ij} = \lambda_i \qquad p \in P,\ i = first(I_p) + N_p, \ldots, last(I_p) \tag{3b}$$

Constraint (3a) ensures that exactly one process $j \in J_i$ is assigned to campaign $i = \mathit{first}(I_P) \ldots \mathit{first}(I_P) + N_p - 1$ in plant $p \in P$. No process is assigned to an optional campaign $i = \mathit{first}(I_P) + N_p \ldots \mathit{last}(I_p)$, if that campaign is not active (i.e. $\lambda_i = 0$). Otherwise, if $\lambda_i = 1$, constraint (3b) is equivalent to (3a).

Duration of campaigns

$$T_i = \sum_{j \in J_i} x_{ij} \cdot \left(\tau_j^f + \tau_j^c \right) + \sum_{j \in J_i} y_{ij} \cdot \tau_j^v \qquad i \in I \tag{4}$$

It is assumed that the various chemical tasks within a cascade are allowed to overlap. Thus, the total duration of a campaign consists of a fixed part which comprises start-up and clean-out times and a variable part which depends on the number of batches within the campaign (see Sect. 3.2). Note that variables T_i can be replaced by the term on the right hand side of the above equation and thus be eliminated from the model formulation.

Size of campaigns

$$\underline{B}_j \cdot x_{ij} \leq y_{ij} \leq \overline{B}_j \cdot x_{ij} \qquad i \in I,\ j \in J_i \tag{5}$$

The number of batches within a campaign must not exceed given lower and upper limits. This constraint also forces the number of batches to zero if a particular chemical process is not assigned to a campaign (i.e. $x_{ij} = 0$). The batch sizes are given for each processing task. As pointed out before, the campaign size limitation is of major importance in the pharmaceutical industry due to legal regulations. Our model formulation easily models this issue.

Resource conflicts

Each campaign requires a specific configuration of equipment units depending of the recipe of the particular chemical process. Within a plant, different campaigns may be operated in parallel, if no *resource conflicts* occur, i.e. campaigns do not use the same equipment unit. After a campaign has been completed, the production equipment is cleaned and may be reconfigured for another type of process.

$$z_{il} \geq \sum_{j \in J_u} x_{ij} + \sum_{j \in J_u} x_{i-l,j} - 1$$
$$p \in P,\ u \in UC_p,\ l = 1, \ldots, MC_p - 1,\ i \in I_p \,|\, i - l \geq first(I_p) \tag{6}$$

The binary variable z_{il} is forced to one, if campaign i and its predecessor $i - l$ use the same equipment unit, indicating that no overlapping of these campaigns is allowed.

Sequence of campaigns

$$ts_{i+1} + T_{i+1} \geq ts_i + T_i \qquad p \in P,\ i = first(I_p), \ldots, last(I_p) - 1 \tag{7}$$

Campaigns are scheduled according to their pre-defined order. The above constraint ensures that campaign $i+1$ finishes after its predecessor campaign i is completed.

$$ts_i \geq ts_{i-l} + T_{i-l} - (1 - z_{il}) \cdot M$$
$$p \in P,\ l = 1, \ldots, MC_p - 1,\ i \in I_p \,|\, i - l \geq first(I_p) \tag{8}$$

In case, no resource conflicts occur ($z_{il} = 0$), campaign i is allowed to start before its predecessor $i - l$ is completed.

$$ts_i \geq ts_{i-MC_p} + T_{i-MC_p} \quad p \in P,\ i \in I_p \,|\, i - MC_p \geq first(I_p) \tag{9}$$

This constraint ensures that no more than MC_p campaigns are carried out concurrently. In practice, MC_p should reflect the number of equipment configurations in a plant, which can be operated concurrently.

Material flow between campaigns

$$\sum_{m \in IR_j} q_{kim} \leq \sum_{j \in JO_k} \beta_{jk} \cdot b_j \cdot y_{ij} \qquad i \in I,\ k \in K \backslash KF \tag{10}$$

The amount of product k transferred from campaign i to the various downstream campaigns $m \in IR_i$ must not exceed the output of product k available from the supplying campaign i. It should be noted that only a single process $j \in J_i$ can be assigned to a particular campaign i. Thus, only a single variable y_{ij} is allowed to take a positive value while the others will be equal to zero. In the above formulation of constraint (10), it is assumed that no initial inventories of intermediate products exist. However, dummy campaigns can be introduced in order to reflect initial inventories.

$$\sum_{i \in IS_m} q_{kim} \geq \sum_{j \in JI_k} \alpha_{jk} \cdot b_j \cdot y_{mj} \qquad m \in I,\ k \in K \backslash KF \tag{11}$$

The amount of feeds of type $k \in K$ required to operate process j in campaign m must have been transferred from the upstream campaigns $i \in^{IS_m}$. Note that index combinations of variables q_{kim} in constraints (10) and (11), respectively, only comprise feasible material flows of product k between campaigns i and m according to the underlying MFP.

$$\pi_{im} \leq 1 + ((ts_m - ts_i - T_i - a_{im} + \sum_{j \in J_i} \tau_j^c \cdot x_{ij})/H) \qquad i \in I,\ m \in IR_i \tag{12}$$

$$q_{kim} \leq \pi_{im} \cdot M \qquad i \in I, m \in IR_i,\ k \in K \backslash KF \tag{13}$$

Constraint (12) enforces the binary variable π_{im} to zero, if the upstream campaign i does not precede the downstream campaign m taking the required transit time a_{im} into account. In this case, no transfer of material takes place, i.e. variables q_{kim} are forced to zero in constraint (13).

Time windows

$$\underline{TS}_i \le ts_i \le \overline{TS}_i \qquad i \in I \tag{14}$$

Each campaign must be started within a pre-defined time window.

Heaviside function: finish events

$$f_{it} \le 1 + ((t - ts_i - T_i + \sum_{j \in J_i} \tau_j^c \cdot x_{ij})/H)$$
$$i \in IF,\ t = \underline{TS}_i, \ldots, \overline{TS}_i \tag{15}$$

$$f_{it} \ge ((t - ts_i - T_i + \sum_{j \in J_i} \tau_j^c \cdot x_{ij}) + 1)/H$$
$$i \in IF,\ t = \underline{TS}_i, \ldots, \overline{TS}_i \tag{16}$$

Heaviside variables indicate if the corresponding campaign i has been finished up to a particular period t. In the proposed model formulation these variables are only used for final products and the pre-defined time window $t = \underline{TS}_i \ldots \overline{TS}_i$.

Quantities of final products

$$p_{ikt} \ge \sum_{j \in JO_k | j \in J_i} \beta_{jk} \cdot b_j \cdot y_{ij} - (1 - f_{it}) \cdot M$$
$$i \in IF,\ k \in KF, t = \underline{TS}_i, \ldots, \overline{TS}_i \tag{17}$$

$$p_{ikt} \le \sum_{j \in JO_k | j \in J_i} \beta_{jk} \cdot b_j \cdot y_{ij}$$
$$i \in IF,\ k \in KF, t = \underline{TS}_i, \ldots, \overline{TS}_i \tag{18}$$

$$p_{ikt} \le f_{it} \cdot M \qquad i \in IF, j \in J_i,\ k \in K_j, t = \underline{TS}_i, \ldots, \overline{TS}_i \tag{19}$$

Variables p_{ikt} indicate the quantities of final product k (produced in campaign i), which have been produced up to period t. In case, campaign i has not been completed up to period t (i.e. $f_{it} = 0$), constraint (17) is not binding and variables p_{ikt} are enforced to zero in constraint (19). It should be further noted that, only for a single process j, variables y_{ij} are allowed to take a positive value, while y_{ij} equals zero for the remaining processes which are not assigned to campaign i. Since constraints (17) and (18) are binding only for the effective type of process (provided that the corresponding campaign has been completed, i.e. $f_{it} = 1$), variables p_{ikt} exactly measure the quantities of the corresponding final products. In our modelling approach, variables p_{ikt} are defined for final products only.

Final product requirements

$$\sum_{i \in IF} \sum_{s=\underline{TS}_i \ldots t} p_{iks} - v_{kt} = d_{kt} \qquad k \in KF, t \in T \tag{20}$$

Variables v_{kt} indicate the excess of cumulative production over cumulative net requirements of final products in period t.

5 Solution procedure

For small-sized supply networks, the MILP model presented in the previous section can be solved within reasonable computational time. Complex supply networks, however, like the one considered in this paper, can only be solved using an integrated approach, which is made up of three subsequent stages: (1) aggregation, (2) solving the actual MILP model using rolling time windows for demand disaggregation, and (3) a post-optimisation stage (see Fig. 8). The motivation of our approach is to provide a practical tool for a complex case of supply network planning in the pharmaceutical industry. It allows near-optimal solutions to be determined even for exceptionally large real life problem instances.

Aggregation

In the *aggregation* stage, processes are aggregated into cascades (see Sect. 3.2), sub-plants are derived from configurations of equipment units (see Sect. 3.3), and global demand figures are considered which will be disaggregated in the subsequent stage of the solution procedure.

Demand disaggregation

In the second stage of the solution procedure, the MILP model of Section 4 is solved using the basic material flow pattern (cf. MFP 1 generated by the algorithm indicated in Figure 6). In the initial solution step of the MILP model, total demand for each final product must be satisfied until the end of the planning horizon disregarding detailed monthly demand figures. We then freeze part of the decision variables in the solution obtained and in the next solution step resolve the model taking not only the total demand figures but also the cumulative demand until the second last period into account. The procedure is repeated by successively including a demand constraint for one further period until the first period is reached. As a result, a smoothed campaign schedule that satisfies detailed monthly demand figures is obtained. In this solution procedure, a pre-defined CPU time limit and a practical MIP gap tolerance are imposed in order to achieve campaign schedules with reasonable computational effort. Details of the solution procedure can be stated as follows.

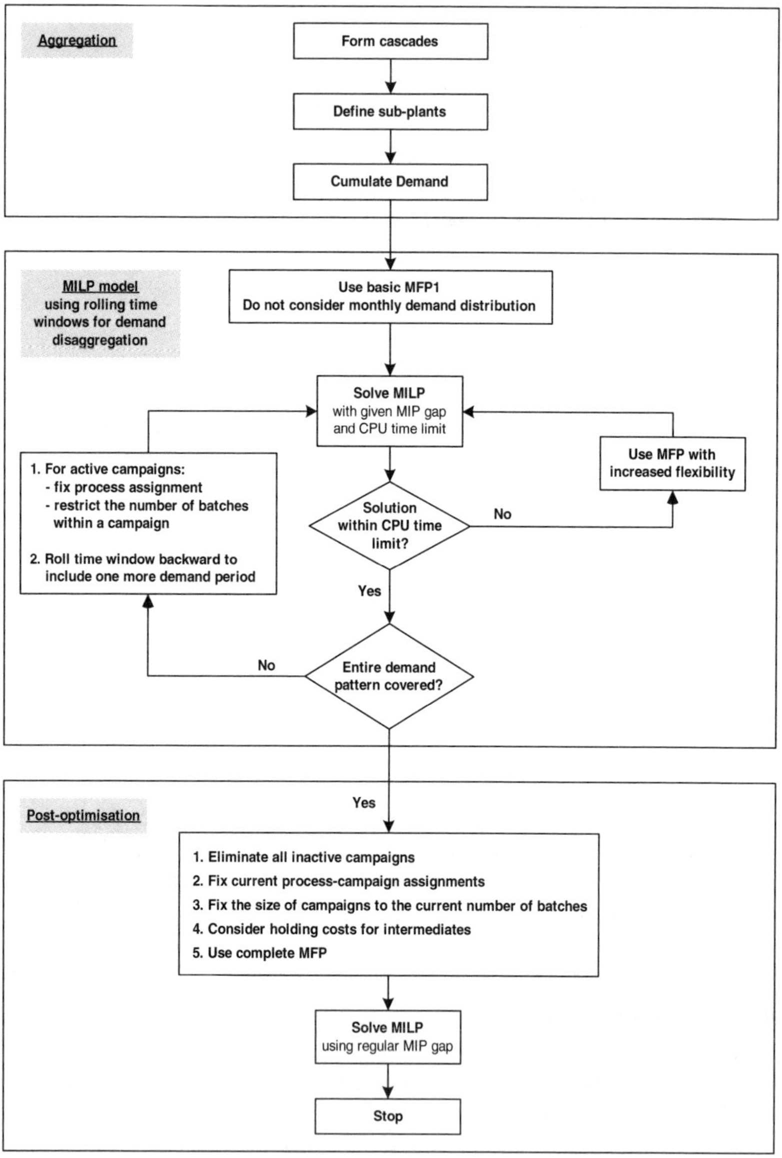

Fig. 8. Iterative procedure for large problem instances

- After each step of the solution procedure, the process assignment of active campaigns is fixed. Explicitly, if $x_{ij} = 1$ is obtained, then x_{ij} is fixed to this value in all subsequent steps. Otherwise, if $x_{ij} = 0$, then x_{ij} remains variable, i.e. $x_{ij} \in \{0, 1\}$.
- Moreover the number of batches y_{ij} of active campaigns is limited to a sufficiently large interval $[y_{ij} - \Delta^-, y_{ij} + \Delta^+]$ in the subsequent solution step.

- An additional demand period is included in the current demand pattern, i.e. the cumulative demand up to the last period not yet included in the model must be satisfied for all final products.

Based on these additional constraints, the MILP model is generated for the next iteration of the solution procedure. The procedure continues until a solution for the entire horizon has been obtained. In our numerical experiments, we observed the following general behaviour of the solution procedure. Initially, campaigns of large size result as the solver seeks to reduce costs for start-up and clean-out operations. The more detailed the demand pattern becomes in the course of the subsequent steps, the more due dates have to be respected. It may thus be impossible to satisfy the demand on time without splitting campaigns and assigning the particular process also to a formerly inactive campaign. Therefore, the interval $[y_{ij} - \Delta^- , y_{ij} + \Delta^+]$ defining the feasible size of a campaign, i.e. the corresponding number of batches, should allow for a large reduction and only a minor increase, i.e. $\Delta^- > \Delta^+$ should be chosen. In case no globally feasible solution is found within the pre-defined CPU time limit and the MIP gap tolerance allowed, a more flexible MFP, i.e. one which includes additional supply links between campaigns (e.g. MFP 2 of Figure 6), is employed and the solution procedure is restarted.

Post-optimisation

The solution approach concludes with a *post-optimisation* stage, which is required for two reasons. First, inventory holding costs of intermediates are not explicitly considered in the objective function of the MILP model. They are only accounted for in the generation of the MFPs. Second, idle campaigns which may result from the underlying material flow pattern and the termination of the optimisation run due the CPU time limit are eliminated. Prior to post-optimisation, all inactive campaigns are removed and the campaign sizes obtained are fixed to the current number of batches. To further reduce inventory holding costs for intermediates, the corresponding campaigns are right-shifted along the time axis. In the post-optimisation stage, the complete MFP linking all campaigns in the schedule is considered. Since all process-campaign assignments have already been determined in the previous stage of the procedure, no particular CPU time limit and no reduced MIP gap tolerance need to be imposed.

6 Numerical results

The industrial case study problem introduced in Section 1 is used to evaluate the practicability of the MILP model and the related solution procedure presented in the previous sections. Table 1 summarizes the main characteristics of the problem instance considered. The complete set of the case study data is available from the authors on request. Because of the necessary lead time for the production of intermediates, demand for the eight final products is given for months 9 to 18 of the 18-months planning horizon (see Table 2). Due to plant holidays, no demand is defined for month 13.

Table 1. Characteristics of the case study

Raw materials, intermediates	29
Final products	8
Chemical processes	44
Plants	8
Equipment configurations	30
Production stages	up to 8
Processing time, clean-out time, transport time	fractional
Planning horizon	18 months

Table 3 summarizes the number of variables and constraints in the MILP model formulation of Section 4. The first row indicates the size of the model for the original supply network based on a complete pattern of material flows. We experienced that not even a feasible solution to this problem could be determined within several hours of CPU time. Thus, the size of the MILP model had to be reduced using the aggregation techniques outlined in Section 3. As can be seen from the second row in Table 3, the number of variables and constraints is considerably reduced, above all the number of integer variables, which primarily contribute to the complexity of the optimisation problem. By introducing sub-plants, an even more compact model formulation can be achieved through further eliminating a great number of continuous decision variables as well as constraints (see the third row of Table 3). It should be noted that no loss of optimality is incurred by applying these aggregation techniques.

As pointed out in Section 3.4, the expertise of the human planner can be exploited in order to reduce the number of material flows allowed between campaigns in the various plants in a controlled manner. Table 4 shows the impact of different material flow patterns on the model size for the case study example. To derive the various MFPs, the algorithm of Figure 6 was used. Although the number of constraints and variables does not decrease considerably with increased MFP simplicity, our computational experiments revealed that CPU times could significantly be reduced by the use of simplified MFPs. Thus, this aggregation technique is also considered as very effective.

ILOG OPL Studio 3.1 was used to implement the supply network planning model presented in Section 4. The MILP-based solution procedure as well as the post-optimisation step proposed in Section 5 were realised using OPL's script language. To solve the MILP models embedded in these procedures, the standard optimisation software CPLEX 7.0 was applied. All calculations were carried out on a PC with a 1 GHz AMD Athlon processor and 512 MB RAM.

In our numerical investigation, the MILP model based on the aggregation of cascades und sub-plants was applied. Since demand figures are defined for 10 months (see Table 2), a corresponding number of solution steps was required to determine a campaign schedule using the MILP-based solution procedure. Table 5 indicates the required CPU time for each step. As mentioned before, an exact

Table 2. Demand of final products in kg

Product	Month									
	9	10	11	12	13	14	15	16	17	18
s1	1988.1	1988.1	1988.1	1988.1	0.0	1988.1	1988.1	1988.1	1988.1	1988.1
s2	95.4	0.0	0.0	0.0	0.0	95.4	0.0	0.0	0.0	95.4
s3	477.2	620.4	601.3	229.1	0.0	534.5	572.7	324.5	295.9	773.1
s4	505.8	1245.5	634.7	634.7	0.0	1083.5	553.6	625.1	1074.1	544.0
s6	55738.9	55738.9	55738.9	55738.9	0.0	55738.9	55738.9	55674.1	55674.1	55674.1
s10	28.6	133.6	95.4	95.4	0.0	19.1	181.3	19.1	95.4	160.2
s12	12400.0	0.0	0.0	0.0	0.0	0.0	12400.0	0.0	0.0	0
s26	3254.6	697.2	2662.8	3483.6	0.0	2844.2	2920.7	3703.1	1689.3	1813.4

Table 3. Number of variables and constraints for different degrees of aggregation (complete pattern of material flow assumed)

	No. of variables		No. of constraints
	integer	total	
Original supply network	20,492	28,498	36,637
Aggregate supply network with cascades	11,707	19,246	35,324
Aggregate supply network with cascades and sub-plants	11,509	13,317	11,476

Table 4. Number of variables and constraints for different material flow patterns (aggregate supply network with cascades and sub-plants assumed)

	No. of variables		No. of constraints
	integer	total	
MFP 1	11,509	13,085	11,056
MFP 2	11,509	13,152	11,178
Complete MFP	11,509	13,317	11,476

Table 5. Overview of the MILP-based solution procedure

	Step 1	Step 2	Step 3	Step 4	Step 5	Step 6	Step 7	Step 8	Step 9	Step 10
CPU time [s]	202.4	145.6	27.8	16.6	43.9	40.1	89.3	39.5	94.4	202.3
MFP	MFP1	MFP1	MFP1	MFP1	MFP1	MFP1	MFP1	MFP1	MFP1	MFP2
MIP gap	$> 5\%$	4.4%	3.0%	4.9%	4.6%	4.0%	1.4%	5.0%	5.0%	$> 5\%$

optimal solution to the optimisation problem at hand is extremely hard to obtain. Therefore, we decided to terminate the optimisation run in each step, once the relative MIP gap was smaller than 5% or the CPU time exceeded 200 seconds. Moreover, we defined parameter values of $\Delta^- = 15$ and $\Delta^+ = 5$ for adjusting the number of batches within a campaign in the course of the stepwise solution procedure.

The MILP-based solution procedure determines a complete campaign schedule within 15 minutes of CPU time. Apart from two steps, the CLPEX solver terminated after the relative MIP gap tolerance of 5% was reached. Contrary to the findings of other studies, which also used MILP models based on a continuous representation of time (cf. Shah (1998), Schilling and Pantelides (1996)), our aggregate model formulation appears to provide fast convergence of the MILP-based solution procedure. Only steps 1 and 10, which were terminated after the CPU time limit was reached, exhibit MIP gaps $> 5\%$. Throughout the solution steps, the basic material flow pattern MFP 1 was used. In step 10, however, additional supply links had to be added. Therefore, the algorithm switched to the more flexible MFP 2.

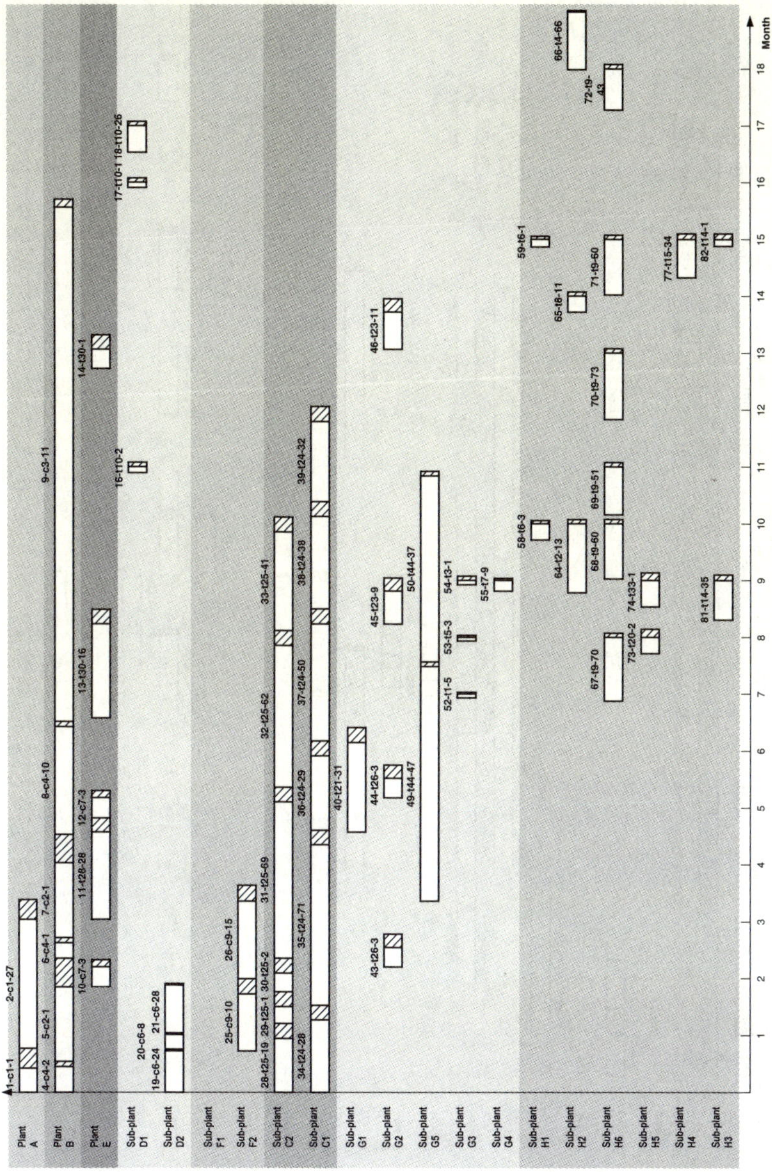

Fig. 9. Campaign schedule before post-optimisation

The final campaign schedule obtained is shown in Figure 9. Horizontal sections in the Gantt chart represent the sub-plants. Campaigns are depicted as white bars. Their attached label identifies the various campaigns. For instance "1c11", which is the label of the first campaign in plant A, indicates campaign No. 1 performing task/cascade c1, of which only one single batch is produced. (For the coding of

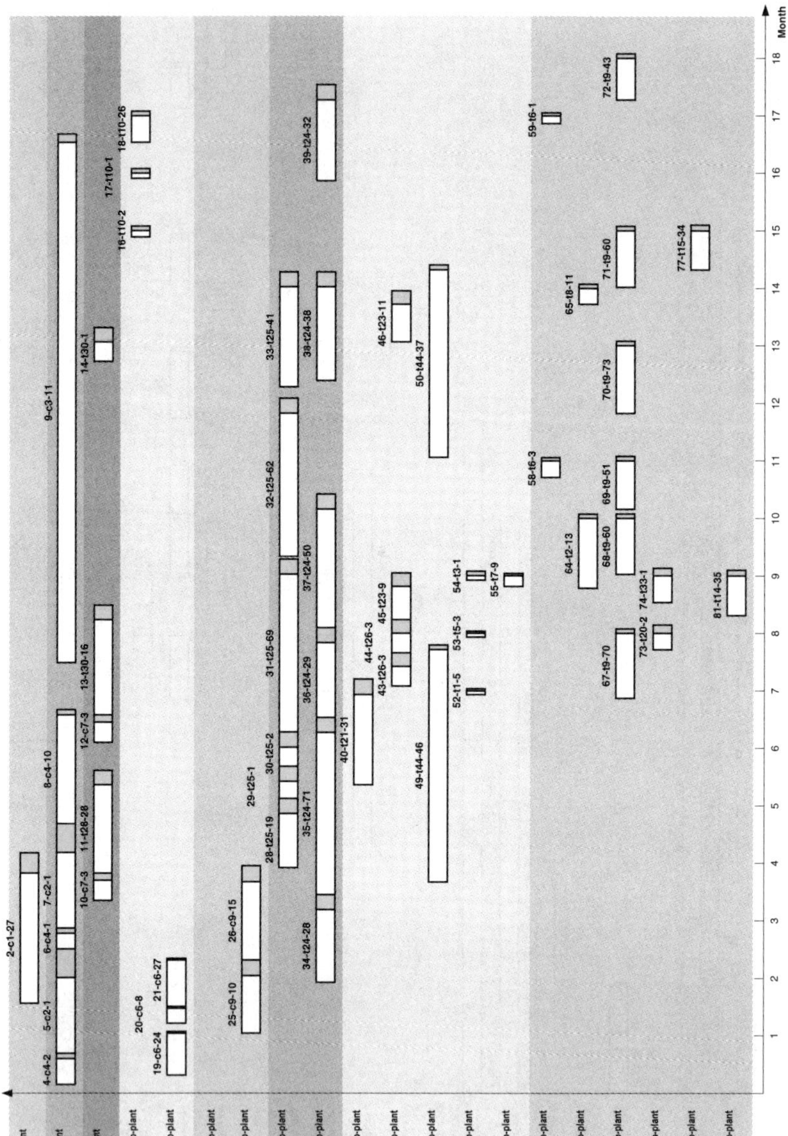

Fig. 10. Campaign schedule after post-optimisation

tasks/cascades we refer to Fig. 4). A closer look at the campaign IDs reveals that the numbering increments from 2 in plant A to 4 in plant B indicating that the third campaign in plant A is inactive. Clean-out operations after each campaign are displayed as hatched rectangles. Apparently, campaigns producing final products are carried out close to the due dates of the final products. However, some of the campaigns producing intermediates start earlier than necessary. This is due to the MILP model formulation, which merely guarantees the timely supply of downstream campaigns. The stock levels of the intermediates, however, are not accounted for in the objective function.

Figure 10 shows the campaign schedule after the post-optimisation stage. 13.6 seconds of CPU time were required to right-shift the campaigns and to reduce inventory holding costs for intermediates accordingly. In this optimisation run, the relative MIP gap was set to 0.001%. A comparison of Figures 9 and 10 reveals further improvements. In the final schedule, the three campaigns 1, 66, and 82 of the initial campaign schedule are eliminated. At the same time, the network of material flows is re-optimised using the complete MFP, i.e. each campaign at an upstream plant is allowed to supply any other campaign at a downstream plant.

A further analysis of Figure 10 shows that the available production capacity is fully utilised only in plants B and C. In the real-life industrial application, the plants also produce additional "local" products, which are not incorporated into the multi-plant supply network considered in the case study. According to the company's policy, their production is separately scheduled after decisions on the size and the timing of the network-based campaigns have been made. The generated campaign schedule may therefore be used as a basis for making capacity reservations at the local plants. In the industrial application, priority is given to the network products, which provide a significantly higher contribution to the company's profit than the numerous "local" products.

7 Concluding remarks

Our research was motivated by a case of supply network planning in the pharmaceutical industry. In the industrial application considered, active ingredients are produced which require numerous processing stages in multiple sites located in different countries. A major characteristic of the highly regulated pharmaceutical industry is the need of extensive cleaning of the equipment units because of rigorous purity requirements. Due to the clean-out times involved when changing to another product type or chemical process, production is typically carried out in campaign mode. Facing cumulative lead times of up to two years, the co-ordination of campaign schedules at different production stages in the various plants is of eminent importance.

To solve the corresponding campaign scheduling problem, a three stage solution procedure is proposed. In the first stage, four different aggregation strategies are employed in order to reduce the complexity of the underlying optimisation problem. These include the aggregation of processing tasks into so-called cascades and the derivation of sub-plants from independently operating equipment configurations. Moreover, human expertise can be used to reduce the large number of supply links arising from the shipments of intermediates. Unrealistic supply links between campaigns in different plants are ruled out in a controlled manner. Finally, total demand is disaggregated into monthly demand figures.

In the second stage of the solution procedure, a series of MILP models is solved, each of which considers detailed demand figures for a limited time window. For this purpose, we present a novel MILP model formulation for supply network planning based on a continuous representation of time. There are two major advances compared to former model formulations known from the literature. First, the proposed continuous time representation allows a much more precise representation of time compared to model formulations, which are based on discrete time periods. For the problem at hand, we also avoid the definition of a large and dense time grid, which would result in prohibitive computational times due to the large number of related binary variables. Second, the model formulation provides opportunities to reduce the size of the model, e.g. by letting the user pre-define the number of campaigns per plant and allow a maximum number of optional campaigns or restrict the feasible types of processes per campaign. To our knowledge, the modelling approach outlined in this paper is the first which uses a continuous time representation for a complex industrial application of supply network planning in the chemical-pharmaceutical industry.

In the actual campaign scheduling stage, the MILP model comprising the entire supply network is solved. In each step of the solution procedure, some of the process-campaign assignments are fixed and an additional time window comprising detailed monthly demand figures is included, until the final solution to the overall supply network planning problem is obtained. During the last stage of the approach, this solution is improved by merging and right-shifting campaigns by taking into account holding costs for intermediate products. For the case study from industry, we demonstrate the efficiency of the aggregation techniques. Even for this extremely complex case study problem, a solution is generated within 15 minutes using a common personal computer. We are currently working on the development of a computer-based decision support tool, which permits the interactive graphical representation of the network and the automatic generation of the optimisation model. Another issue of future research is the development of a hierarchical modelling framework, which combines the approach presented in this paper with detailed scheduling models for local plants.

References

Bassett MH, Dave P, Doyle III FJ, Kudva GK, Pekny FJ, Reklaitis GV, Subrahmanyam S, Miller DL, Zentner MG (1996a) Perspectives on model based integration of process operations. Computers and Chemical Engineering 20: 821–844

Bassett MH, Pekny FJ, Reklaitis GV (1996b) Decomposition techniques for the solution of large-scale scheduling problems. AIChE Journal 42: 3373–3387

Berning G, Brandenburg M, Gürsoy K, Mehta V, Tölle FJ (2002) An integrated system solution for supply chain optimization in the chemical process industry. OR Spectrum 24: 371–402

Blömer F (1999) Produktionsplanung und –steuerung in der chemischen Industrie. Gabler – Deutscher Universitäts Verlag, Wiesbaden (in German)

Blömer F, Günther HO (1998) Scheduling of a multi-product batch process in the chemical industry. Computers in Industry 36: 245–259

Blömer F, Günther HO (2000) LP-based heuristics for scheduling chemical batch processes. International Journal of Production Research 38: 1029–1051

Dimitriadis, AD, Shah N, Pantelides CC (1997) RTN based rolling horizon algorithms for medium-term scheduling of multipurpose plants. Computers and Chemical Engineering S21: S1061–S1066

Grossmann IE, Quesada I, Raman R, Voudouris VT (1996) Mixed-integer optimization techniques for the design and scheduling of batch processes. In: Reklaitis V, Sunol AK, Rippin DWT, and Hortacsu O (eds) Batch processing systems engineering, pp 451–494. Springer, Berlin Heidelberg New York

Grunow M, Günther HO, Lehmann M (2002) Campaign planning for multi-stage batch processes in the chemical industry. OR Spectrum 24: 281–314

Hui CW, Gupta, A (2000) A novel MILP formulation for short term-scheduling of multistage multi-product batch plants. Computers and Chemical Engineering 24: 1611–1617

Hui CW, Gupta A, van der Meulen HAJ (2000) A novel MILP formulation for short term-scheduling of multi-stage multi-product batch plants with sequence dependent constraints. Computers and Chemical Engineering 24: 2705–2717

Iyer RR, Grossmann IE (1998) A bilevel decomposition algorithm for long range planning of process networks. Industrial and Chemical Engineering Research 37: 474–481

Kallrath J (2002a) Combined strategic and operational planning – A MILP success story in chemical industry. OR Spectrum 24: 315–341

Kallrath J (2002b) Planning and scheduling in the process industry. OR Spectrum 24: 219–250

Kim SB, Lee HK, Lee IB, Lee ES, Lee B (2000) Scheduling of non-sequential multipurpose batch processes under finite intermediate storage policy. Computers and Chemical Engineering 24: 1603–1610

Kondili E, Pantelides CC, Sargent RWH (1993) A general algorithm for short-term scheduling of batch operations – I. MILP formulation. Computers and Chemical Engineering 17: 211–227

Lin X, Floudas CA (2001) Design, synthesis and scheduling of multipurpose batch plants via an effective continuous-time formulation. Computers and Chemical Engineering 25: 665–674

Liu ML, Sahinidis NV (1995) Computational trends and effects of approximations in an MILP model for process planning. Industrial and Chemical Engineering Research 34: 1662–1673

Maravelias CT Grossmann IE (2001) Simultaneous planning for new product development and batch manufacturing facilities. Industrial and Chemical Engineering Research 40: 6147–6164

McDonald CM, Karimi IA (1997) Planning and scheduling of parallel semicontinuous processes. 1. Production planning. Industrial and Chemical Engineering Research 36: 2691–2700

Mendez CA, Cerda J (2000) Optimal scheduling of a resource-constrained multiproduct batch plant supplying intermediates to nearby end-product facilities. Computers and Chemical Engineering 24: 369–376

Mendez CA, Cerda J (2002) An MILP-based approach to the short-term scheduling of make-and-pack continuous production plants. OR Spectrum 24: 403–430

Mendez CA, Henning GP, Cerda J (2000) Optimal scheduling of batch plants satisfying multiple product orders with different due-dates. Computers and Chemical Engineering 24: 2223–2245

Mendez CA, Henning GP, Cerda J (2001) A MILP continuous time approach to short-term scheduling of resource-constrained multistage flowshop batch facilities. Computers and Chemical Engineering 25: 701–711

Mockus L, Reklaitis GV (1997) Mathematical programming formulation for scheduling of batch operations based on nonuniform time discretization. Computers and Chemical Engineering 21: 1147–1156

Neumann K, Schwindt C, Trautmann N (2002) Advanced production scheduling for batch plants in process industries. OR Spectrum 24: 251–279

Norton LC, Grossmann IE (1994) Strategic planning model for complete process flexibility. Industrial and Chemical Engineering Research 33: 69–76

Orcun S, Altinel IK, Hortacsu Ö (2001) General continuous time models for production planning and scheduling of batch processing plants: mixed integer linear programming formulations and computational issues. Computers and Chemical Engineering 25: 371–389

Oxe G (1997) Reducing overcapacity in chemical plants by linear programming. European Journal of Operational Research 97: 337–347

Papageorgiou LG, Rotstein GE, Shah N (2001) Strategic supply chain optimisation for the pharmaceutical industries. Industrial and Chemical Engineering Research 40: 275–286

Papageorgiou LG, Pantelides CC (1996a) Optimal campaign planning/scheduling of multipurpose batch/semicontinuous plants. 1. Mathematical formulation. Industrial and Chemical Engineering Research 35: 488–509

Papageorgiou LG, Pantelides CC (1996b) Optimal campaign planning/scheduling of multipurpose batch/semicontinuous plants. 2. A mathematical decomposition approach. Industrial and Chemical Engineering Research 35: 510–529

Pinto JM, Grossmann IE (1994) Optimal cyclic scheduling of multistage continuous multiproduct plants. Computers and Chemical Engineering 18: 797–816

Pinto JM, Türkay A., Bolio B, Grossmann IE (1998) STBS: A continuous time MILP optimization for short-term scheduling of batch plants. Computers and Chemical Engineering 22: 1297–1308

Sahinidis NV, Grossmann IE (1991) Reformulation of multiperiod MILP models for planning and scheduling of chemical processes. Computers and Chemical Engineering 15: 255–272

Sahinidis NV, Grossmann IE, Fornari RE, Chathrathi M (1989) Optimization model for long range planning in the chemical industry. Computers and Chemical Engineering 13: 1049–1063

Schilling G, Pantelides CC (1996) A simple continuous time process scheduling formulation and a novel solution algorithm. Computers and Chemical Engineering S20, S1221–S1226

Schwindt C, Trautmann N (2000) Batch scheduling in process industries: an application of resource-constrained project scheduling. OR Spektrum 22: 501–524

Shah N (1998) Single- and multisite planning and scheduling: Current status and future challenges. AIChE Symposium Series No. 320 94: 75-90

Timpe CH (2002) Solving planning and scheduling problems with combined integer and constraint programming. OR Spectrum 24: 431-448

Timpe CH and Kallrath J (2000) Optimal planning in large multi-site production networks. European Journal of Operational Research 126: 422–435

Wilkinson SJ, Shah N, Pantelides, CC (1995) Aggregate modelling of multipurpose batch plant operation. Computers and Chemical Engineering S19: S583–S588

Wilkinson SJ, Cortier A, Shah N, Pantelides, CC (1996) Integrated production and distribution scheduling on a Europe-wide basis. Computers and Chemical Engineering S20: S1275–S1280

Zentner MG, Peckny JF, Reklaitis, GV, Gupta JND (1994) Practical considerations in using model based optimization for the scheduling and planning of batch/semicontinous processes. Journal of Process Control 4: 259–280

Zentner MG, Reklaitis GV (1996) An interval-based mathematical model for the scheduling of resource-constrained batch chemical processes. In: Reklaitis V, Sunol AK, Rippin DWT, Hortacsu O (eds) Batch processing systems engineering, pp. 779–807. Springer, Berlin Heidelberg New York

An MILP-based approach to the short-term scheduling of make-and-pack continuous production plants*

Carlos A. Méndez and Jaime Cerdá

INTEC (UNL – CONICET), Güemes 3450, 3000 Santa Fe, Argentina
(e-mail: jcerda@intec.unl.edu.ar)

Abstract. Many fast moving consumers good manufacturing companies produce a moderate number of intermediates that are combined in many different ways to generate an enormous variety of end products. To do that, such companies usually run continuous production plants in a make-to-stock environment. The process structure includes a fabrication area yielding basic intermediates that are stocked in a large middle storage space, and a packing sector where finished products usually comprising several intermediates are manufactured. Intermediates all undergo the same sequence of processing stages and the production of any campaign is sequentially allocated to an ordered set of end products. An MILP continuous time scheduling problem formulation handling independently assignment and sequencing decisions and considering sequence-dependent setup times and specific due dates for export orders has been developed. The problem objective is to meet all end-product demands at minimum make-span. The proposed model is able to account for assorted products, multiple campaigns for a particular intermediate even at the same unit and the consecutive allocation of an intermediate campaign to different finished products. Moreover, it can easily embed powerful preordering rules to yielding reduced MILP formulations so as to tackle real-world industrial problems at low computational cost. The approach has been successfully applied to large-scale industrial examples.

Key words: MILP mathematical model – Continuous production plants – Optimal campaign scheduling

* The authors acknowledge financial support from FONCYT under Grant 14-07004, and from "Universidad Nacional del Litoral" under CAI+D 121.

Correspondence to: J. Cerdá

1 Introduction

Most of the publications on short-term scheduling have been focused on multiproduct batch facilities (Kondili et al., 1993a; Pinto and Grossmann, 1995; Cerdá et al., 1997; Ierapetritou and Floudas, 1998a; Hui and Gupta, 2000; Méndez et al., 2000, 2001). Recent contributions on the subject can even assign production stages to different plants and develop a scheduling approach within a multi-plant supply network environment (Grunow et al., 2002). However, manufacturing firms often run continuous processes to perform some production steps yielding intermediates and/or final products. Unfortunately, a few scheduling methodologies for multiproduct continuous facilities have been reported in the last decade. Kallrath (2002) published a comprehensive overview of the current state-of-the-art of planning and scheduling problems in the chemical process industry. An early contribution on short-term scheduling of multiproduct continuous facilities was due to Kondili et al. (1993b). These authors presented an MILP scheduling algorithm for a single-stage energy-intensive multiproduct plant with multiple continuous production lines. It was based on the notion of time slots with variable lengths and the partitioning of the time horizon into multiple periods at the deadlines of the required deliveries. Schilling et al. (1994) addressed the short-term scheduling of a single-stage multiproduct plant with multiple continuous processors using time-slots and the state-task-network (STN) representation of the process structure. They allowed non-zero setup times, continuous and point demands and, in addition, inventory, changeover and energy costs were considered. Likewise recent methodologies, the scheduling horizon was divided into multiple time periods with point demands at period ends. Constraints forbidding a slot from overlapping more than a single period were included. Schilling and Pantelides (1996) extended the scope of the previous work by using the resource-task-network (RTN) framework. They observed that the continuous time formulation for multiproduct semicontinuous facilities exhibited a poor computational performance. To overcome such a problem, they proposed a special branch-and-bound scheme that starts branching on the continuous slot lengths to get a tighter bound and then switches to the normal branching on the binary variables. Zhang and Sargent (1994) and (1998) presented a unified MINLP model for the problem of determining the optimal operation of a mixed production facility, involving batch and continuous processes. The approach was later extended to deal with various types of operational restrictions, such as task precedence and changeover constraints. Additionally, nonlinearities introduced by allowing variable processing rates were eliminated through an exact linearization technique, so that the problem could be modeled as a large-scale MILP. Karimi and McDonald (1997) proposed a couple of slot-based MILP formulations to deal with the scheduling of single-stage multiproduct facilities involving no storage requirements and multiple product demands at specified due dates. Ierapetritou and Floudas (1998b) developed a novel continuous-time formulation for short-term scheduling of mixed production facilities that utilizes the STN representation of the production network and the notion of task and unit event times. The approach is capable of handling limited storage capacity by defining storage operations and cleanup times without incorporating any further task. However, none of the previous work can easily consider order due dates since they divide the scheduling horizon into multiple periods and production campaigns can spread over two or more periods. As a result, a significant increase in binary variables and constraints normally occurs. Méndez and Cerdá (2002) extended the MILP continuous-time scheduling framework for multistage batch facilities previously proposed by Méndez et al. (2001) to also deal with multiproduct continuous plants. The approach can account for sequence-dependent changeovers and limited intermediate storage without defining any additional task. It uses neither time-slots nor events for units and tasks but the basic concept of direct/non-direct campaign precedence at any equipment item. When applied to the scheduling of a large-scale make-and-pack continuous industrial plant, a better schedule was found in less CPU time than previous approaches despite running a much lower number of production campaigns. Test examples found in the literature for comparing the performance of different scheduling approaches generally assume (i) simultaneous supply of intermediates to the packing lines directly or through storage silos and (ii) end products requiring just a single intermediate.

This paper introduces a new MILP continuous time formulation for the scheduling of a make-and-pack continuous industrial facility producing a large number of assorted products, each one containing various intermediates in different proportions. Several parallel continuous processors run at both the production and packaging stages, and an unlimited storage space for intermediates and end products is assumed to be available. The production of any intermediate campaign is sequentially allocated to an ordered set of end products and the amount assigned to each one is stored in a different cluster of containers and conveyed to the packing section. Unlike the approach of Méndez and Cerdá (2002), the proposed formulation takes advantage of the unlimited storage assumption to allow the treatment of assorted products, multiple runs for any intermediate even at the same unit and the consecutive allocation of a production campaign to different finished products. Similar to our previous approach, it permits to consider order due dates and sequence-dependent changeovers. Moreover, it can easily embed powerful preordering rules to tackle real-world industrial problems at low computational cost. Several publications on the scheduling of multiproduct batch facilities have already considered the use of preordering rules (Pinto and Grossmann 1995; Cerdá et al., 1997; Méndez et al., 2000). However, by using the notion of direct/non-direct campaign/run precedence, the proposed MILP approach is able to incorporate different preordering rules and limit the application of them to just a subset of intermediate/end campaigns at the continuous production/packing lines.

2 The structure of a make-and-pack continuous facility

A candy production plant will be used to illustrate the typical structure of a make-and-pack continuous facility. As described by Henning and Cerdá (2000), a candy production plant makes a wide variety of final products in a T-shaped processing structure consisting of two major sectors: production and packaging (see Fig. 1). In the fabrication area, different types of wrapped candies (*intermediates*) are produced in several continuous lines or units running in parallel. Processing rates can vary with both the type of wrapped candy being made and the assigned production

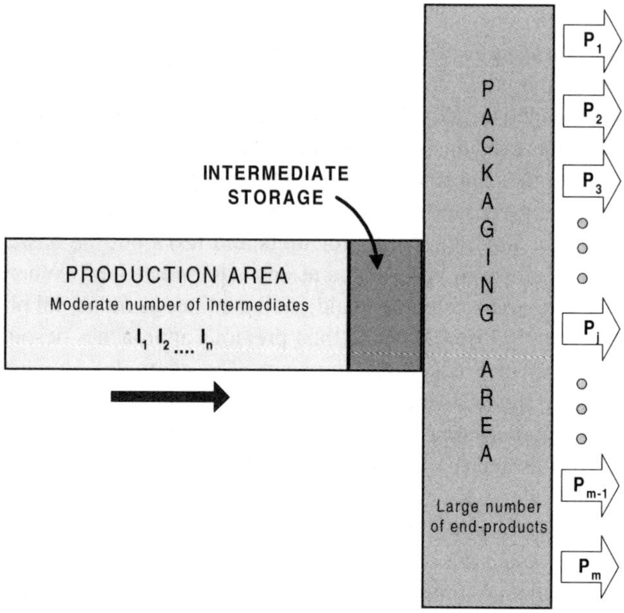

Fig. 1. Schematic representation of a make-and-pack continuous production facility

line. Intermediates $i \in I$ are then temporarily stocked in a rather large storage space located in-between both sectors from which they are fed to the packing area. In the middle storage space, intermediates are stored in different sets of containers, each one holding the required amount of an intermediate $i \in I$ for a particular end product $p \in P$. In the packaging sector, several units are run in parallel to obtain a high number of end products $p \in P$ by packaging different amounts of one or several types of wrapped candies in many different ways and combinations. Each final product p consists of a number of identical bags/cans all manually packed in cardboard boxes. Two end products differ in at least one of the following features: (i) type of packaging; (ii) types of intermediates contained in each bag/can, (iii) proportions of the intermediates included in a bag/can, (iv) total weight of wrapped candies contained in a bag/can, (v) number of bags/cans in a cardboard box. Packing rates may significantly change from one to another end product. Moreover, they are generally larger than the intermediate production rates, in the sense that it takes more time to produce the amount of the unique intermediate contained in a single-flavor end product than to packing it. In other words, packing campaigns are generally shorter than the related production runs supplying the required intermediates. As a result, the packing lines could remain idle over some portion of the time horizon. To get a better use of the packing capacity, external orders involving wrapped candies coming from neighboring plants are also handled in the packing section. Such external orders have to be considered at the time of developing the short-term schedule for the whole facility.

The candy manufacturing plant works 24 hours each day, five days a week from Monday to Friday in a make-to-stock environment, i.e. nearly all production orders are placed to replenish end product inventories. Just a few ones are export orders to be sent overseas. Such orders involving special products for foreign markets should be ready at specified due dates. The other ones must be completed before the end of the scheduling horizon. Every week, the plant scheduler has to develop a production schedule for both fabrication and packing areas based on (i) the warehouse and exporting orders to be fulfilled, each one specifying codes, amounts and sometimes due dates of the required end-products; (ii) packing orders from neighboring plants to be completed at given due dates; (iii) the status of the plant at the beginning of the week, including lines/units operating in each sector and their ready times, and (iv) the length of the time horizon, usually just a single week.

2.1 Operating mode at the production area

It is a typical production policy in a candy plant to produce every week the set of required intermediates in proper amounts to just meet the demands of the end products to be packed within the scheduling period. Storage of intermediates over the weekend is avoided to guarantee the quality of the final product. Consequently, no intermediate inventory is available neither at the beginning nor at the end of the week. To establish the weekly production targets for the fabrication sector, the warehouse orders involving one or several end products should be exploded to determine the set of intermediates and the amount of each one to be made. Since a number of end products may share some types of wrapped candies, the requirements of the same intermediate by different end products are added up to determine its total demand over the time horizon.

To reduce changeover time and intermediate losses, usually a low number of production campaigns is run for each required intermediate. Switching intermediates in a production line often implies a sudden change in flavor and/or color. Then, the intermediate output at the early stage of a new production campaign does not meet flavor/color specifications and must be discarded. Running a few campaigns for each required intermediate is the preferred option to minimize changeover losses. Sometimes, however, such a production strategy may lead to long campaigns delaying the production in the packing sector. In that case, several runs are usually accomplished at the same or different processing units to avoid an increase in the overall make-span. In industry, the duration of a production run must never be lower than a specified minimum length. Therefore, running many campaigns for a particular intermediate very often becomes an unfeasible option. On the other hand, only a single packing campaign for every finished product is just run. This is so mostly because (a) high packaging rates usually lead to rather short campaigns and (b) changeover operations requiring time and especially labor costs must be avoided.

2.2 The influence of assorted products on the scheduling task

Assorted end products contain several types of wrapped intermediates that are produced at different processing lines in the fabrication area (see Fig. 2). If not

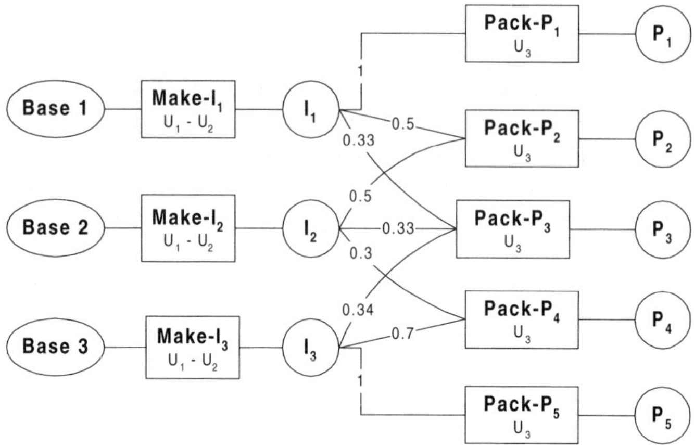

Fig. 2. State-task network representation for Example 1

all the necessary intermediates are available in the right proportions in the storage space, then the packing of an assorted product cannot be started. Furthermore, in case some of the required intermediates are produced much earlier than the remainder, the WIP inventory level may increase undesirably.

Usually, the output of a production campaign for a certain intermediate is sequentially allocated to several end products requiring the same type of wrapped candy and stored in a different set of containers. Therefore, a production campaign for intermediate $i \in I$ can be regarded as composed of a sequence of production runs (i, p), each one yielding the same intermediate $i \in I$ to meet the demand of a different end product $p \in P_i$ requiring i. In the middle storage space, the output of a production run (i, p) remains in the assigned cluster of containers until the packing of end product p begins. Not only the plant scheduler should decide on: (a) the allocation of continuous processing units $j \in J$ to required intermediates $i \in I$ and (b) the ordering of the intermediate campaigns accomplished at the same processing line but also (c) the matching of intermediate campaign outputs to end products and (d) the arrangement of production runs (i, p), each one tied to a different end product $p \in P_i$, within an intermediate campaign. Because such scheduling choices are handled in an MILP-based approach through binary variables, the additional decisions (c) and (d) could significantly increase the problem size and the computational requirements. Matching production runs to end products means to specify the set of final products and the requirement of each one to be satisfied by a particular intermediate campaign. In turn, the arrangement of production runs (i, p) within a campaign for intermediate $i \in I$ provides the order in which the end product requirements will be fulfilled. This is important because the packaging of an end product demanding a certain intermediate can be started before the supplying campaign has been completed. In addition, it facilitates the logistics at the storage space by providing the information on the ordered list of end products to which a production campaign is allocated as well as the amount of intermediate assigned to each one.

In this paper, the best sequence of production runs (i, p) accomplished at the same processing line $j \in J_i$ is sought. Often, those ones producing the same intermediate i are consecutively run to minimize changeover time and cost, thus giving rise to a longer campaign. In the general case, however, two or more campaigns yielding intermediate $i \in I$ can be accomplished at the same or different production lines. A characteristic arrangement of production runs (i, p) within an ith-intermediate campaign is usually observed at near-optimal schedules. If a pair of runs (i, p) and (i, p') involving the same intermediate i are accomplished at the same production line $j \in J_i$ and the end product p is packed before p', then production run (i, p) is usually accomplished earlier to keep the stock of intermediates considerably lower.

2.3 The average work-in-process inventory level

Dealing with assorted end products in a make-and-pack continuous facility really makes the scheduling task much harder for the necessary synchronization in the operation of the two major plant sectors. In fact, the packing of an end product is mostly made in parallel with the campaigns supplying the intermediates but fairly out-of-phase to guarantee that the operation must never be stopped because of intermediate shortages. This also implies that the output of the production lines at the fabrication area should be properly coordinated to speed-up the packing of end products by simultaneously providing all the required intermediates. In this way, the intermediates are packed as soon as they are available in the storage space and, therefore, their average lifetime is so minimized. Consequently, there is nearly always enough storage space for intermediates coming from the production area and the storage constraint is never active at the optimum. Therefore, an unlimited storage space is a fair assumption. Unfortunately, such synchronization on the proposed schedules has a modeling cost to be paid in terms of 0–1 variables and linear constraints.

2.4 Schedule performance measures

Usually, an optimal schedule for the whole plant permits to reach the weekly output targets at both sections almost at the same time. This is so because the WIP inventory level remains low over the entire scheduling horizon, even at the time the last intermediate campaign has just ended. In contrast, a poor coordination of the production/packing operations builds up a large inventory of intermediates that often reaches its peak when the production of wrapped candies is completed and the operations at the fabrication area are stopped. In such a case, extra shifts are needed to run the packing lines on Saturdays so as to transform the remaining WIP inventory into final products. As a result, the overall make-span also increases.

Therefore, an improvement in the production schedule generally brings about not only a sizable reduction in labor and changeover costs but also a significant decrease in both the average WIP inventory level and the overall make-span. These performance measures are usually adopted at a candy production plant to judge the quality of the proposed schedule. Since the proposed approach can handle sequence-dependent changeover times and, in addition, labor and WIP inventory carrying costs are closely related to the make-span, the problem objective will aim to just minimizing the overall make-span. By doing so, the other performance measures are mostly optimized.

3 Problem statement

Given a make-and-pack continuous production facility comprising:

- A fabrication center producing a moderate number of intermediates.
- A packing area where different combinations of intermediates are packed to yield a large number of assorted finished products.
- A storage space of unlimited capacity receiving intermediates from the fabrication center and supplying them to the packing area.
- A set of make-to-stock orders for final products to be satisfied before the end of the time horizon and some export/packing orders to be shipped at certain due dates.
- A set of continuous processing lines/packing units running in parallel at each center.
- A specified time horizon, usually one-week long.

The problem goal is to determine:

- The set of production campaigns for each intermediate/end product to be accomplished at any continuous line/unit at the fabrication/packing center.
- The run length and the starting time for each production/packing campaign.
- The ordered list of end products $p \in P_i$ allocated to each campaign producing intermediate $i \in I$, where the first appearing in the list is the first served; i.e. the sequence of production runs (i, p) at each processing unit.
- The run length and the starting time for every production run (i, p) being accomplished to meet the requirements of end-product $p \in P_i$.

So as to minimize the overall make-span while strictly meeting the specified due dates for exporting and packing orders. At the minimum make-span schedule, production and packing tasks are almost completed at the same time through a perfect synchronization of the operations in both plant sectors. In other words, there is nearly no stock of intermediates to be packed at the time the fabrication center has achieved the weekly production target.

4 Major assumptions

To derive the problem mathematical formulation, the following assumptions have been made:

- All problem data are known with certainty.

- Every final product demand can be regarded as a point or discrete demand to be shipped either to inventory before the end of the scheduling period or to customers at certain due dates.
- Every final product is obtained by packing the appropriate types of intermediates in the right proportions. End-product structures are given.
- Production and packing centers both use continuous processes with multiple lines/units running in parallel.
- A single packing campaign for every end product is at most allowed.
- Each production line/packing unit can make one or several intermediate/end products.
- The output of an intermediate-producing campaign can be sequentially assigned to several end products.
- A storage space of unlimited capacity between fabrication and production areas is available.
- The duration of a production run (i, p) assigned to a particular end product should never be lower than the specified minimum run time. The minimum run length for a campaign involving several consecutive production runs of the same intermediate is not directly considered in the proposed MILP model.
- Changeover times at production/packing units are sequence-dependent.
- Inventory of intermediates is available neither at the beginning nor at the end of the scheduling period.

5 The mathematical model

One of the major features of the proposed mathematical model is the handling of allocation and sequencing decisions through separate sets of 0–1 variables. Thus, Y_{ipj} is an allocation variable that is equal to 1 whenever production run (i, p) is assigned to processing line $j \in J_i$. On the other hand, the sequencing variable $X_{ipi'p'}$ establishes the relative ordering of production runs (i, p) and (i', p'). If both runs have been allocated to the same production line, then $X_{ipi'p'} = 1$ denotes that run (i, p) is a direct or a non-direct predecessor of run (i', p'). In contrast, if $X_{ipi'p'} = 0$, then production run (i', p') is started earlier, i.e. (i', p') precedes (i, p). In this way, a single 0–1 variable is necessary to determine the relative ordering of any pair of runs if they have been allocated to the same unit. If not, the value of $X_{ipi'p'}$ is meaningless.

5.1 Notation

Sets

I	intermediates
I_p	intermediates required for end product p
P	end products
P_i	end products requiring intermediate i
J	continuous processing units

J_i, J_p available lines to run campaigns producing intermediate i or end product p

$J_{ii'}, J_{pp'}$ available units to run campaigns for intermediates (i, i') or end products (p, p')

Parameters

d_p demand of end product $p \in P$

dd_p due date for end product $p \in P$

l_{ipj}^{min} minimum run length of any production run (i, p) producing intermediate i for final product p at unit j

M a very large number

ρ_{pi} quantity of intermediate i required per unit amount of end product p

r_{ij}, r_{pj} production rate of intermediate i or end product p at unit j

ru_j ready time of unit j

ro_i release time for intermediate i

$\tau_{ii'j}$ changeover time between consecutive campaigns producing intermediates i, i' at unit j

Variables

CI_{ipj} completion time for production run (i, p) at line $j \in J_i$

CP_p completion time for packing campaign yielding final product p

LI_{ipj} length of production run yielding intermediate i for product p at line $j \in J_i$

LP_{pj} length of campaign producing final product p at unit $j \in J_p$

MK make-span

$S_{pp'}$ binary variable denoting that the campaign for end product p is run before ($S_{pp'} = 1$) or after ($S_{pp'} = 0$) another one producing end product p'

W_{pj} binary variable denoting that a campaign for final product p is run in unit $j \in J_p$

$X_{ipi'p'}$ binary variable denoting that a production run yielding intermediate i for product p is started before ($X_{ipi'p'} = 1$) or after ($X_{ipi'p'} = 0$) another one producing intermediate i' for product p'at the same production line

Y_{ipj} binary variable denoting that a production run yielding intermediate i for product p is run at line $j \in J_i$

$Z_{ii'}$ binary variable denoting that a campaign producing intermediate i is started before ($Z_{ii'} = 1$) or after ($Z_{ii'} = 0$) another one yielding intermediate i' at the same production line.

5.2 Problem constraints

- *A single packing campaign for every end product $p \in P$ can at most be run.*

Since the packing rate is relatively high and changeover operations involve labor costs, a single packing campaign for each end product is at most accomplished.

$$\sum_{j \in Jp} Wpj = 1 \;\; \forall p \in P \tag{1}$$

- *Minimum run length for the production run (i,p)*
  The duration of a production run (i, p) assigned to a particular end product p should never be lower than a specified minimum run time.

$$lipj^{\min}\, Yipj \;\leq LI\, ipj \;\leq\; (dp\; \rho pi/rij\,)\, Yipj \;\; \forall\, i \in I, p \in Pi, j \in Ji \tag{2}$$

- *Length of a packing campaign yielding end product p∈ P*
  Since d_p is the required amount of product p and r_{pj} is the packing rate at unit $j \in J_p$, then the length of the packing campaign will be (d_p/r_{pj}) if product p is allocated to unit j $(W_{pj} = 1)$.

$$LPpj = (dp/rpj)\, Wpj \;\; \forall p \in P, j \in Jp \tag{3}$$

- *Starting times for intermediate/end product runs*
  The starting time of a production run (i, p) yielding intermediate i for end-product p at unit j, i.e. $(CI_{ipj} - LI_{ipj})$, can never be lower than either the ready time of the allocated unit j or the release time of intermediate i.

$$CIipj - LIipj \geq Max\,[ruj, roi] \;\; \forall i \in I, p \in Pi, j \in Ji \tag{4}$$

$$CPp - LPpj \geq Max\,[ruj, rop] \;\; \forall p \in P, j \in Jp \tag{5}$$

- *Enough amount of intermediate i ∈ I_p should be provided to meet the requirement of any end product p ∈ P_i through one or more production runs.*
  If d_p is the demand of end product p to be satisfied and ρ_{pi} is the quantity of intermediate i required per unit amount of p, then $(d_p\rho_{pi})$ is the overall amount of i to be allocated to p. In turn, the output of a production run (i, p) at line $j \in J_i$ is obtained by multiplying the processing rate r_{ij} by the run length LI_{ipj}. Therefore,

$$\sum_{j \in Ji} rij\; LIipj = dp\; \rho pi \;\; \forall i \in I, p \in Pi \tag{6}$$

- *A production run (i, p) supplying intermediate i ∈ I_p to end product p must never begin later than the packing campaign for product p and will never finish after completing such a campaign.*
  Since $(CI_{ipj} - LI_{ipj})$ is the starting time of production run (i, p) and $(CP_p - \Sigma_{j'} LP_{pj'})$ is the time at which the single packing campaign for p begins, then:

$$CPp - \sum_{j' \in Jp} LPpj' \geq CIipj - LIipj \;\; \forall i \in I, p \in P, j \in Ji \tag{7}$$

$$CPp \geq CI\, ipj \;\; \forall i \in I, p \in Pi, j \in Ji \tag{8}$$

- *Sequencing constraints on production runs* (i, p)
  Sequencing constraints must be imposed on a pair of production runs (i, p) and (i', p') only if they are accomplished at the same unit $j \in J_{ii'}$, i.e. $Y_{ipj} = Y_{i'p'j} = 1$. If run (i, p) is started before run (i', p'), i.e. $X_{ipi'p'} = 1$, then constraint (9) holds to ensure that (i', p') begins after completing both the run (i, p) and the subsequent changeover operation. In such a case, the other sequencing constraint (10) becomes redundant. Instead, if run (i, p) is begun later ($X_{ipi'p'} = 0$), then constraint (10) now holds to prevent from starting (i, p) before ending (i', p') and, in addition, Eqn. (9) becomes redundant. Otherwise $Y_{ipj} + Y_{i'p'j} < 2$ and, consequently, constraints (9) and (10) are both redundant and the value of the sequencing variable $X_{ipi'p'}$ is meaningless. In this way, the number of sequencing variables is cut by half compared to other scheduling approaches since a single 0–1 variable is required to define the relative ordering of any pair of intermediate production runs (i, p) and (i', p').

$$CIi'p'j - LIi'p'j \geq CIipj + \tau ii'j - M\,(1 - Xipi'p')$$
$$-M\,(2 - Yipj - Yi'p'j)$$
$$\forall i, i' \in I, p \in Pi, p' \in Pi', j \in Jii' : (i < i') \text{ or } (i = i' \text{ and } p < p') \quad (9)$$

$$CIipj - LIipj \geq CIi'p'j + \tau i'ij - MXipi'p' -$$
$$M\,(2 - Yipj - Yi'p'j)$$
$$\forall i, i' \in I, p \in Pi, p' \in Pi', j \in Jii' : (i < i') \text{ or } (i = i' \text{ and } p < p') \quad (10)$$

- *Sequencing constraints on end-product packing campaigns*
  A similar idea is used to define the sequencing constraints (11) and (12) to be imposed on any pair of packing campaigns. In this way, the number of sequencing variables $S_{pp'}$ is also cut by half.

$$CPp' - LPp'j \geq CPp + \tau pp'j - M\,(1 - Spp')$$
$$-M\,(2 - Wpj - Wp'j)$$
$$\forall p, p' \in P, j \in Jpp' \; : \; p < p' \quad (11)$$

$$CPp - LPpj \geq CPp' + \tau p'pj - MSpp'$$
$$-M\,(2 - Wpj - Wp'j)$$
$$\forall p, p' \in P, j \in Jpp' \; : \; p < p' \quad (12)$$

- *Due date constraints*
  The packing campaign yielding product p required at certain due date dd_p must never be completed after dd_p. Most of the end products are made to replenish inventories. The due date for them is the end of the scheduling horizon.

$$CPp \leq ddp \;\; \forall p \in P \quad (13)$$

- *Make-span definition*
  The schedule make-span is the least upper bound on the completion time of any packing campaign.

$$CPp \leq MK \;\; \forall p \in P \quad (14)$$

5.3 The objective function

The problem objective is to minimize the overall time needed to complete all the required end products, i.e. the overall make-span. Therefore, the proposed mathematical formulation aims to optimize the objective function (15), while satisfying all problem constraints (1)–(14). In this way, the short-term scheduling of a make-and-pack continuous manufacturing plant producing a wide range of end products from a moderate number of intermediates can be modeled as an MILP. Objective function (15) is generally applied when production orders for refilling warehouse inventories are to be satisfied. If instead a large portion of the overall demand comes from customer orders to be fulfilled at given due dates, then the objective function (16) seeking to minimize the total order earliness is mostly used. Since constraint (13) ensures that the tardiness for every customer order will be equal to zero at the optimum, then the objective pursued by (16) is to complete all the production orders just in time.

$$\text{Min } MK \tag{15}$$

$$\text{Min } \sum_{p \in P} (ddp - CPp) \tag{16}$$

As shown in a next section, preordering rules can effectively be applied to get a reduced MILP problem formulation still providing a very good schedule in a short CPU time for large-scale industrial problems. Application of preordering rules permits to eliminate a significant number of sequencing variables $(S_{pp'}, X_{ipi'p'})$ and constraints.

6 Embedding preordering rules in the mathematical formulation

The size of the proposed problem formulation shows an exponential growth with the number of production/packing units and the range of intermediates and final products to be manufactured. Therefore, the short-term scheduling of a real-world make-and-pack continuous industrial facility could hardly be solved to optimality. However, different preordering rules for intermediate and end product campaigns can easily be embedded in the problem formulation to significantly shrink the solution space by excluding many options rarely adopted at good schedules. Since binary and continuous variables related to such options are very likely to be zero at the optimum, then they can be deleted from the mathematical model. Despite that, the so-reduced problem formulation could still provide near-optimal schedules at very reasonable CPU times. An overview summarizing the equations belonging to the different solution strategies presented in this section is included in Table 1.

6.1 Implicit preordering rules for intermediate runs

As mentioned before, a characteristic campaign-sequencing pattern is usually observed at near-optimal schedules. If a pair of production runs (i, p) and (i, p') are

Table 1. Summary of the proposed solution strategies

Solution strategy	Equations
Without preordering rules	1–14
FPFS rule	1–8; 9[a]–10[a]; 9'–10'; 11–14
G/FPFS rule	1–8; 9[b]–10[b]; 9'–10'; 9"–10"; 11–14
SIPT rule	1–10; 13–14; 17
SIPT & FPFS rules	1–8; 9[a]–10[a]; 13–14; 17–18
SIPT & G/FPFS rules	1–8; 9[b]–10[b]; 13–14; 17–19
Z-FPFS approach	1–8; 9'–10'; 11–14; 20–21
SIPT & Z-FPFS approach	1–8; 13–14; 17–18; 20–21

[a] Only for any pair of runs producing different intermediates (i, i').
[b] Only for any pair of runs producing different intermediates (i, i') for the same end product p.

accomplished at the same production line and the end product p is packed before p' $(S_{pp'} = 1)$, then production run (i, p) is started earlier to keep the WIP inventory level much lower, i.e. $(X_{ip,ip'} = 1)$. In other words, production runs involving the same intermediate for different end products and accomplished at the same processing line are arranged according to the so-called "first-packed-first-served (FPFS)"-implicit preordering rule. The term "implicit" is used because the preordering rule is not directly applied to any pair of intermediate production runs (i, p) and (i, p'). By assuming such a sequencing pattern, therefore, the 0–1 variables $X_{ip,ip'}$ can be substituted by the variables $S_{pp'}$ and, therefore, deleted from the problem formulation. In this way, a significant saving of sequencing variables $X_{ip,ip'}$ can be achieved and a near-optimal schedule is still found. The sequencing constraints (9)–(10) on such type of production runs (with a common intermediate for two different end products) take the following form:

$$CIip'j - LIip'j \geq CIipj - M\,(1 - Spp') - M\,(2 - Yipj - Yip'j)$$
$$\forall i \in I, p \in Pi,\ p' \in Pi, j \in Ji : p < p' \qquad (9')$$

$$CIipj - LIipj \geq CIip'j - MSpp' - M\,(2 - Yipj - Yip'j)$$
$$\forall i \in I, p \in Pi,\ p' \in Pi, j \in Ji : p < p' \qquad (10')$$

Therefore, the FPFS-implicit preordering rule permits to arrange production runs (i, p) yielding the same intermediate i for different end products at the same processing line through constraints (9')–(10').

A more important reduction in the number of sequencing variables and constraints can be achieved by applying the FPFS implicit rule to any pair of runs (i, p) and (i', p'), even if they produce different intermediates $(i \neq i')$ for different products $(p \neq p')$ at the same production line. In other words, run (i, p) should precede run (i', p') if end product p is packed before p'. It can be labeled as the General FPFS-implicit preordering rule, i.e. the G/FPFS-rule. Obviously, some deterioration in the objective function may occur. However, finding a good feasible schedule for real world industrial problems at low computational cost is still a worthy goal for the scheduler. In this case, sequencing constraints (9)–(10) are to be replaced by,

$$CIi'p'j - LIi'p'j \geq CIipj + \tau ii'j - M\,(1 - Spp')$$
$$-M\,(2 - Yipj - Yi'p'j) \qquad (9'')$$
$$\forall i, i' \in I, p \in Pi, p' \in Pi', j \in Jii' : (i < i' \text{ and } p \neq p')$$

$$CIipj - LIipj \geq CIi'p'j + \tau i'ij - MSpp' \qquad (10'')$$
$$-M\,(2 - Yipj - Yi'p'j)$$
$$\forall i, i' \in I, p \in Pi, p' \in Pi', j \in Jii' : (i < i' \text{ and } p \neq p')$$

Constraints (9)–(10) still apply to arrange any pair of runs (i, p) and (i', p) producing different intermediates $(i \neq i', ord\ i < ord\ i')$ for the same product p at the same production line.

6.2 Explicit preordering rules for end products

As explained before, intermediates are immediately packed as soon as they are available in the middle storage space when the plant is well operated. However, packaging units will be steadily working only if proper types of intermediates in the right amounts are continually available in the storage space to start the processing of another end product. This can be achieved by always packing first at any unit the end product requiring intermediates that are processed faster. In other words, end products are to be ordered at any packing unit by increasing processing times of their required intermediates, i.e. the shortest intermediate processing time (SIPT)-rule. A characteristic processing time for the set of intermediates required by a particular end product p (MPT_p) can be defined as follows:

$$MPTp = \sum_{i \in Ip} \left(dp\ \rho pi / \sum_{j \in Ji} rij \right)$$

MPT_p usually is, but not always, a good lower bound on the overall processing time of the intermediates allocated to end product p. Therefore, the proposed explicit preordering rule states that end products $p \in P$ must be arranged at each packing unit by increasing MTP_p. If end products p and p' can be packed in the same unit and the inequality $MTP_p < MTP_{p'}$ holds, then $S_{pp'} = 1$. As a result, the constraint (11) takes the following simpler form,

$$CPp' - LPp'j \geq CPp + \tau pp'j - M\,(2 - Wpj - Wp'j)$$
$$\forall p, p' \in P, j \in Jpp' : MTPp < MTPp' \qquad (17)$$

and the redundant constraint (12) can be excluded from the problem formulation. If at the same time the FPFS-implicit rule is applied, then the sequencing constraints

(9')–(10') on production runs yielding the same intermediate for different end products reduce to a single constraint set given by:

$$CIip'j - LIip'j \geq CIipj - M\,(2 - Yipj - Yip'j)$$
$$\forall i \in I, p \in Pi, p' \in Pi, j \in Ji : MPTp < MPTp' \quad (18)$$

In this way, the combined use of the SIPT-rule for end products and the FPFS-rule for intermediate runs significantly reduces the number of sequencing variables and constraints. On the other hand, if the general G/FPFS implicit rule is instead enforced together with the SIPT-rule, then the sequencing constraints (9")–(10") on runs producing different intermediates for different end products are to be replaced by,

$$CIi'p'j - LIi'p'j \geq CIipj + \tau ii'j - M\,(2 - Yipj - Yi'p'j)$$
$$\forall i, i' \in I, p \in Pi, p' \in Pi', j \in Jii' : (i \neq i' \text{ and } MPTp < MPTp') \quad (19)$$

and a larger saving in binary variables and constraints is so achieved.

When the end products are to be delivered at specified due dates, the SIPT-rule can be replaced by the earliest-due-date (EDD) rule to arrange the end product campaigns by increasing due dates at every packing line. At the same time, the objective function (16) minimizing the total order earliness is to be applied. When the setup time $\tau_{pp'j}$ at the packing lines for a pair of end products (p, p') is important and highly dependent on the selected campaign sequence, then the ordering of such end product campaigns must still remain as a model decision. In other words, the sequencing variable $S_{pp'}$ should stay in the problem formulation. In such a case, the EDD rule will be successively applied to every pair of end product campaigns, different from the ones involving important sequence-dependent changeovers, to establish their relative ordering beforehand.

6.3 An alternative MILP problem formulation

As mentioned before, the combined use of the SIPT-rule for end products and the G/FPFS-rule for intermediate runs significantly reduces the number of sequencing variables and constraints. However, at the same time, it usually increases the number of campaigns producing the same intermediate and, therefore, the total number of campaigns. For problem instances involving significant sequence dependent changeovers, the use of such preordering rules may become unattractive because of the additional setup costs. To limit the number of changeovers and, at the same time, reduce the model complexity, a new binary variable $Z_{ii'}$ should be introduced in the proposed MILP formulation to control the sequencing of intermediate campaigns at every production line, where a campaign for intermediate i can be regarded as a series of consecutive production runs (i, p) involving the same intermediate i for different end products. By including $Z_{ii'}$ in the problem formulation, however, a single campaign for each intermediate i can at most be run at every available production line $j \in J_i$. Moreover, if campaigns of any two intermediates are produced on different lines, they must be produced in the same relative order on all these lines. On the other hand, it is better to keep applying the FPFS rule for ordering the production runs (i, p) and (i, p') involving the same intermediate i for different end products p and p', i.e. constraints (9') and (10') still hold.

The decision variable $Z_{ii'}$ will be equal to one if production campaigns for intermediates i and i' are run at the same unit and the intermediate i is produced before i'. If both are produced at the same line but the intermediate i' is produced earlier than i, then $Z_{ii'} = 0$. Finally, if both campaigns are run at different production lines, then $Z_{ii'}$ will be meaningless. The variable $Z_{ii'}$ can be used not only to choose the best sequence of intermediate campaigns but also the ordering of production runs (i, p) and (i', p'), involving different intermediates $(i \neq i')$ for similar or different end products, at the same production line.

Therefore, the new sequencing constraints for ordering production runs (i, p) and (i', p') at the same production line in terms of the new binary variable $Z_{ii'}$ are the following:

$$\begin{aligned} &CIi'p'j - LIi'p'j \geq CIipj + \tau ii'j - M\,(1 - Zii') \\ &-M\,(2 - Yipj - Yi'p'j) \\ &\quad \forall i, i' \in I, p \in Pi, p' \in Pi', j \in Jii' : (i < i') \end{aligned} \tag{20}$$

$$\begin{aligned} &CIipj - LIipj \geq CIi'p'j + \tau i'ij - MZii' \\ &-M\,(2 - Yipj - Yi'p'j) \\ &\quad \forall i, i' \in I, p \in Pi, p' \in Pi', j \in Jii' : (i < i') \end{aligned} \tag{21}$$

In short, for problem instances involving significant sequence dependent changeovers at the production lines, it is better to solve an MILP problem formulation including constraints (1)–(8), (9')–(10'), (11)–(14), (20)–(21) and the objective function (15) or (16). Such a formulation comes from using both the FPFS-rule and the set of binary variables $Z_{ii'}$, i.e. the Z/FPFS approach. Indeed, the Z/FPFS can be applied to any problem instance whether or not it involves significant changeovers at the production lines. Its major advantage is to provide a better control on the number of intermediate campaigns being run. If, additionally, the SIPT-rule is applied to sequence end product packing campaigns (i.e., the SIPT & Z/FPFS approach), then constraints (9')–(10') are to be replaced by Eqn. (18) and constraints (11)–(12) by Eqn. (17).

7 Results and discussion

A small-size example has first been solved to illustrate the special features of the short-term scheduling problem being tackled and the type of results provided by the proposed MILP scheduling approach. In addition, the best production schedules and CPU requirements with and without preordering rules were compared to assess the rule performance. Next, the approach was applied to a large-scale industrial case study twice. First, it was assumed that final product requirements are mostly due for refilling warehouse inventories at the end of the week. Afterwards, the problem was revisited but this time the make-and-pack continuous facility have been properly scheduled to timely satisfy a set of customer orders at the promised due dates. In

Table 2. Processing unit data for Example 1

Intermediate	Rate (tn/h)		Changeover time (h)		
	U_1	U_2	I_1	I_2	I_3
I_1	12	10.5		1.5	2.3
I_2	10	8	1.2		1.7
I_3	13	11	1.3	2.0	

any case, it was assumed unit ready times and release times for intermediates all cqual to zcro. Thc proposcd MILP problcm formulations wcrc solved on a 1 GIIz Pentium III PC using ILOG OPL Studio 3.1 (ILOG OPL Studio, 2000) together with the embedded CPLEX Mixed-Integer Optimizer 7.0 release (ILOG CPLEX, 2000).

7.1 Example 1

Five assorted products (P_1–P_5) are to be produced from three different intermediates (I_1–I_3) at a make-and-pack continuous industrial facility during next week. Intermediates are made from three different base materials available as required. The plant is operated on a closed-shop mode from Monday to Friday, three shifts a day. Figure 2 shows the STN representation of the process network. Thus, product P_2 requires intermediates I_1 and I_2 in equal proportions (50:50) while every unit amount of P_3 includes similar quantities of the three intermediates (I_1–I_3). Two parallel processing units (U_1–U_2) producing intermediates at the production area, a single packaging unit U_3 and an unlimited storage space for intermediates and final products are available. Since intermediates are unstable, no WIP inventory should remain at the end of the week. Product-dependent processing rates, intermediates that can be processed and sequence-dependent setup times for every production line at the making stage are all given in Table 2. In turn, Table 3 shows the bill of materials, demands (in tons) and product-dependent processing rates for the packing units. The bill of materials specifies the amount of each intermediate needed to pack one hundred tons of end products. A minimum run length of 8 h was adopted for every production run (i, p) at every continuous unit.

Table 3. Packing unit data for Example 1

Product	Demand	Rate (tn/h)	Product composition		
	(tn)	U_3	I_1	I_2	I_3
P_1	750	30	100		
P_2	350	25	50	50	
P_3	280	20	33	33	34
P_4	400	18		30	70
P_5	600	23			100

Table 4. Model sizes and computational requirements for Example 1

Solution strategy	Binary vars, cont. vars, rows	Objective function	CPU time[a]	Iterations	Nodes
Without preordering rules	64, 47, 278	108.99	47.79	189076	30521
FPFS rule	55, 47, 278	108.99	16.69	67276	9354
G/FPFS rule	33, 47, 278	108.99	10.93	45162	6231
SIPT & G/FPFS rules	23, 47, 206	111.26	0.22	294	43
Partial SIPT & FPFS rules[b]	26, 47, 225	114.26	0.44	1045	163
Z/FPFS approach[b]	31, 47, 278	108.99	5.65	30167	3837

[a] Seconds on Pentium III PC (1 GHz) with ILOG 3.1 / CPLEX 7.0
[b] With changeovers at the packing lines.

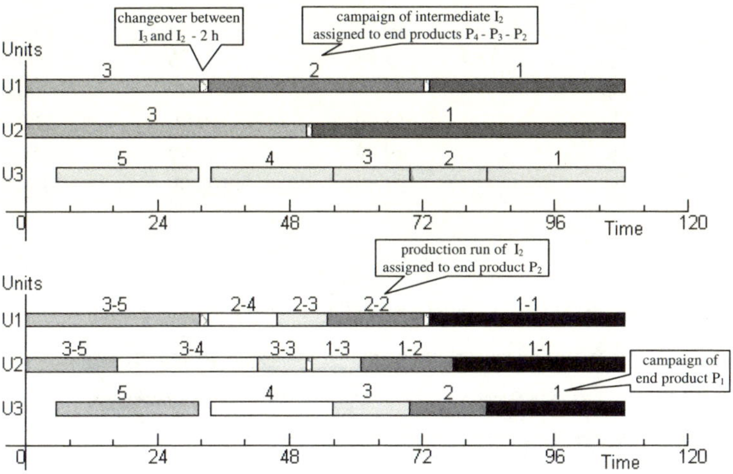

Fig. 3. Optimal production schedule for Example 1

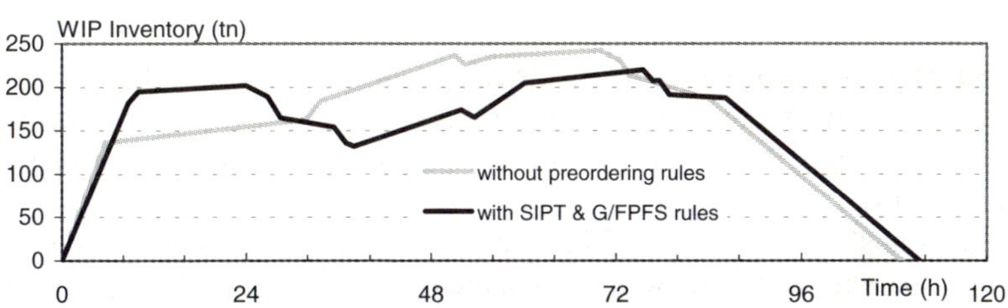

Fig. 4. Evolution of intermediate WIP inventory for Example 1

7.1.1 Example 1 (without sequence dependent setup times at the packing lines)

Example 1 can be formulated as an MILP comprising 64 binary variables, 47 continuous variables and 278 linear constraints (see Table 4). The best schedule shown in both Figure 3 and Table 5 was found in 47.79 s. Figure 3 shows the optimal schedule twice; first, in terms of intermediate campaigns without specifying the

Table 5. Optimal schedule for Example 1

Unit	Intermediate/ product	Starting time (h)	Completion time (h)	Production (tn)
U_1	I_3 - P_5	0.00	31.76	412.94
	I_2 - P_4	33.76	45.76	120.00
	I_2 - P_3	45.76	55.00	92.40
	I_2 - P_2	55.00	72.50	175.00
	I_1 - P_1	73.70	108.99	423.39
U_2	I_3 - P_5	0.00	17.01	187.06
	I_3 - P_4	17.01	42.46	280.00
	I_3 - P_3	42.46	51.11	95.20
	I_1 - P_3	52.41	61.21	92.40
	I_1 - P_2	61.21	77.88	175.00
	I_1 - P_1	77.88	108.99	326.61
U_3	P_5	5.68	31.76	600.00
	P_4	33.76	55.99	400.00
	P_3	55.99	69.99	280.00
	P_2	69.99	83.99	350.00
	P_1	83.99	108.99	750.00

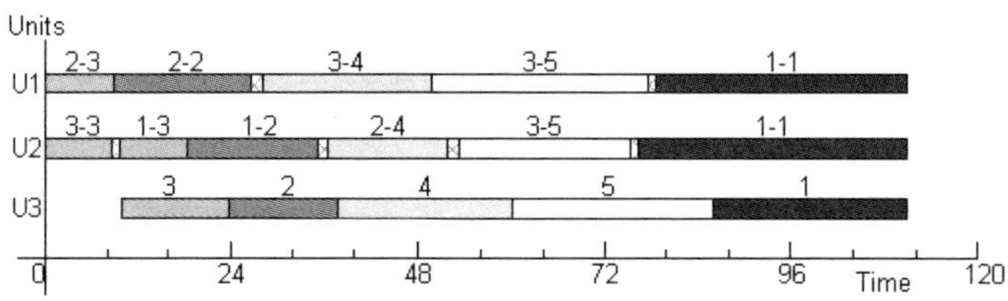

Fig. 5. Best schedule for Example 1 (with SIPT & G/FPFS rules)

ordered set of final products to which the campaign output is sequentially assigned, and then, in terms of the series of constituent production runs (i, p) comprised by every campaign. By applying the FPFS-implicit rule and therefore substituting constraints (9)–(10) by equations (9')–(10') for ordering the runs producing the same intermediate for different end products, the optimal schedule described in Figure 3 is still found but at lower computational cost. The number of 0-1 variables drops from 64 to 55 (see Table 4). At the optimum, a total of five intermediate campaigns are accomplished at two production stage by consecutively running those ones producing the same intermediate for different end products. In this way, changeover time and losses are also minimized. Since a long production campaign would be required to meet the overall demand of some intermediates like I_3 and I_1, then shorter parallel runs for such intermediates are accomplished at two production lines, in order to prevent potential bottlenecks in the packing stage. Moreover, production and packing operations are completed at the same time t= 108.99 h, i.e. eleven hours before the end of the scheduling period. Figure 4 describes the evolution of the intermediate inventory over the scheduling period. The overall WIP inventory level reaches a peak amounting to 242 ton, which is even less than the quantity of intermediates (280 ton) required by product P_3 featuring the lowest demand.

To evaluate the performance of the proposed preordering rule for end products, Example 1 was again solved but this time embedding the Shortest Intermediate Processing Time (SIPT) rule together with the G/FPFS implicit rule in the MILP formulation. End product campaigns have been arranged at the packing lines by decreasing values of the characteristic processing time MPT_p, i.e. $(P_3, P_2, P_4, P_5, P_1)$. In this way, the near-optimal production schedule depicted in Figure 5 was found. The overall make-span slightly rises to 111.26 h due to the rise in the changeover time resulting from running three more intermediate production runs. However, the CPU time is reduced by a factor of 217 with regards to the rigorous model (see Table 4).

7.1.2 Example 1 (with sequence dependent setup times at the packing lines)

So far, setup times at the packing lines were assumed to be negligible. Let us now suppose that finite sequence dependent setup times are just given for the pair of end products $P_2 - P_4$, $P_4 - P_5$ and $P_2 - P_5$. When they are consecutively processed in such an order, the setup time is equal to 3 h. If the processing ordering is reversed, the setup time is decreased by half, i.e. the setup time will decrease to 1.5 h. Example 1 was solved again but this time the relative ordering of the packing campaigns for each pair of end products with finite sequence dependent changeovers was still a model decision. In other words, binary variables $S_{P2,P4}$, $S_{P4,P5}$ and $S_{P2,P5}$ were included in the problem formulation. For any other pair of end products, the SIPT-rule was still applied to establish the relative ordering of the packing campaigns beforehand. In this way, the production schedule shown in Figure 6a was found. The number of binary variables just increases by three, while the makespan rises by only 3 h (see Table 4). This is achieved by changing the sequence of end product campaigns at the packing line. Instead of sequencing the end product campaigns by increasing MPT_p, i.e. $(P_3, P_2, P_4, P_5, P_1)$ to get an overall setup time of 6 h, the alternative sequence $(P_3, P_5, P_4, P_2, P_1)$ is chosen with a total setup time of 3 h.

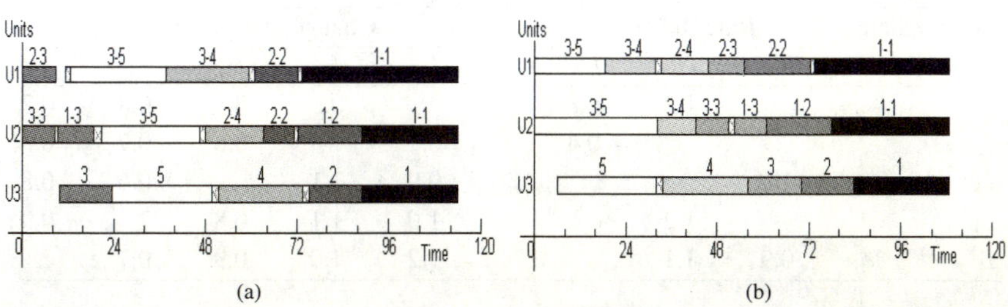

Fig. 6 a,b. Best schedules for Example 1with changeovers at the packing line **a** using partial SIPT & G/FPFS rules, **b** using Z/FPFS approach

A greater improvement is attained if the SIPT-rule is no longer applied and the Z/FPFS approach is used to solve the modified Example 1 with finite sequence dependent setup times. In this case, all the sequencing variables $S_{pp'}$ for any pair of end products remain in the model and the binary variables $Z_{ii'}$ for ordering intermediate campaigns are additionally considered. The best schedule found is given in Figure 6b. The schedule make-span drops by approximately 5 h from 114.26 h to 108.99 h. Moreover, the production schedule provided by the Z/FPFS approach features the same sequence of packing campaigns and the same make-span as the best solution found for Example 1 without finite changeover times (see Figs. 3 and 6a). When compared with the production schedule shown in Figure 6a, the number of intermediate campaigns drops from 9 to 5 and the number of setups decreases by a half at both the production and the packing areas. Such an improvement is achieved at very low computational cost.The number of 0-1 variables only increases by 5 and the CPU time amounts to 5.65 s (see Table 4).

Table 6. Product demand data for Examples 2 and 3

Product	Demand	Due date	Product composition					Product	Demand	Due date	Product composition				
	(tn)	(example 3)	I_1	I_2	I_3	I_4	I_5		(tn)	(example 3)	I_1	I_2	I_3	I_4	I_5
P_1	13.5	72	50	50				P_{10}	20.5	56	20	20	20	20	20
P_2	15	96		100				P_{11}	20	120	25		25	25	25
P_3	22	80			20	40	40	P_{12}	21	112			25	75	
P_4	18	56	40			60		P_{13}	19.5	96				100	
P_5	17.5	120		33		33	34	P_{14}	18	64	33	33			34
P_6	18	80			100			P_{15}	16	32		30			70
P_7	15	32	100					P_{16}	16	120	50		50		
P_8	14	72				100		P_{17}	25	48[a]					
P_9	14.5	64					100	P_{18}	30	56[a]					

[a] Due dates for packing orders from neighboring plants are also considered at Example 2.

Table 7. Processing unit data for Examples 2 and 3

Intermediate	Rate (tn/h)			Changeover time (h)				
	U_1	U_2	U_3	I_1	I_2	I_3	I_4	I_5
I_1		1.2	0.8		0.5	0.8	0.8	0.7
I_2			0.4	0.7		0.6	0.7	0.8
I_3	1.0			0.9	1.1		0.6	0.8
I_4		1.2		1.0	1.1	0.8		0.9
I_5	0.9	1.1		1.2	1.0	0.9	0.7	

Table 8. Packing unit data for Examples 2 and 3

Product	Setup time (h) - rate (tn/h)		Product	Setup time (h) - rate (tn/h)	
	U_4	U_5		U_4	U_5
P_1	0.4 – 5.0	0.5 – 6.0	P_{10}		0.9 – 4.5
P_2		0.4 – 2.7	P_{11}	0.5 – 3.5	
P_3		0.6 – 2.0	P_{12}	0.5 – 2.0	
P_4	0.5 – 1.5		P_{13}		0.55 – 5.0
P_5	0.6 – 3.3		P_{14}	0.7 – 4.0	
P_6	0.4 – 1.5		P_{15}		0.4 – 2.0
P_7		0.55 – 4.0	P_{16}	0.8 – 2.6	
P_8	0.9 – 3.5	0.8 – 2.5	P_{17}		1.0 – 1.6
P_9		0.6 – 5.0	P_{18}	0.9 – 2.3	

7.2 Example 2

Example 2 deals with a large-scale industrial problem. Specified amounts of sixteen end products (P_1–P_{16}) for refilling warehouse inventories are to be produced from five intermediates (I_1–I_5) at a make-and-pack continuous industrial plant over a scheduling period of 120 h. Moreover, two packaging orders from neighboring plants (P_{17}–P_{18}) are also to be considered. The plant consists of three parallel continuous processing lines (U_1–U_3) producing intermediates at the fabrication area and two packaging units (U_4–U_5) working simultaneously at the packing stage. The plant is run from Monday to Friday, three shifts a day. Data for Example 2 are included in Tables 6–8. Intermediate production targets for the fabrication area are chosen so as to exactly meet end product requirements and no WIP inventory remains at the end of the scheduling horizon. Moreover, no initial inventory of intermediates is available and a minimum run length of 3 h for every run (i, p) in all units was adopted.

By applying the proposed scheduling approach without any preordering rule, the problem can be formulated as an MILP involving 502 binary variables, 137 continuous variables and 1306 linear constraints. The size of the model still remains too large if either the FPFS or the G/FPFS-implicit preordering rule is only applied. Since the problem cannot be solved to optimality in a reasonable CPU time, the SIPT-preordering rule for end products together with either the FPFS or the G/FPFS-implicit rule for intermediates were next embedded in the MILP formulation. By enforcing both the SIPT & FPFS rules, the number of binary variables diminishes to 331 and the production schedule shown in Figure 7 was found in 633.78 s (see Table 9). Packing operations are completed at time $t = 112.63$, i.e. 7.37 hours before the end of the time horizon. By analyzing Figure 7, an effective coordination of the short-term schedules for the parallel production lines and the packing units can be observed. The average level of work-in-process inventory is reduced as much as possible. Runs producing the same intermediate for different end products were grouped together to avoid setup times. In this way, a total of eight intermediate campaigns is accomplished, i.e. three above the least possible number of campaigns. Moreover, packing orders (P_{17}, P_{18}) were completed just at their due dates.

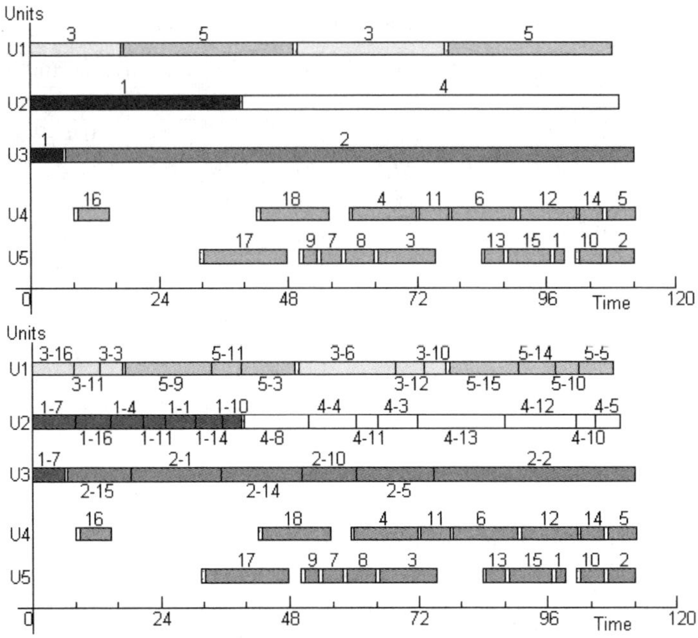

Fig. 7. Best solution for Example 2 (with SIPT & FPFS rules)

To further reduce the problem size and make the relative gap tolerance for the branch-and-bound search equal to zero, the SIPT-preordering rule for end products together with the Z/FPFS approach was applied to again solve Example 2. In this manner, the number of 0-1 variables decreases by a half (see Table 9). The best solution depicted in Figure 8 stands for a near-optimal production schedule featuring an overall make-span of 112.63 h, just equal to the value exhibited by the schedule found by using the SIPT & FPFS preordering rules (see Table 9). Surprisingly, the best schedule was found in a very low CPU time as large as 2.74 s. When compared with the production schedule shown in Figure 7, the number of campaigns decreases from 8 to 6, and the number of setups drops from 5 to 3.

Table 9. Model sizes and computational requirements for Examples 2 and 3

Example	Solution strategy	Binary vars, cont. vars, rows	Objective function	CPU time[a]	Iterations	Nodes
2	SIPT & FPFS rules	331, 137, 1092	112.63	633.78[b]	890581	125125
2	SIPT & G/FPFS rules	102, 137, 863	115.51	50.20	112983	16738
2	SIPT & Z/FPFS approach	153, 137, 1092	112.63	2.74	2990	324
3	EDD & G/FPFS rules	84, 136, 844	66.50	5.88	15656	2479

[a] Seconds on Pentium III PC (1 GHz) with ILOG 3.1/CPLEX 7.0.

[b] Relative gap tolerance: 0.03.

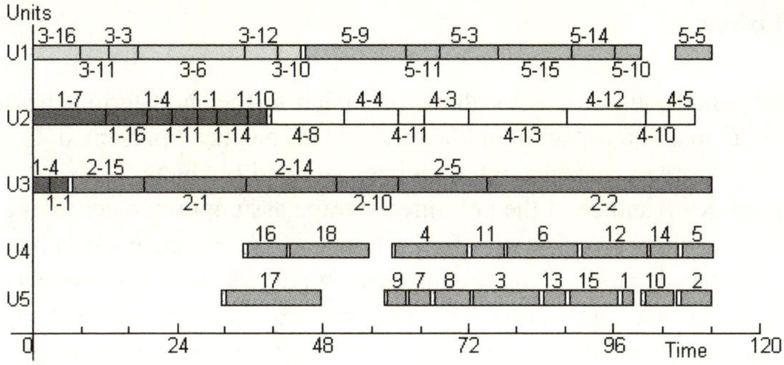

Fig. 8. Best solution for Example 2 (with SIPT & Z/FPFSapproach)

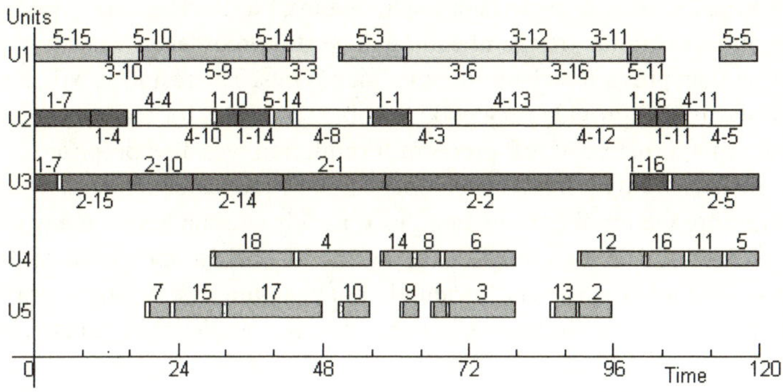

Fig. 9. Best schedule for Example 3 (with EDD & G/FPFS rules)

7.3 Example 3

Example 2 is revisited but this time a large fraction of end product demands comes from export orders to be shipped at certain due dates. Only three products (P_5, P_{11}, P_{16}) are made for inventory and, additionally, two packing orders from neighboring plants ($P_{17} - P_{18}$) are also to be considered. The proposed approach does not require to dividing the scheduling horizon into multiple periods at the order due dates. Instead, it just includes the specified due date as an upper bound for the product completion time. By using the objective function (16) and embedding both the "earliest due date (EDD)"-preordering rule for end products together with the G/FPFS-rule for intermediates in the proposed MILP formulation, Example 3 can be modeled through a reduced MILP model involving 84 binary variables and 844 linear constraints. In this way, the production schedule shown in Figure 9 was found in 5.88 s (see Table 9). Convergence to the optimum is apparently much faster when the EDD-rule and the minimum total earliness goal are both applied. The overall earliness amounts to 66.5 h, i.e. an average earliness of 3.7 h for each final product. From Figure 9, it follows that the handling of due dates produces a sizable increase in the number of intermediate campaigns in order to meet customer orders in time.

8 Conclusions

An MILP continuous-time algorithmic approach to the short-term scheduling of a continuous make-and-pack manufacturing plant has been presented. It assumes an unlimited storage space for intermediates and end products. The proposed formulation takes advantage of the unlimited storage assumption to consider assorted products, multiple production runs for each intermediate even at the same unit and the consecutive allocation of every campaign output to several end products. Moreover, it permits to account for order due dates and sequence-dependent changeovers. If necessary, the proposed approach can freely insert idle time at the beginning, in-between campaigns or at the end of the scheduling horizon. By describing the campaign sequence at every unit through the full set of (direct/non-direct) predecessors for any campaign, the proposed approach achieves a much greater saving in 0-1 variables and constraints from embedding powerful preordering rules in the MILP formulation than previous scheduling methodologies. The resulting MILP reduced formulation can still provide very good feasible schedules for real-world industrial problems. An alternative MILP problem formulation was also proposed to handle problem instances involving considerably large sequence dependent setup times. By constraining the number of campaigns for each intermediate i at any available production line $j \in J_i$ to a single one, the model variant is capable of finding a production schedule with a very low number of changeovers. A more complete and thorough analysis of the limitations arising from the single intermediate campaign assumption and the effectiveness of the proposed alternative formulation will be included in future publications. Three examples were solved to illustrate the computational performance of the MILP approach with and without preordering rules.

References

Cerdá J, Henning GP, Grossmann IE (1997) A mixed-integer linear programming model for short-term scheduling of single-stage multiproduct batch plants with parallel lines. Industrial & Engineering Chemistry Research 36: 1695–1707

Grunow M, Günther HO, Lehmann M (2002) Campaign planning for multi-stage batch processes in the chemical industry. OR Spectrum 24: 281–314

Henning GP, Cerdá J (2000) Knowledge-based predictive and reactive scheduling in industrial environments. Computers and Chemical Engineering 24: 2315–2338

Hui C-W, Gupta A (2000) A novel formulation for short-term scheduling of multistage multi-product batch plants. Computers and Chemical Engineering 24: 1611–1617

Ierapetritou, M.G., Floudas, C. (1998a). Effective continuous-time formulation for the short-term scheduling. 1. Multipurpose batch processes. Industrial & Engineering Chemistry Research 37: 4341–4359

Ierapetritou MG, Floudas CA (1998b) Effective continuous-time formulation for short-term scheduling. 2. Continuous and semicontinuous processes. Industrial & Engineering Chemistry Research 37: 4360–4374

ILOG OPL Studio 3.1 (2000) User's manual. ILOG S.A. France

ILOG CPLEX 7.0 (2000) User's manual. ILOG S.A. France

Kallrath J (2002) Planning and scheduling in the process industry. OR Spectrum 24: 219–250

Karimi IA, McDonald CM (1997) Planning and scheduling of parallel semicontinuous processes. 2. Short-term scheduling. Industrial & Engineering Chemistry Research 36: 2701–2714

Kondili E, Pantelides CC, Sargent RWH (1993a) A general algorithm for short-term scheduling of batch operations – I. MILP Formulation. Computers and Chemical Engineering 17: 211–227

Kondili E, Shah N, Pantelides CC (1993b) Production planning for the rational use of energy in multiproduct continuous plants. Computers and Chemical Engineering 17: S123–S128

Méndez CA, Henning GP, Cerdá J (2000) Optimal scheduling of batch plants satisfying multiple product orders with different due dates. Computers and Chemical Engineering 24: 2223–2245

Méndez CA, Henning GP, Cerdá J (2001) A MILP continuous time approach to short-term scheduling of resource-constrained multistage flowshop batch facilities. Computers and Chemical Engineering 25: 701–711

Méndez CA, Cerdá J (2002) An MILP continuous-time formulation for short-term scheduling of multiproduct continuous facilities. Computers and Chemical Engineering 26: 687–695

Pinto JM, Grossmann IE (1995) A continuous time mixed integer linear programming model for short term scheduling of multistage batch plants. Industrial & Engineering Chemistry Research 34: 3037–3051

Schilling G, Pantelides CC (1996) A simple continuous-time process scheduling formulation and a novel solution algorithm. Computers and Chemical Engineering 20: S1221–S1226

Schilling G, Pineau Y, Pantelides CC, Shah N (1994) Optimal scheduling of multipurpose continuous plants. Presented at AIChe Annual Meeting. San Francisco, CA, Paper 233i

Zhang X, Sargent RWH (1994) The optimal operation of mixed production facilities - A general formulation and some approaches for the solution. Fifth International Symposium on Process Systems Engineering, pp 171–178

Zhang X, Sargent RWH (1998) The optimal operation of mixed production facilities - Extensions and improvements. Computers and Chemical Engineering 22: 1287–1295

Long term planning of inter-company energy supply concepts

W. Fichtner, M. Wietschel, and O. Rentz

Institute for Industrial Production (IIP), University of Karlsruhe,
Hertzstraße 16, 76187 Karlsruhe, Germany
(e-mail: wolf.fichtner@wiwi.uni-karlsruhe.de)

Abstract. In order to assess the economic and ecological effects of inter-company energy supply concepts, an optimising model integrating investment and long-term production planning has been developed. The model represents the energy production of companies on a very disaggregated level taking into account different resources, energy carriers and production processes. This model has been applied to five industrial companies to analyse the economic and ecological implications resulting from investments in inter-company energy supply concepts. Existing technologies as well as possible investment options have been assessed by a techno-economic analysis taking into account company specific circumstances. In addition, the new German combined heat and power law has been modelled in detail, because the supplementary payments legally established in this law may effect energy go supply concepts in a considerable way. Computational results show the installation of a combined cycle power plant adapted to the specific conditions of the case at hand would be the most promising option to fulfil the future energy demand of the companies involved.

Key words: Investment planning – Long-term production planning – Inter-company energy supply concepts – Mixed-integer programming – Combined heat and power

1 Introduction

Since the end of the sixties economists have been concerned with the physical-quantitative analysis of material and energy flows. Based on scientific laws, e.g. the second law of thermodynamics, they came to the insight that, in a closed thermodynamic system, each energy and material transformation involves a loss of workable

Correspondence to: W. Fichtner

energy and material (see Georgescu-Roegen [19] and the critical remarks of Ayres [3]). This lead to the call for a conservation of limited, non-renewable resources. In the late eighties, the protection of natural resources and the reduction of emissions were identified as essential prerequisites of a sustainable development. According to the Brundtland report [34], a sustainable development is a development that meets the needs of the present without compromising the ability of future generations to meet their own demands.

Due to the fact that, in the field of industrial production, investment options resulting in a more sustainable development will only be realised if they are in accordance with a company's economic objectives, political decision makers will have to change framework conditions, e.g. by introducing new policy instruments, in order to stimulate investments in sustainable development. To encourage sustainable development in the industrial sector, even in the short term, economically favourable investments have to be identified which also have a positive social and environmental impact.

Within their own production processes, most companies have already realised economically favourable options. Investment options realised by several companies in co-operation, however, can open up new opportunities which have not yet been comprehensively realised. Generally speaking, inter-company co-operations for ecological reasons are rather seldom [24]. In analogy to the integral concept of supply chain management [31], economic advantages can be gained by realising an inter-company optimum instead of an optimum within each of the companies involved. Moreover, inter-company investment options could help to realise ecological benefits, e.g. if they are used to interconnect the energy and material flows of different companies. Since the energy supply of industrial companies is a major source of emissions as well as an important prerequisite for economic development, inter-company energy supply concepts present a promising approach for companies to contribute to sustainable development. Such co-operative concepts could help to overcome the problem that recycling energy flows (e.g. waste heat) in production processes is not always economically and/or technically feasible for an individual company. Another important reason for the analysis of inter-company co-operations lies within the deregulation of energy markets, which has lead to a situation in which environment-friendly technologies such as Combined Heat and Power (CHP) units are often driven out of the market by cheaper technologies. Therefore, in January 2002, the German Parliament passed a new CHP law establishing supplementary payments on each kWh electricity produced in cogeneration, i.e. the simultaneous production of both useable heat or steam and electricity, and fed into the public grid in order to support the operation of CHP plants as one mechanism to reduce the current level of greenhouse gas emissions. Due to the fact that this new law only subsidizes existing CHP installations and not the installation of new ones, there is only one possibility to open up new CHP potentials and still profit from the new law: expanding existing CHP potentials e.g. by linking the energy demand of different companies located in the same area resulting in regional energy supply concepts.

Whereas some theoretical and practical work on inter-company material-flow management exists, e.g. in the field of industrial ecology [12], especially with regard to eco-industrial parks [10], there has hardly been any practical experience in inter-company energy-flow management between production companies until now (see e.g. [25]). Due to the fact that linking energy flows between companies requires the development of specific hardware such as steam pipes, requiring considerable investments, possible investors are faced with a need for decision support tools for a simultaneous planning of investment and long term production. Up to now, the practical use of models combining investment and long-term production planning [9] has been very limited. Due to favourable peculiarities the energy sector is one of the few areas where these models have been used even though only in national energy planning. Because of this aggregated view existing energy models for long-term planning do not consider the details of company-specific energy systems (e.g. their requirements for protection against the breakdown of single boilers or generators as well as the company specific load profiles).

Therefore, the objective of this paper is to present a model for the analysis of the economic and ecological effects of inter-company energy supply concepts under the new framework conditions of liberalised energy markets. The theoretical exposition is illustrated by the application of the model to the inter-company energy supply network "Karlsruher Rheinhafen". Due to problems with regard to the measurement of social criteria, the social dimension of an investment project is not considered.

The remainder of this paper is organised as follows. The next section provides an overview of requirements for a model of techno-economic analyses of inter-company energy supply concepts. Section 3 describes the developed model which combines investment and long term production planning of inter-company energy supply concepts. In Section 4 this model is used to identify economic and ecological implications resulting from an investment in an inter-company energy supply concept. Finally, conclusions are drawn and the prospects and recommendations for future research are given in Section 5.

2 The analysis of inter-company energy supply concepts

Due to the long technical lifetime of energetic units, investment planning in the energy sector typically takes place at the strategic planning level. In order to analyse such investments, static or time-adjusted investment appraisals could be used (possibly enlarged by socio-environmental considerations, [4]). However, pure investment planning with an exogenously given output of the new energy supply unit not taking into account different modes of operation (e.g. different full load operating hours or different ratios between heat and electricity) would not be adequate. Not considering the output of an energetic unit as a decision variable of the investment decision could lead to wrong investment decisions because the profitability of a new energy supply concept strongly depends on its utilisation and the corresponding payments e.g. for primary energy carriers. Furthermore, in order to minimise the emissions resulting from energy production, the utilisation of the energetic unit has to be a decision variable. The utilisation, in turn, depends on the energetic units already available, the technology of the new energetic unit, the primary energy carriers (especially if fluctuating energies are to be used) and the demand profiles to be met by the new energetic unit. Furthermore, interdependencies arising from specific energetic units which have to be operated due to their simultaneous integration into the production process (e.g. CO-boilers in refineries) have to be taken into account when determining the utilisation of a new energy supply concept. Different approaches based on linear optimisation have been developed to solve the interdependence problem in the assessment of investment projects [9]. The application of these methods has focused on the manufacturing industry, but due to the complexity of the problem the models have hardly gained any practical relevance so far. Compared to the situation in manufacturing companies there are some peculiarities in the energy sector (e.g. electricity and heat can hardly be stored) making it easier to consider interdependencies between investment and long term production planning [33].

Since the two oil crises in the seventies, a growing number of models of energy systems have been developed and used in the energy sector. The conceptual structure of energy models can be found in Hafkamp [22]; however in [26] and [13] different energy models are presented. The investment planning of electric utilities in particular has received considerable attention and many formulations of the problem have been proposed over the last decades (see e.g. [1,8,23]). Therefore, the use of mathematical programming in the context of energy planning is a well-known field of research [18], but is still of great interest (see e.g. [21] and [30]), especially due to the new situation in liberalised energy markets. However, existing electricity models for strategic planning can hardly be used to model the details of the energy system of a production company. Regarding technical parameters, a methodology to analyse energy supply concepts on a company level has to take into account technical details of energetic units, for instance minimum load requirements. In addition, company-specific demands regarding protection against the breakdown of energetic units have to be modelled. In many companies, back-up capacities (cold reserve and warm reserve) are required because an interruption of the process heat supply would lead to high monetary losses or even to irreversible damage to production units. It has already been mentioned that economic profitability is a prerequisite for the realisation of inter-company energy supply concepts. A model for the analysis of energy supply concepts must therefore be able to take into account company-specific requirements, e.g. the requested return on investment. Due to stringent investment requirements in many companies existing energetic units are regularly overhauled and/or retrofitted to keep them operational for longer than initially intended. To determine the optimal time for an investment in new capacities, a model must therefore be capable of taking into account increasing maintenance costs.

The integration of possible inter-company energy supply concepts into the energy planning approach of production companies considerably enlarges the decision areas of these companies. In addition to investments in boilers, turbines etc. realised independently by the companies, various forms of energy supply concepts (including different alternatives of energy transportation) realised in co-operations with possible partners – such as the development of new power plants – have to be taken into account. A model to analyse inter-company energy supply concepts has to be able to represent the production of different qualities of steam with regard to temperature and pressure, since each company tends to have an own standard.

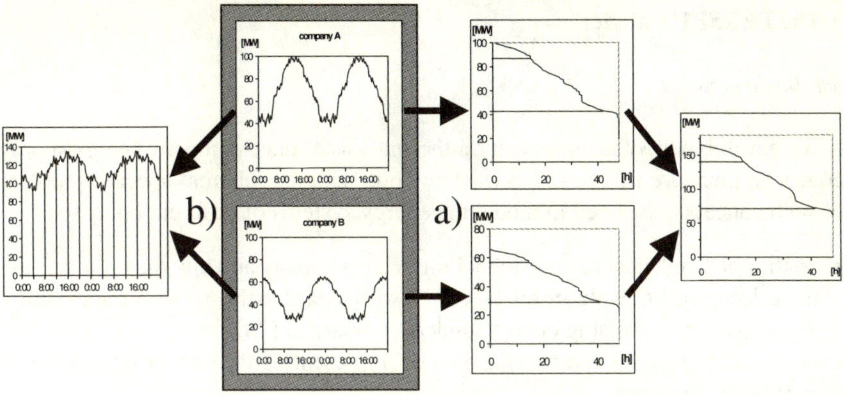

Fig. 1. Different possibilities for modelling the load

In addition it must be possible to identify which units of a company should not be operated at full-load in order to be able to compensate for a breakdown of an unit in one of the other companies of the network. Furthermore, in order to assess the effects of inter-company energy concepts, e.g. due to the combination of different load profiles, it is indispensable to represent the required electricity and heat of the companies relating to their machines and appliances. There are two basic possibilities to represent the load of a system. Firstly, load curves can be used. A load curve is a curve of power versus time showing the level of load for each point of time covered. The horizontal axis is time and the vertical axis is load (MW). In contrast to this procedure, the majority of existing energy models for strategic planning does not represent the load profile directly as a time series but rely on a simplified representation of the load curves by load duration curves giving the load level as a function of the duration of its occurrence. A load duration curve is a curve of loads, plotted in descending order of magnitude against time (see Fig. 1). The area under the load duration curve is divided into alternative load ranges, which have to be satisfied by different types of energetic units (e.g. peak load, middle load and base load power plants (shown at the right side of Fig. 1)). Figure 1 illustrates the differences between these two possibilities for modelling the load. For this, exemplary load curves of two companies for two typical working days (grey highlighted in Fig. 1) have been illustrated. Procedure a) in Figure 1 shows the representation of the load curves by two company-specific load duration curves and the corresponding joint load duration curve. The inherent problem of this aggregation is the loss of relevant information, e.g. the combination of the two load duration curves indicates that the demand of both companies would be less than 80 MW for about 15 hours. In reality the joint demand is never smaller than 90 MW, as can be seen from the corresponding load curve for the joint supply of both companies resulting from procedure b).

3 The PERSEUS model

3.1 Basic concept

In the development of an approach for the analysis of inter-company energy supply concepts, there are two possible ways to cope with the inherent specific problem of such concepts, the need to model the energy systems of different companies:

- on the one hand, develop one model for every company and link these models using a decomposition algorithm (see [11], with regard to the use of decomposition algorithms in optimising energy models compare to [28]) or,
- on the other hand, develop a single model representing the energy systems of all companies involved.

Due to the fact that inter-company energy supply concepts can lead to a variety of ways to link energy flows between the companies (e.g. one CHP unit produces different forms of steam for different partners of the network), it seems to be more advantageous to develop one single model representing the energy structures of all companies involved. For this task, the multi-period mixed-integer linear optimisation approach PERSEUS-ICC (Program Package for Emission Reduction Strategies in Energy Use and Supply – Inter-Company Concepts) has been developed [14]. Considering energy conversion technologies, which already exist in the analysed system, as well as future investment options, it is the objective of this model to provide information on how the exogenously given energy demand needed to maintain production in the firms involved should be met in an economically optimal way.

3.2 Assumptions made in the model building phase

In the model building phase some modelling assumptions had to be made in order to construct a suitable model. The most important of these assumptions are described below. In order to adequately model the energy demand resulting from different shift times in the companies, it has been decided that the hours of the day have to be taken into account. Representing load profiles as a time series is recognised as essential to identify the capacity needed and to calculate the amount of electricity produced in cogeneration, which is decisive for the supplementary payment under the German CHP law. Of course it is not possible to represent the energy demand at 8760 hours per year over the whole time horizon of this strategic planning problem, which spans up to 20 years due to the technical life-time of energetic units. Therefore, to represent load profiles within the model, time series of the real daily load curves have to be averaged, e.g. over all working days that have more or less the same load profile, resulting in typical daily load profiles. Demand profiles for each of these typical days of the year are assumed to be identical to all other days of this type. In addition, the load curves for typical days have to be divided into representative time ranges (e.g. night range from 24.00 to 6.00) in a way that permits an adequate representation of the peculiarities of the analysed system, such as shift times or peak hours. A constant demand in each time range is assumed. In order to calculate the annual energy demand, the number of days of the different types in one year has to be multiplied by the demand in the different time ranges of the corresponding days.

Assumptions also had to be made with regard to the efficiencies of energetic units, because the efficiency of an energetic unit normally depends on it's actual utilisation intensity. Furthermore, most energetic units can only be operated if minimum load requirements are fulfilled. To model these non-linear interdependencies between the input and output of energy conversion technologies it has been assumed that the efficiencies of these energy conversion units are fixed in certain utilisation ranges in order to avoid a non-linear optimisation problem. To realise this integer variables have been used to represent the increase of the efficiency in certain steps.

Finally, some assumptions on the target function of the optimisation problem had to be made. Due to reasons of complexity reduction, it was decided to minimise all decision-relevant costs within the whole supply system over the time horizon as a one-dimensional optimisation function. Reducing the planning problem in this way has the disadvantage that the model identifies possible benefits of inter-company energy supply concepts by negating individual objectives and strategies of different companies involved in the network. But in discussions with energy experts of different companies it became obvious that the minimisation of the costs for energy supply is the absolutely dominating strategy on the company level. On the other side a problem has been identified in the fact that the optimum found for the whole network does not have to be optimal for each individual company involved (see also Sect. 5).

3.3 Model description

Let the energy systems of the companies involved consist of a set $U = \{1, \ldots, m\}$ of existing and possible future energy conversion units. Each of these energetic units $u \in U$ has at least one technical process $p \in P$ ($P = \{1, \ldots, n\}$) that stands for a possible mode of operation. The processes represent the transformation of energy input into energy output. Therefore, let $f \in F$ ($F = \{1, \ldots, r\}$) denote the different energy carriers. The model uses a discrete-time approximation in which the time horizon is broken down into periods $t \in T$ ($T = \{0, \ldots, w\}$). Each period $t \in T$ is divided into time intervals $seas \in S$, with the help of which the load curves of typical days are represented. Each time interval represents one time range of a typical day (e. g. the time range from 8.00 to 10.00 at a summer working day) for all days of this type. Therefore, the modelled energy demand in a time interval $seas \in S$ of a period $t \in T$ corresponds to the average energy demand in this time interval of this period multiplied by the number of days of the corresponding type. This exogenously given energy demand $D_{f,t,seas}$ is the driving force of the model, which has to be fulfilled for each energy carrier $f \in F$ in each time interval $seas \in S$ of each period $t \in T$. The discrete variables of this optimisation model are the numbers $newZ_{u,t}$ of energy conversion units $u \in U$ installed in $t \in T$. With the help of the parameter $rest_{u,t,t'}$ the numbers of energy conversion units $u \in U$ installed in $t \in T$ and still available in the period $t' \geq t$ are assigned to $Z_{u,t'}$. Furthermore, due to the fact that a pure investment planning would not be adequate, the process activity levels $PL_{p,t,seas}$ for each process $p \in P$, each period $t \in T$ and each time interval $seas \in S$, the flow levels $Xi_{p,f,t,seas}$ of energy carriers $f \in F$ which are an input of the process $p \in P$, and the flow levels $Xo_{p,f,t,seas}$ of energy carriers $f \in F$ which are an output of the process $p \in P$ are non-negative decision variables of the model. Environmental constraints, e.g. upper bounds $E\max_{k,t}$, can be set for different pollutants or greenhouse gases $k \in K$ in each period $t \in T$. The planning model, which represents a mixed integer programming model with a lot of technological and economical issues modelled on a very detailed level, can be outlined as follows (for more information see the Appendix):

Minimise

$$\sum_{t \in T} \alpha_t \cdot \tag{1}$$
$$\left(\sum_{f \in F} \left[\sum_{seas \in S} \left[\sum_{p \in P} (Xi_{p,f,t,seas} \cdot Cvar_{p,f,t,seas} + Xo_{p,f,t,seas} \cdot Cvar_{p,f,t,seas}) \right] \right] + \sum_{seas \in S} \sum_{p \in P} PL_{p,t,seas} \cdot Cvp_{p,t,seas} + \sum_{u \in U} [Z_{u,t} \cdot Capn_u \cdot (Cfix_{u,t} + Cinv_{u,t})] \right)$$

Subject to

$$Cpo_{u,t} + Z_{u,t} \cdot Capn_u \geq \sum_{p_u} \left(\frac{PL_{p_u,t,seas}}{h_{seas}} \right) \qquad \forall seas \in S; \forall u \in U; \forall t \in T \tag{2}$$

$$\sum_{t=0}^{t'} rest_{u,t,t'} \cdot newZ_{u,t} = Z_{u,t'} \qquad \forall u \in U; \forall t' \in T \tag{3}$$

$$Xi_{p,f,t,seas} = \frac{in_{f,p} \cdot PL_{p,t,seas}}{\eta_{p,t}} \qquad \forall f \in F; \forall p \in P; \forall seas \in S; \forall t \in T \tag{4}$$

$$PL_{p,t,seas} \cdot out_{f,p} = Xo_{p,f,t,seas} \qquad \forall f \in F; \forall p \in P; \forall seas \in S; \forall t \in T \tag{5}$$

$$\sum_{p_{out}} Xo_{p_{out},f,t,seas} \geq D_{f,t,seas} \qquad \forall f \in F; \forall seas \in S; \forall t \in T \tag{6}$$

$$\sum_{seas \in S} \sum_{p \in P} PL_{p,t,seas} \cdot E_{p,k} \leq E \max_{k,t} \qquad \forall k \in K; \forall t \in T \tag{7}$$

$$\sum_{p_u} PL_{p_u,t,seas} \leq ml_{u,t,seas} \qquad \forall u \in U; \forall t \in T; \forall seas \in S \tag{8}$$

$$Cpo_{u,t} + Z_{u,t} \cdot Capn_u \leq capm_{u,t} \qquad \forall u \in U; \forall t \in T \tag{9}$$

$$\frac{\sum\limits_{p_u} PL_{p_u,t,seas}}{h_{seas}} + NT_{u,t,seas} \geq (Cpo_{u,t} + Z_{u,t} \cdot Capn_u) \cdot mika_u \quad \forall u \in U; \forall t \in T; \forall seas \in S \quad (10)$$

$$NG_{u,t,seas} \geq \frac{NT_{u,t,seas}}{ml_{u,t,seas}} \quad \forall u \in U; \forall t \in T; \forall seas \in S \quad (11)$$

$$LT_{u,t,seas} \geq \frac{\sum\limits_{p_u} PL_{p_u,t,seas}}{ml_{u,t,seas}} \quad \forall u \in U; \forall t \in T; \forall seas \in S \quad (12)$$

$$LT_{u,t,seas} = 1 - NG_{u,t,seas} \quad \forall u \in U; \forall t \in T; \forall seas \in S \quad (13)$$

$$NT_{u,t,seas} \leq ml_{u,t,seas} \quad \forall u \in U; \forall t \in T; \forall seas \in S \quad (14)$$

$$PL_{p,t,seas} \in \mathbf{R}^+ \quad \forall p \in P; \forall seas \in S; \forall t \in T \quad (15)$$

$$Xi_{p,f,t,seas}, Xo_{p,f,t,seas} \in \mathbf{R}^+ \quad \forall f \in F; \forall p \in P; \forall seas \in S; \forall t \in T \quad (16)$$

$$NT_{u,t,seas} \in \mathbf{R}^+ \quad \forall u \in U; \forall t \in T; \forall seas \in S \quad (17)$$

$$newZ_{u,t}, Z_{u,t} \in \mathbf{Z}^+ \quad \forall u \in U; \forall t \in T \quad (18)$$

$$LT_{u,t,seas}, NG_{u,t,seas} \in \{0,1\} \quad \forall u \in U; \forall t \in T; \forall seas \in S \quad (19)$$

The objective function (1) minimises the discounted decision-relevant cost of energy supply in order to maintain the production in the companies involved. Decision-relevant cost comprises fuel and transport cost related to energy flows, variable cost depending on the process activities in energy conversion units, fixed cost and investments in new generation capacity distributed as annuities. The explicit consideration of time-dependent changes within parameters of existing energetic units, for instance decreasing plant efficiencies or increasing maintenance costs, is facilitated by integrating annually differentiated parameter values. Using annuities the investment related cost items for periods after the time horizon are not included in the total discounted cost function. This way, the rest value of the capacity after the time horizon is taken into account. With constraints (2) it is ensured that enough capacity is available to satisfy the activity level of the modes of operation of the units. Equations (3) determine the number of new energetic units which have been installed during the time horizon and are still available in period t. Constraints (4) link the activity level of a process with energy flows being input into this process taking into account the efficiency and the share of this energy carrier of the whole input of the process. Constraints (5) connect the activity level of a process with the output energy flows of this process. Constraints (6) guarantee that the energy demand of each company involved is satisfied by corresponding energy flows in each time interval of a period. Emissions constraints are set by using constraints (7). With the help of constraints (8) and (9) the sum of the activity levels of all processes belonging to one unit and the capacity installed can be limited to upper levels. To consider that some energetic units can only be operated if minimum load requirements are fulfilled, constraints (10) to (14) have been implemented. Constraints (10) guarantee that each unit is running at least with its minimum load requirement. Due to the fact that the capacity installed is an optimisation variable, a new continuous variable NT is needed, in order to model the case that capacity is installed but not used in this time interval. In addition, the binary variable NG, taking on a value of one if the variable NT has a value greater than zero (11), and the binary variable LT, taking on a value of one if the sum of the activity levels of the processes within one unit is greater than zero (12), are needed. Constraints (13) in conjunction with the zero-one restrictions on the (indicator) variables NG and LT (19) impose mutual exclusivity. Constraints (14) limit NT to the upper limits also used in constraints (8).

In addition to these balance and capacity restrictions, there are specific features to model technical restrictions for certain technologies, e.g. to limit the share of technologies with fluctuating power supply (such as wind turbines or photovoltaic installations) or to enforce a minimum share of capacities suitable for frequency and voltage regulation purposes. The model allows to exclude the operation of energetic units from specific time intervals in order to facilitate the consideration of different system characteristics. For instance, it has to be ensured that energetic units using solar radiation can not be used during the night. In addition, it has to be taken into account that not all energetic units can change their power output rapidly, necessitating the integration of the rate of change of power output into the model. It has also been modelled that an investment option can only be realised if the investment amortises within the company-specific payoff period. In analogy to the modelling of minimum load requirements binary variables have been used to represent the rising of the efficiency in certain steps.

$$\frac{\sum\limits_{p_u} PL_{p_u,t,seas}}{h_{seas}} \leq \sum_{u' \in U_{UW} / \{u\}} WRES_{u',t,seas} + LAbw \qquad \forall u \in U_{UW}; \forall t \in T; \forall seas \in S \quad (20)$$

$$WRES_{u,t,seas} = (Cpo_{u,t} + Z_{u,t} \cdot Capn_u) - \frac{\sum\limits_{p_u} PL_{p_u,t,seas}}{h_{seas}} - HV_{u,t,seas} \qquad \forall u \in U_{UW}; \forall t \in T; \forall seas \in S \quad (21)$$

$$Cpo_{u,t} + Z_{u,t} \cdot Capn_u = HV_{u,t,seas} + HP_{u,t,seas} \qquad \forall u \in U_{UW}; \forall t \in T; \forall seas \in S \quad (22)$$

$$HV_{u,t,seas} \leq capm_{u,t} \qquad \forall u \in U_{UW}; \forall t \in T; \forall seas \in S \quad (23)$$

$$HP_{u,t,seas} \leq capm_{u,t} \qquad \forall u \in U_{UW}; \forall t \in T; \forall seas \in S \quad (24)$$

$$1 + \frac{\sum\limits_{p_u} PL_{p_u,t,seas}}{ml_{u,t,seas}} > PLG_{u,t,seas} \qquad \forall u \in U_{UW}; \forall t \in T; \forall seas \in S \quad (25)$$

$$HVG_{u,t,seas} \geq \frac{HV_{u,t,seas}}{capm_{u,t}} \qquad \forall u \in U_{UW}; \forall t \in T; \forall seas \in S \quad (26)$$

$$HPG_{u,t,seas} \geq \frac{HP_{u,t,seas}}{capm_{u,t}} \qquad \forall u \in U_{UW}; \forall t \in T; \forall seas \in S \quad (27)$$

$$PLG_{u,t,seas} = HPG_{u,t,seas} \qquad \forall u \in U_{UW}; \forall t \in T; \forall seas \in S \quad (28)$$

$$HVG_{u,t,seas} = 1 - PLG_{u,t,seas} \qquad \forall u \in U_{UW}; \forall t \in T; \forall seas \in S \quad (29)$$

$$WRES_{u,t,seas}, HP_{u,t,seas}, HV_{u,t,seas} \in \mathrm{R}^+ \qquad \forall u \in U_{UW}; \forall t \in T; \forall seas \in S \quad (30)$$

$$PLG_{u,t,seas}, HVG_{u,t,seas}, HPG_{u,t,seas} \in \{0,1\} \qquad \forall u \in U_{UW}; \forall t \in T; \forall seas \in S \quad (31)$$

Constraints (20) ensure that the breakdown of a boiler can be compensated by the available warm reserves and the maximum load rejection that can be handled (a simplified formulation can be found in [16]). With the help of equations (21) the available warm reserve of an energetic unit is determined. Due to the fact that there is no warm reserve if the unit is not used in this time interval the continuous variable HV is needed. If the unit is not used HV is assigned the value of the capacity installed, if the unit is used HV is set to zero. In order to model this assignment to the optimisation variable HV, one more continuous variable HP is needed. Equations (22) guarantee that the installed capacity equals the sum of HV and HP. Furthermore the binary variable HVG, taking on a value of one if the variable HV has a value greater than zero (26), and the binary variable HPG, taking on a value of one if the variable HP has a value greater than zero (27), are needed. Constraints (29) in conjunction with the zero-one restrictions on the variables HVG and HPG (31) impose mutual exclusivity. In addition, it has to be ensured that HP equals the capacity installed only if the unit is used, what can be guaranteed by equations (28) and constraints (25), with the help of which the binary variable PLG is set to zero if the unit is not used. Constraints (23) and (24) limit HV and HP to the upper limits also used in constraints (9).

The model is programmed in the algebraic modelling language GAMS [7]; a data management system based on Microsoft Access is used for storing and maintaining input data [20]. This data management system contains all relevant data on the structure of the energy systems as well as on the characteristics of existing and new energy conversion technologies in the companies involved. The data management system consists of two databases; the first one contains the complete functionality of the system – i. e. forms, queries and code – while the actual data is stored in the second one. In addition to being able to work on different "data"-databases with one "system"-database, this structure facilitates the maintenance and upgrading of the system as existing data-databases can still be used even if enhancements and modifications to the whole system necessitate the implementation of a new version of the system-database. For the analysis of the model results, a module which automates the generation of aggregated tables and charts has been implemented using Microsoft Excel.

4 Application of the model to analyse inter-company energy supply concepts

4.1 The case study

In the project "Analysis of the Inter-Company Energy Supply Concept Karlsruher Rheinhafen", supported by the German Federal Ministry of Education and Research (BMBF), the PERSEUS-ICC model has been applied to evaluate the economic and ecological effects of inter-company energy supply concepts between five energy-intensive production companies (in addition, the municipal utility of Karlsruhe is integrated in this case study). It should be mentioned that the results of this project can only be briefly outlined due to the confidential nature of the data. The companies involved are key-players in the energy sector of Karlsruhe. This can be illustrated by the fact that their electricity demand is equivalent to the demand of about 500.000 average German households. All companies are located in the area near the Rhine harbour of Karlsruhe with a maximum distance of about 4 km between them. Since investments as well as heat losses increase with the length of pipelines needed to link the companies, this close spatial proximity of the sites in question is an important prerequisite.

In a first step, the energy systems of these companies had to be described using technical, economic and environmental parameters. To determine these parameters, an engineering analysis of the real energetic processes in these companies was required to gain insights into the process characteristics. Using this engineering approach, existing configurations were assessed using various technical and environmental parameters showing, e.g. that for most energetic units it is sufficient to model a minimum load requirement and one efficiency if this requirement is fulfilled. In addition, the economic data of the existing units had to be determined, e.g. by interpreting information provided by the cost accounting systems. The analysis shows that up to now, the demand for electricity and heat in these companies has been met partly by auto-production plants and partly by delivery from utilities. Due to the age of the existing boilers and turbines, the companies are faced with the need to redesign their energy supply systems.

To obtain technical data on possible investment options, simple approaches like the utilisation of data base systems with generic data sets or the comparison with already realised projects with similar technical characteristics can't be used because real inter-company projects necessitate significant process modifications or purpose-built units. Therefore, the flow-sheeting simulation program "Aspen Plus" [2], a mathematical simulation program for the calculation of interdependent mass and energy balances of complex process engineering systems, has been applied to develop an optimal adaptation of new energetic units to the specific circumstances of the case at hand (see [16], other applications of flow-sheeting simulation programs can be found in [15]). With the detailed technical information regarding possible future options provided by this flow-sheeting simulation program, investments in each major piece of equipment have been estimated using cost estimating charts combined with current price quotations. Furthermore, investments in components besides the processing equipment have been approximated by using individual plant component factors (see [6] with regard to the data and [17] with regard to the methodologies). Finally, fixed and variable costs have been calculated based on the determined mass and energy balances.

4.2 Inter-company energy supply concepts

In the framework of the case study "Karlsruher Rheinhafen", the following possible inter-company energy supply concepts have been analysed:

- different natural gas fired Combined Cycle Power Plants ($CCPP$) for electricity and heat production,
- joint purchase on spot and forward markets,
- diverting unused energy flows from one company into another company to meet the demand for low temperature heat,
- the use of geothermal energy for the production of electricity and process steam and
- the installation of a combined heat and power plant fired with biomass.

The different combined cycle power plants have been designed, on the one hand, to meet the steam demand of one specific company and the electricity demand of different partners, and on the other hand, to meet the electricity and heat / steam demand of all companies involved. Table 1 shows some of the technical and economic parameters of the different natural gas fired combined cycle power plants, which have been calculated using Aspen Plus and are input into PERSEUS-ICC. Whereas $CCPP3$ is used to fulfil the electricity and steam demand of all companies involved, $CCPP1$ and $CCPP2$ have been designed with less capacity in order to satisfy the total thermal demand of one of two companies with considerable auto-production (in the following these companies are called company A and B).

On the demand side, the future development of energy demand in the different companies was estimated to stay nearly constant at the level of the year 2000. In addition, an adequate level of aggregation of the daily load curves had to be found. Therefore, data on the energy demand of the companies at every hour in the year 2000 was analysed and clustered. In this context different possibilities of characterising the energy demand of the companies using different representative days and time intervals were considered. Due to the fact that in most of the companies involved the load curves hardly vary between working days and the weekend, the characterisation of the power and heat demand of the companies by load curves for one typical day was seen to be sufficient. This typical daily load has been approximated using 5 different time intervals in order to represent base load hours (8.00 p.m. to 8.00 a.m.) and peak load hours (8.00 a.m. to 8.00 p.m.) as well as the different shift times in the companies.

Finally, some energy-economic framework assumptions had to be made including projections of primary energy carrier prices and legal framework conditions. It was assumed that the price for natural gas will increase from 1.34 Eurocent/kWh in 2004, to 1.43 Eurocent/kWh in 2008 and 1.52 Eurocent/kWh in 2012 to 1.61 Eurocent/kWh in the period starting 2016. In addition, the new German CHP law

Table 1. Technical and economic parameters of different combined cycle power plants

Plant	Satisfaction of electricity demand of which firms	Satisfaction of thermal demand of which firms	Capacity (MWel)	Total efficiency (%)	Electrical efficiency (%)	Invest ment (€/kW)	Fixed cost (€/kW)
$CCPP\,1$	firm A fully other firms partly	firm A	102	81	38	910	28
$CCPP\,2$	firm B fully other firms partlly	firm B	109	83	33	965	28
$CCPP\,3$	all firms fully small surplus of feed-in electricity		164	85	33	875	22

has been represented in detail. This law regulates the feed-in tariffs for electricity deliveries from certain types of CHP units into the public grid. On top of the market price for these deliveries, the operators of the CHP units are entitled to obtain supplementary payments for each kWh of electricity produced in cogeneration (most CHP units can also be used to produce only electricity (condensation mode)). In accordance with the German CHP law, it was assumed that the feed-in supplement for electricity from modernised combined heat and power plants will decline from 1.74 Eurocent/kWh in 2002 to 1.59 Eurocent/kWh in 2010, whereas the support for electricity produced in old CHP plants will decline from 1.53 Eurocent/kWh in 2002 to 0.97 Eurocent/kWh in 2006 and then be phased out. Finally, the development of the market price for electricity was assumed to increase from 2.64 Eurocent/kWh in 2004 to 2.76 Eurocent/kWh in 2008, 2.88 Eurocent/kWh in 2012 and about 3 Eurocent/kWh in 2016. This price can be obtained for electricity exports into the public grid by the operators of CHP units; on the other hand, this electricity price plus a transmission fee has to be paid for electricity imports from the net.

It should be mentioned that the whole modelling procedure, including estimates for the energy-economic framework assumptions, required a multi-disciplinary analysis in which specialists in different fields contributed to data provision. Close cooperation with the experts in the companies involved was also an important prerequisite.

Currently, the data management system for the inter-company energy supply network "Karlsruher Rheinhafen" consists of about 150 technologies. Due to the fact that most of the constraints of the mathematical model have to be evaluated for each time interval of each period of the time horizon, the resulting model is very large. Considering five time periods (4 year-steps from 2000 to 2020), the model consists of approximately 40,000 continuous variables, 3,000 integer variables and 32,000 equations with 140,000 non-zero elements. The planning model has been implemented on a Pentium III PC with 128 MB RAM and 500 MHz clock pulse.

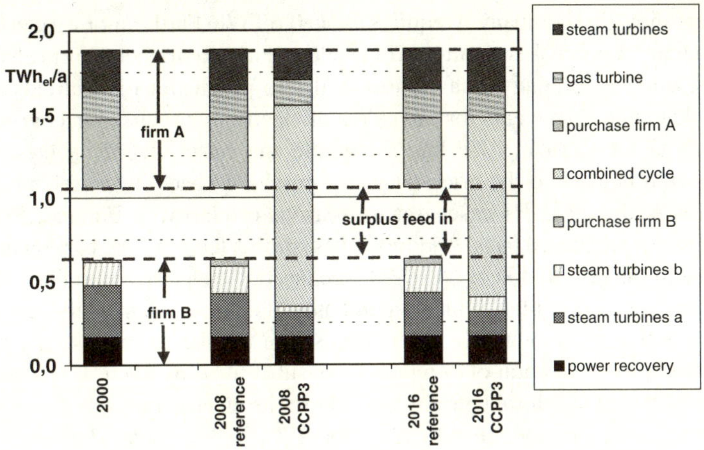

Fig. 2. Operation of energetic units in the companies A and B

To solve the mixed-integer linear optimisation problems, the commercial solver CPLEX 7.0 was used. Depending on the chosen scenario, the identification of the exact solutions required extensive computational times (up to eight hours), but due to the fact that an interactive use of this model for long-term planning is not required no fast heuristic methods have been used until now.

In order to validate the model the energetic situation of the companies involved in the year 2000 has been optimised, showing only minor differences (under 1%) between the model results and the real energy flows and the corresponding costs. In a next step the model has been used to analyse future energy supply concepts. If inter-company energy supply concepts are excluded (reference case), the model results indicate that the companies will continue to meet their electricity and heat demand partly by auto-production and partly by purchase from utilities. Due to the pay-back-times required there will only be smaller investments, e.g. for peak load boilers. As the new CHP law also introduces a feed-in supplement for electricity from old CHP plants (but with lower price supplements compared to modern CHP plants), all electricity produced in cogeneration is fed in and at least partly purchased back leading to a different utilisation of existing units in the different companies compared to the actual utilisation. If investments in inter-company energy supply concepts can also be chosen, the model results show that the installation of the combined cycle power plant producing heat and power for all the partners involved ($CCPP3$ in Table 1) is the minimum cost option. Compared to the reference case, the discounted decision-relevant cost of energy supply needed to maintain production in the companies can be reduced by about 10 percent. In Figure 2, the model results are shown for the generation and supply of electricity (not of the different kinds of heat) in the periods starting in 2000, 2008 and 2016 for the two companies with considerable auto-production. The bars on the left represent the reference case and the bars on the right correspond to the inter-company energy concept $CCPP3$. The upper parts of the bars for the reference case indicate that in company A about 0.8 TWh electricity are needed per year, whereas the lower parts of these bars show that company B requires about 0.6 TWh. Furthermore, it can be seen that company A purchases about half of its electricity requirements from electricity markets, whereas company B has almost total auto-production of electricity. Due to the supplementary payments for each kWh of electricity produced in cogeneration according to the German CHP law, the whole amount of electricity produced in cogeneration is fed into the grid and partly purchased back in the reference case as well as in the $CCPP3$ case. But for reasons of clarity, in Figure 2, only the electricity not purchased back is addressed as surplus feed-in. The bars on the right show that, if the installation of $CCPP3$ is realised, the electricity produced by this combined cycle power plant replaces purchases of company A to a large extent and auto-production of company B by the two steam turbines, whereas only a small part of the auto-production of company A is substituted by $CCPP3$. In addition about 0.4 TWh of surplus electricity are sold to the electricity markets per year.

Furthermore, the model results show that $CCPP3$ is operated almost in continuous load (base load) to produce electricity and heat. Therefore, the long-term production plan for this inter-company energy supply concept has only small variations over the time of day, leading to a high utilisation of the new energetic unit. Remaining peaks are met by existing boilers and turbines. In addition, the model results illustrate that the use of the combined cycle power plant, compared to the autonomous development without inter-company cooperation to link energy flows, results in to a reduction of the network's contribution to environmental problems. For instance, the Global Warming Potential of the emitted greenhouse gases decreases by roughly 21 %. This emission reduction would be in line with the obligations of the Kyoto Protocol, in which Germany (the European Community) committed itself to reducing emissions of greenhouse gases by 21 % (8%) during the period 2008-2012 in comparison with 1990. Besides this reduction of CO_2, the inter-company energy supply concept would also result in considerable improvements in other ecological problem areas such as the protection of natural resources or the reduction of pollutants like SO_2.

If Option $CCPP3$ is excluded from the optimisation, the second best solution identified is the separate installation of $CCPP1$ and $CCPP2$. In the first years of the time horizon the differences in the discounted decision-relevant cost of energy supply between the installation of $CCPP3$ and the installation of $CCPP1$ and $CCPP2$ are about 3 % (these savings might seem rather small, but it has to be taken into account that the discounted decision-relevant costs amount to several millions of Euros). In the years after 2010, when the feed-in supplement for electricity from modernised combined heat and power plants will be phased out, the differences are much larger. This is due to the fact that the installation of $CCPP1$ and $CCPP2$ can only compensate the economies of scale of $CCPP3$ by feeding much more electricity into the net and obtaining the supplementary payments for each of these kWh.

4.3 Sensitivity analysis in different scenarios

In a next step a sensitivity analysis has been performed as a critical review of the obtained results. With the help of this sensitivity analysis, it has been analysed by

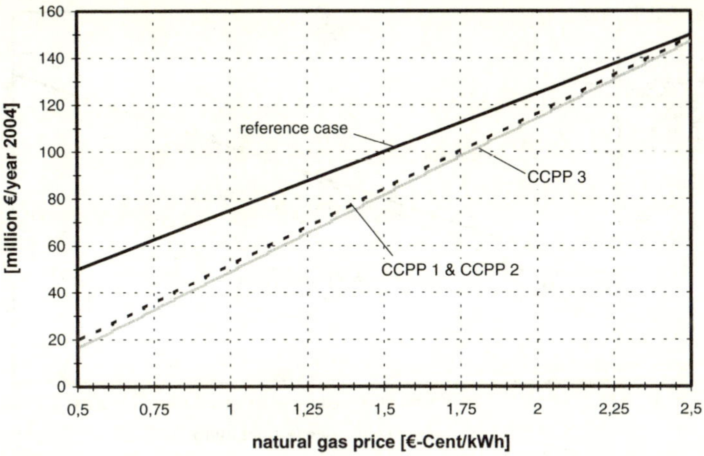

Fig. 3. Sensitivity analysis in the reference scenario

how much the gas price can increase without altering the results obtained with the initial set of gas prices, all other parameters being kept constant. Therefore, the gas prices in the different years of the time horizon have been altered by certain percentages compared to the originally assumed prices in the corresponding years. In order to be able to represent the results of this sensitivity analysis graphically the following explanations focus on the year 2004. The sensitivity analysis shows that $CCPP3$ remains the optimal solution as long as the gas prices do not double. This would correspond to a gas price of about 2.5 Eurocent /kWh in the year 2004 (see Fig. 3), indicating that the original solution is quite stable (in the year 2001 gas prices ranged from 1.17 to 1.52 Eurocent / kWh in Germany). For reasons of clearity, only the costs for energy supply for $CCPP3$, the second best solution comprising the separate installation of $CCPP1$ and $CCPP2$, and the reference case are shown in Figure 3.

Due to the fact that a decision for a new energy supply concept would affect the energy systems of the companies involved for several decades, possible repercussions of changing framework conditions have to be taken into account. For this reason, uncertainties have been considered by evaluating different scenarios. In a first scenario, the investments in gas turbines have been raised by 10 $\%$ due to the increasing demand for gas turbines. The analysis shows that with current gas prices, $CCPP3$ would still be the optimal solution. But under these conditions the sensitivity analysis indicates that $CCPP3$ would only lead to the lowest energy supply costs as long as the gas price remains below 2 Eurocent /kWh in the year 2004 (see Fig. 4). Higher prices for natural gas would now result in the reference case being the minimum cost solution.

In another scenario electricity demand has been increased by 9.3 percent, steam demand by 7.4 percent in the periods from the year 2012 due to the assumption that a new paper machine would be needed in one of the production companies involved. In this scenario, the model results show that the installation of $CCPP3$ is still the most cost-efficient investment in order to satisfy the energy demand of

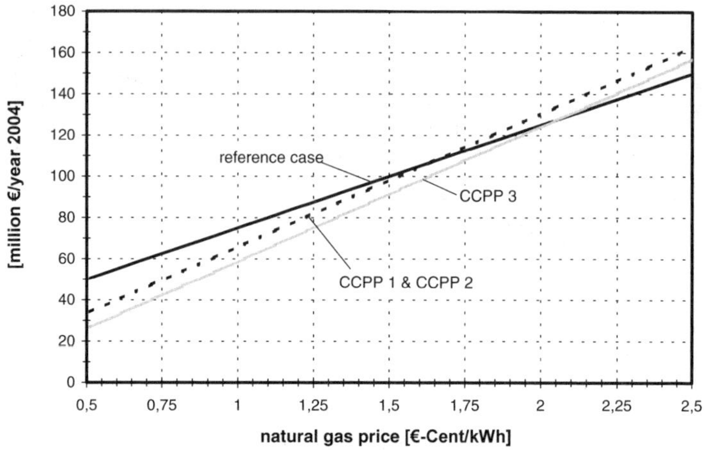

Fig. 4. Sensitivity analysis in the scenario with increased investments in gas turbines

the companies involved. In a third scenario it has been assumed that a greenhouse gas emission trading scheme within the European Community will be implemented in order to meet the obligations of the European Union under the Kyoto Protocol. In this scenario an equilibrium price of about 20 € per ton of carbon dioxide equivalent [5] has been assumed, increasing the economic profitability of the analysed inter-company energy supply concept $CCPP3$.

In a final scenario, the support for electricity produced in combined heat and power plants and fed into the grid (German Combined Heat and Power Act) has been ignored, leading to a different optimal solution. Under this assumption, the model shows smaller investments realised independently by the individual companies and the installation of $CCPP2$, which is used to meet the total electricity and steam demand of company B, to be the minimum cost configuration. Here, the relatively low market prices for electricity and the investments in steam pipes to link the heat flows between the companies can not be overcompensated by supplementary payments on the large amount of electricity produced in cogeneration, which results from bundling the energy demand of different companies. Compared to the optimal solution in this scenario, the installation of the inter-company energy supply concept $CCPP3$ would result in an increase of the discounted decision-relevant cost of about 1.2 percent, demonstrating the significance of the German CHP law for the profitability of inter-company energy supply concepts.

5 Conclusions and recommendations for future research

To analyse the economic and ecological effects of inter-company energy supply concepts, a pure investment planning is not sufficient. This is due to the fact that future payments connected with a new inter-company energy supply concept strongly depend on the utilisation of these energetic units. In turn, this utilisation depends on the energetic units available in the companies of the network. Furthermore, it is necessary to represent realistically the existing energetic units in the companies as well as possible investment options which requires that various supply side and demand side details are taken into account.

In this paper, we have presented an approach for long-term production and investment planning of inter-company energy supply concepts considering the new conditions of liberalised energy markets. A mixed-integer linear optimisation model has been developed which represents all the necessary technical, economic and ecological parameters of the energy systems of the companies and the inter-company energy supply concepts. This model has been used to identify economic and ecological implications of inter-company energy supply concepts in a network of five industrial companies located in the area near the Rhine harbour of Karlsruhe. Under the chosen framework assumptions of this specific case, the model results illustrate that the installation of a custom-designed combined cycle power plant would be the most promising solution to meet the energy demand needed to maintain production in the companies involved. This strategy would result in considerable cost savings and environmental benefits compared to the optimal strategies realised independently by the individual companies including sub-optimal system performance.

In addition a sensitivity analysis and a scenario analysis have shown that with the exception assuming an improbably high increase of the gas price, the installation of this combined cycle power plant is the minimum cost solution as long as there are supplementary payments on each kWh electricity produced in cogeneration and fed into the public grid in accordance to the new German CHP law. Furthermore, the results show that this law will lead to a new utilisation of existing CHP units in industrial companies because it also introduces a feed-in supplement for electricity from old CHP plants. More generally the case study indicates that a simultaneous planning of investments and long-term production on the basis of a mixed integer programming model can generate solutions which are of relevance for practical applications if the most important requirements resulting from technological and economic issues are modelled on an adequate level.

In order to be able to consider uncertainties in inter-company energy supply concepts (e.g. bankruptcy of one of the partners, variation of the feed-in supplement for electricity according to the German CHP law), this deterministic optimisation approach is to be enhanced by multi-periodic stochastic programming methods [29]. With the help of these methods, it is possible to determine hedging strategies against the occurrence of events changing the assumed energy-economic framework assumptions for the inter-company concept. By assigning probabilities to possible scenarios, a strategy can be elaborated which is not necessarily "optimal" for any specific case, but which can be adjusted to different outcomes of the political discussion without sacrificing competitiveness. With the help of this procedure robust solutions are to be identified in order to accommodate future internal and external uncertainties. In this context it seems to be of great importance to consider changing parameters of the German CHP law over time due to the dominating role of the supplementary payments on each kWh electricity produced in cogeneration in the cost effective solution.

Another weakness of the developed model is the central, one-dimensional optimisation which assumes the same target function for all participants of the inter-company energy supply concept. The model identifies possible economic and environmental benefits of inter-company energy supply concepts by determining the global optimum for the whole network instead of the optimum for each company. For this reason, the decisions reached are not necessarily the same as the decisions from the point of view of individual players. For example, the model will identify the installation of a new combined heat and power plant as the optimal solution, even if this leads to higher costs for one partner company compared to its reference case. In order to realise the overall benefits, the companies placed at a disadvantage would have to be compensated by the other partners. In the specific case of inter-company energy supply concepts, the problem of assigning costs and benefits to the different partners is even more complex due to the fact that most of these concepts produce electricity and process heat/steam in cogeneration. Due to this simultaneous production, it is not possible to assign costs to the joint products on the basis of the respective expenses only. Therefore, it is the subject of further research to tackle this problem of assignment using game theoretic approaches, which have been successfully used for similar problems in other areas [27]. Besides this add-on solution, another future research topic is the transformation of the optimisation model to a multi-agent system as a construct of software agents, representing each partner of the co-operation by a single software agent [32]. As a central idea in the paradigm of multi-agent systems the role concept will be used as the enabler for different players to pursue their individual objectives and strategies, in order to avoid an all-encompassing target function negating alternative goals for different participants of the network.

Appendix

Indices:

(If the same index is required several times in one equation or in the objective function, the index will be marked by a dash.)

t := time period
f := energy carrier
$seas$:= time interval
p := process; standing for one operation mode of a unit
k := emission
u := unit; a unit is a real physical capacity of a special energy conversion technology,
p_u := process belonging to the unit u
p_{out} := process producing final energy to satisfy demand

Sets of indices

S := time intervals
T := time periods
P := processes
U := units

U_{UW} := units for which a protection against a short-term breakdown has to be taken into account
F := energy carriers
K := emissions

Parameters

α_t := discount factor in period t
$D_{f,t,seas}$:= demand for energy carrier f at time interval seas in period t
$in_{f,p}$:= input share of energy carrier f in process p
$out_{f,p}$:= output share of energy carrier f in process p
$out_{f,p}$:= efficiency of process p in period t
$E_{p,k}$:= emission factor of pollutant/greenhouse gas k for process p
$Emax_{k,t}$:= upper limit of total emissions of pollutant / greenhouse gas k in period t
$ml_{u,t,seas}$:= upper limit of the sum of activity levels of all processes belonging to a unit u at time interval seas in period t
$capm_{u,t}$:= upper limit of the capacity u installed in period t
$rest_{u,t,t'}$:= 1 if an unit, with a given lifetime, built in t is still available in t'; 0 else
$mika_u$:= share of the capacity that has to be reached to operate this unit u
$LAbw$:= the maximum rejection of the load which does not damage the production system
$Cvar_{p,f,t,seas}$:= specific variable cost for energy carrier f flowing into / out of process p at time interval $seas$ in period t
$Cvp_{p,t,seas}$:= specific variable cost of the process p at time interval $seas$ in period t
$Cfix_{u,t}$:= specific fixed cost of unit u in period t
$Cinv_{u,t}$:= specific investments of unit u allocated as annuity to period t
h_{seas} := hours of the time interval $seas$ (hours of the time range of the corresponding typical day multiplied by the number of these days)
$Capn_u$:= capacity of unit u
$Cpo_{u,t}$:= in period t still remaining capacity of an unit u installed before the time horizon

Non-negative optimisation variables

$PL_{p,t,seas}$:= activity level of process p at time interval $seas$ in period t
$Xi_{p,f,t,seas}$:= flow of energy carrier f into a process p at time interval seas in period t
$Xo_{p,f,t,seas}$:= flow of energy carrier f out of a process p at time interval seas in period t
$WRES_{u,t,seas}$:= back-up capacity of unit u at time interval $seas$ in period t (auxiliary variable, $WRES_{u,t,seas}$ is a result of $Z_{u,t}$, $HV_{u,t,seas}$ and $\sum_{pu} PL_{pu,t,seas}$)

Discrete optimisation variables

$newZ_{u,t}$:= numbers of energetic units u installed in period t
$Z_{u,t}$:= numbers of energetic units u, installed during the time horizon; still available in period t (auxiliary variable, $Z_{u,t}$ is a result of $newZ_{u,t}$)

Auxiliary variables to model minimum load requirements

Non-negative optimisation variables

$NT_{u,t,seas}$:= variable to model the case that an existing u is not used at the time interval $seas$ in period t

Binary optimisation variables

$NG_{u,t,seas}$:= variable taking on a value of one if the variable $NT_{u,t,seas}$ has a value greater than zero and a value of zero if $NT_{u,t,seas}$=0

$LT_{u,t,seas}$:= variable taking on a value of one if the sum of the activity levels of the processes within the unit u is greater than zero at time interval $seas$ in period t, else the value is zero

Auxiliary variables to model the protection against a short-term breakdown

Positive optimisation variables

$HV_{u,t,seas}$:= variable taking on the value of the capacity installed if this unit u is not used at time interval $seas$ in period t

$HP_{u,t,seas}$:= variable taking on the value of the capacity installed if this unit u is used at time interval seas in period t

Binary optimisation variables

$PLG_{u,t,seas}$:= variable taking on a value of zero if the unit u is not used at time interval $seas$ in period t

$HVG_{u,t,seas}$:= variable taking on a value of one if the variable $HV_{u,t,seas}$ has a value greater than zero

$HPG_{u,t,seas}$:= variable taking on a value of one if the variable $HP_{u,t,seas}$ has a value greater than zero

References

Anderson D (1972) Models for determining least-cost investments in electricity supply. Bell Journal of Economics: 3: 267–299

Aspen (1994) Aspen plus™getting started. Harvard University Press, Cambridge, MA

Ayres RU (1999) The second law, the fourth law, recycling and limits to growth. Ecological Economics 29: 473–483

Bardouille P, Koubsky J (2000) Incorporating sustainable development considerations into the energy sector decision making: Malmö Flintränen district heating facility case study. Energy Policy 28: 689–711

Blok K, de Jager D, Hendriks C (2001) Economic evaluation of sectoral emission reduction objectives for climate change. Study on behalf of the European Commission, http://europa.eu.int/comm/ environment/enveco/climate_change/sectoral_objectives.htm

Boeddicker D (1997) Thermodynamische und energiewirtschaftliche Bewertung eines Kombikraftswerks mit integrierter Kohlevergasung und CO_2-Rezyklierung. VDI-Verlag, Düsseldorf

Brooke A, Kendrick D, Meeraus A (1988) GAMS – A user's guide. Scientific Press, Redwood City

Caramanis MC (1983) Electricity generation expansion planning in the eighties: requirements and available analysis tools, energy models and studies, pp 541–562. North-Holland, Amsterdam New York Oxford

Charnes A, Cooper W, Miller H (1959) Application of linear programming to financial budgeting and the costing of funds. The Journal of Business 32: 20–46

Cote RP, Cohen-Rosenthal E (1998) Designing eco-industrial parks: a synthesis of some experiences. Journal of Cleaner Production 6: 181–188

Dantzig GB, Wolfe P (1960) Decomposition principle for linear programme. Operation Research 48: 101–111

Erkman S (1997) Industrial ecology: an historical view. Journal of Cleaner Production 5 (1–2): 1–10

Forum für Energiemodelle und Energiewirtschaftliche Systemanalysen (1999) Energiemodelle zum Klimaschutz in Deutschland: Strukturelle und gesamtwirtschaftliche Auswirkungen aus nationaler Perspektive. Physica-Verlag, Heidelberg

Fichtner W, Frank M, Rentz O (2000) Entwicklung eines Instrumentariums zur Analyse ökonomischer und ökologischer Potentiale bei der Vernetzung von Energieflüssen. In: Fleischmann B, Lasch R, Derigas U, Domschke W, Rieder U (eds) Gesellschaft für Operations Research e.V. (GOR). Operations Research Proceedings 2000, pp 439–444. Springer, Berlin Heidelberg New York

Futterer E, Munsch M (1990) Flow-Sheeting-Programme für die Prozesssimulation. Chem-Ing-Tech 62 (1): 9–16

Frank M, Göbelt M, Fichtner W, Rentz O (2001) Optimierende Energie- und Stoffflussmodelle für die strategische Planung von betriebsübergreifenden Energieversorgungssystemen im liberalisierten Energiemarkt. In: VDI-GET (eds) Betriebliches Energiemanagement 2001. VDI-Verlag, Düsseldorf

Garrett D (1989) Chemical engineering economics. Van Nostrand Reinhold, New York

Gately (1970) Investment planning for the electric power industry: an integer programming approach. Research Report 7035, Dept. of Economics, University of Western Ontario

Georgescu-Roegen N (1971) The entropy law and the economic process. Harvard University Press, Cambridge, MA

Göbelt M, Fichtner W, Wietschel M, Rentz O (2000) Development and application of an optimising energy and material flow model for strategic planning of energy utilities. Conference Proceedings of the 5th Conference of Asian-Pacific Operational Research Societies within IFORS, Singapore

Gonzalez-Monroy L, Cordoba A (2002) Financial costs and environment impact optimization of the energy supply systems. International Journal of Energy Research 26 (1): 27–44

Hafkamp WA (1984) Economic-environmental modelling in a national-regional system. North-Holland, Amsterdam New York Oxford

International Atomic Energy Agency et al. (1995) Computer tools for comparative assessment of electricity generation options and strategies. DECADES Project Series Publication, Vienna

Kaluza B, Blecker Th, Bischof Ch (1999) Networks - a cooperative approach to environmental management. University of Klagenfurt, Institute for Economics

Korhonen J, Savolainen I (2001) Cleaner energy production in industrial recycling networks. Eco-Management and Auditing 8 (3): 144–153

Lev B (ed) (1983) Energy models and studies. North-Holland, Amsterdam New York Oxford

Nouweland A, Borm P, Brouwers W, Bruinderink R, Tijs S (1996) A game theoretic approach to problems in telecommunication. Management Science 42 (2): 35–47

Nurminski E, Balabanov T (1983) Decomposition of a large-scale energy model. Large Scale System 4: 295–308

Prekopa A (1995) Stochastic programming. Kluwer, Dordrecht

Song YH (ed) (1999) Modern optimisation techniques in power systems. Kluwer, Dordrecht

Stadtler H, Kilger C (2000) Supply chain management and advanced planning. Springer, Berlin Heidelberg New York

Weinhardt Ch, Zuber P, Göbelt M, Fichtner W, Wietschel M, Rentz O (2000) A competition-model for the electricity sector: from an or approach to a multi-agent system. In: Urban Ch (ed) Workshop 2000 – Agent-based simulation, pp 63–68. SCS Europe, Ghent
Wietschel M (2000) Produktion und Energie: Planung und Steuerung industrieller Energie- und Stoffströme. Habilitationsschrift, TH Karlsruhe. Lang, Frankfurt a. M. et al.
World Commission on Environment and Development (1987) Our common future. Oxford University Press, Oxford

Business case Océ: Reverse logistic network re-design for copiers

Fallstudie Océ: Redesign eines Recyclingnetzwerks für Photokopierer

H.R. Krikke[1], A. van Harten[2], P.C. Schuur[2]

[1] Erasmus University, School of Management Studies, MTI, P.O.Box 1738, NL-3000 DR Rotterdam, The Netherlands (e-mail: H.Krikke@fac.fbk.eur.nl)
[2] University of Twente, Technology and Management, OMST, P.O.Box 217, NL-7500 AE Enschede, The Netherlands

Received: 9 June 1998 / Accepted: 18 November 1998

Abstract. The introduction of extended producer responsibility forces Original Equipment Manufacturers to set up a logistic network for take back, processing and recovery of discarded products. In this paper, we discuss a business case study carried out at Océ, a copier firm in Venlo (NL). It concerns the installment of remanufacturing processes. There is a choice from two locations in Venlo (NL) and one in Prague (Czech Republic), where assignments are subjected to managerial constraints. The study is meant to verify whether the strategic decision of Océ to move remanufacturing activities to the Czech Republic is also economically feasible. We limit ourselves to an optimisation of the HV02-machine network. We follow our general approach, in which we first determine *how* return products are processed (recovery strategy) and subsequently optimise the *reverse logistic network design*. We optimise on total operational costs over all possibilities and also compare three pregiven managerial solutions (=network designs) with a Mixed Integer Linear Programming model. Differences in economic costs appear to be very small, hence installing recovery activities in Prague for the HV02-machine must be well motivated from a strategic point of view. Moreover, we argue that besides cost minimisation, Océ should include performance indicators, such as JIT, reliability, in logistic optimisation to support its quality oriented business strategy. In addition, we discuss aspects regarding specific modelling elements in this case situation, the definition of cost functions, the possibility of optimising the forward and reverse logistic network and the use of LP- versus MILP-models in this kind of situations.

Correspondence to: H.R. Krikke

Zusammenfassung. Die Einführung einer erweiterten Produktverantwortung zwingt die Hersteller von Wirtschaftsgütern dazu, ein logistisches Netzwerk für die Rücknahme, Aufarbeitung und Verwertung von gebrauchten Produkten aufzubauen. In der vorliegenden Arbeit wird über eine Fallstudie bei Océ, einem Hersteller von Kopierern in Venlo (NL), berichtet. Es wird ein Standortplanungsproblem betrachtet, bei dem unter Beachtung verschiedener unternehmerischer Restriktionen eine Entscheidung über den Aufbau von Kapazitäten zur Durchführung von Recyclingaktivitäten an Standorten in Venlo (NL) und Prag (Tschechien) getroffen werden muß. Mit Hilfe der Studie soll geprüft werden, ob die Entscheidung von Océ zur Verlagerung von Recyclingaktivitäten nach Tschechien auch ökonomisch zweckmäßig ist. Hierbei wird die Betrachtung auf die Optimierung des Recyclingnetzwerks für eine bestimmte Produktgruppe (HV02-Geräte) beschränkt. Zunächst wird die Recyclingstrategie festgelegt, aus der sich ergibt, wie die zurückgenommenen Produkte aufgearbeitet und verwertet werden, bevor anschließend eine Optimierung der Konfiguration des logistischen Netzwerks zum Recycling erfolgt. Das Ziel besteht in der Minimierung der gesamten operativen Kosten für alle möglichen Netzwerkkonfigurationen mit Hilfe eines Modells der gemischt-ganzzahligen linearen Programmierung (MILP), wobei zusätzlich drei vom Management vorgegebene Konfigurationen zum Vergleich durchgerechnet werden. Auch das generelle Optimum wird kalkuliert. Es ergibt sich, daß die Kostenunterschiede der verschiedenen Lösungen sehr gering sind, so daß der Aufbau von Recyclingkapazitäten für die HV02-Geräte in Prag im wesentlichen nur durch strategische Gesichtspunkte begründet werden kann. Zusätzliche Aspekte, die in dieser Arbeit angesprochen werden, betreffen spezifische Elemente der Modellierung des vorliegenden Falles, die Festlegung von Kostenfunktionen, die Möglichkeit der Optimierung von integrierten logistischen Netzwerken für Produktion und Recycling sowie die Anwendung von LP- vs. MILP-Modellen für die genannten Planungssituationen.

Key words: Reverse logistics – Remanufacturing – Copying machines – Océ – Location-allocation – Case-study – Mixed integer linear programming

Schlüsselwörter: Kreislauflogistik – Wiederverwendung – Aufarbeitung – Kopierer – Océ – Standortplanung – Fallstudie – Gemischt-ganzzahlige lineare Programmierung

0 Introduction

Traditionally manufacturers only took back discarded products and components selectively, if at all. Products were usually returned to the **O**riginal **E**quipment **M**anufacturer (OEM), due to contractual obligations (lease products), technical failure, etc. However, the growing public interest in environmental issues causes customer demand for recycling and the implementation of new government policies. These new policies aim at the closure of material flows as part of **I**ntegral **C**hain **M**anagement (ICM). One of the main political instruments used is *extended producer responsibility*, which makes OEMs formally responsible for take back and recovery of their 'own' discarded products. This will confront many OEMs with large volumes of unwanted return products. A new management field, Product

Recovery Management (PRM), emerges. PRM is defined as the "management of all used and discarded products, components and materials for which a manufacturing company is legally, contractually, or otherwise responsible" (cf. Thierry et al., 1995). PRM includes the set-up and management of a reverse logistic system.

As one of its sub-aspects, one must choose efficient locations and capacities for recovery processes and optimise good flows in the system. This class of problems is generally known in the OR-literature as facility location or location-allocation problems.

General approach

In our studies, we have the following approach. Firstly, supply and demand are balanced in a predetermined **recovery strategy**, see Krikke et al. (1998b). In these recovery strategies, decision rules are formulated on the degree of disassembly and the assignment of **R**ecovery and **D**isposal (RD-) options to the product or its released components. In some RD-options, the product or component is reused or disposed of. In other options, it is broken down to the material level to facilitate material recycling. Secondly, now that we have determined the magnitudes of good flows from supply to demand via various processing steps, we determine the **reverse logistic network design**, i.e., optimal locations and good flows. To enable the implementation of the recovery strategies for the return products, a reverse logistic system has to provide the capacity necessary for these strategies. In other words, in our approach there are two major inputs for the reverse logistic network design: (i) the recovery strategies applied to return flows, represented by a *processing graph* for each product, and (ii) available logistic systems that can *potentially* be used in the reverse logistic network, represented in a *transportation graph*. The processing graphs and the transportation graph are combined into a *network graph*. A network graph reflects the maximal reverse logistic network that can be realised. The final design will be a sub-graph of the network graph, namely the one with the lowest overall costs providing sufficient capacity. This final design is to be found by an optimisation model, e.g. a MILP-model. For brevity, we omit an extensive discussion on the use of processing - and transportation graphs. For this, we refer to Krikke (1998a).

Literature

A review of location-allocation models in a traditional production-distribution context can be found in Domschke and Krispin (1997). The use of location-allocation models in reverse logistics is described in a number of studies, mostly related to cases. Below, we give a review of these models, based on Krikke (1998a).

Caruso et al. (1993) consider an Urban Solid Waste Management System (USWMS). They develop a location-allocation model to find the number and locations of the processing plants, given the locations of the waste generators and landfills. For each processing plant, the technology – incineration, composting or recycling –, the amount of waste processed as well as the allocation of service users (waste sources) and landfills (waste sinks) are determined. No more than one facility may be located in one geographic zone and there are maximum capacities for all facilities and landfills. The model is single period and has a multi-criteria objective function, with components for economic cost, waste of resources and ecological impact. Efficient heuristics are developed to solve the problem. The model was applied in a case study for the region of Lombardy (Italy).

Ossenbruggen and Ossenbruggen (1992) describe a computer package for solid waste management (SWAP) based on LP-modelling. The model describes a waste management district as a network, where nodes represent waste sources, intermediary (capacitated) processing facilities and destinations (sinks) on given locations. Sources, sinks and intermediary stations can be of multiple (technology) types. Decision variables are the amount of waste to be processed by each facility and the magnitude of flows between the facilities. Implicitly, the processing paths are determined, where a flow can be split into sub-streams for different processing. Constraints follow from technically allowed processing sequences and capacity limitations. The algorithm finds a cost optimal solution, where the cost function only includes variable costs per waste unit, e.g. kg. These unit costs incorporate tipping fees, shipping costs and revenues from reuse.

Pugh (1993), describes the HARBINGER model, which gives decision support in the long term waste management planning of a city or county. The waste management system involves collection, transportation, treatment and disposal or reuse of a communities waste stream. These systems tend to be very complicated, which explains the need for mathematical analysis. The heart of HARBINGER lies in the multi-period allocation sub-model, which determines the cost-optimal assignment of waste flows from the sources to treatment and disposal facilities on given locations, within constraints set by the user (e.g. for capacity). Optimisation occurs on least cost. Other sub-models of HARBINGER are used to specify the input for the allocation sub-model and for post-optimality analysis. Unfortunately, the model description is not very detailed.

In a study of Marks (1969), the problem of selecting transfer stations is considered. Waste is generated at discrete sources and from there routed via intermediary transfer stations to discrete sinks, representing the disposal locations. The sinks have a demand that varies between a lower and upper bound reflecting minimal throughput requirements and maximum capacities of these disposal locations. At the intermediary transfer stations activities like transfer, packing and sorting can take place. The transfer stations can be located at a number of locations, where capacities are restricted. Each opened location has a fixed cost and linear processing cost. Also transportation cost between sources, intermediary transfer stations and sinks are linear. A Branch & Bound algorithm, using an out-of-kilter algorithm at the nodes, is developed to find the solution with the least overall cost for the MILP-problem.

Gottinger (1988) develops a similar, but more extensive regional management model. The model is concerned with the number, location and capacity (expansion) of both intermediary transfer stations and the ultimate disposal locations (sinks) as well as the routing from discrete waste sources through the system to the sinks. There is one type of transfer station and one type of disposal facility. For both types of facilities a set of potential and *existing* locations is given. The concave cost functions are approximated by linear segments, whereby one segment is represented by a pseudo-facility. Each pseudo-facility has a fixed cost and linear processing cost, in compliance with the cost function of the corresponding real life facility, within the capacity range covered by the pseudo-facility. Only one pseudo-facility per location can be opened. Existing locations have a restricted (current) capacity, potential new locations have infinite capacity. In addition, source locations and magnitude of waste flows generated and (linear) transportation costs are given. The aim is to minimise overall cost. A B&B procedure, very similar to the one of Marks (1969), is used for optimisation. Some variations of the model are described, for which special purpose algorithms are developed. The general model is applied in a case study for the Munich Metropolitan Area.

Spengler et al. (1997) develop a MILP-model for the recycling of industrial by-products in German steel industry. The model is based on the multi-level warehouse location problem and modified for this case study. It has to be determined which locations will be opened and how flows are routed from the sources through the intermediary facilities to the sinks. The model is multi-stage and multi-product, while it is allowed to transfer sub-streams of interim products from one intermediary facility to another in various ways, before delivering it at a sink. A sink can be either a reuse or a disposal location. Facilities can be installed at a set of potential locations and at different capacity levels, with corresponding fixed and variable processing cost. The type of processes to be installed at the intermediary facilities also have to be determined, hence the processing graph is not given in advance. Maximum facility capacities are restricted and transportation costs between locations are linear. While the amounts of waste generated at the sources are fixed, the demand at the sinks is flexible within a range. This range is set by the minimal required throughput and the maximum capacity of the sink.

Barros et al. (1998) present a MILP-model to determine an optimal network for the recycling of sand. In this real life case, sieved sand is coming from construction works, which represent the sources. The sand is delivered at a regional depot, where it is sorted in three quality classes. The first two classes, clean and half clean sand, are stored at the regional depot in order to be reused. The dirty sand is cleaned at a treatment facility, where it is also subsequently stored as clean sand. Both the clean and the half clean sand can be reused in new projects, which represent the sinks. Supply and demand are fixed for the respectively three and two qualities of sand. It has to be determined at which locations regional - and treatment centres must be opened, where locations can be picked from a pregiven set of potential locations. Also the capacities of the facilities and the routing through the system have to be determined, where capacities of both facilities are restricted. Opening a facility incurs a fixed and variable linear processing cost, transportation costs are also linear. The model used is a multi-level capacitated warehouse location model, for which heuristic algorithms are developed.

Thierry (1997) describes the underlying LP-model of a Decision Support System REVLOG, which supports the design of a network for both forward and return flows. Given the locations of warehouses and (re-) manufacturing plants, good flows are optimised. Trade offs are made with respect to the amount of new production versus remanufacturing, the amount of remanufacturing versus disposal, local versus central testing and direct distribution versus distribution via intermediary warehouses. All facilities are capacitated, transportation links are not. The aim is to maximise net profits, i.e., the net difference between sales from new and remanufactured products minus the variable processing and transportation costs. Supply and market demand are given, but not all market demand needs to be met. If applicable, lost sales are accounted for.

In Kooi et al. (1996), a MILP-model is developed for the set-up of a multi-echelon reverse logistic network with given and balanced supply and demand. The heart of the model lies in the concept of the *processing route*. A processing route is a physical implementation (or path) of processes needed to realise a recovery or disposal (RD-)option. Now, the clue to the problem is to generate a set of possible processing routes for each RD-option. If a processing route is used then costs are incurred, i.e., volume related (=variable) processing costs of every facility in the processing route and volume related transportation costs between the facilities (from the first to the last facility on the route). A facility must be opened, if at least one processing route is chosen that 'passes' through this particular facility. This incurs a certain fixed cost, which are added to the variable costs. If multiple facilities are opened at one location, facility investment costs are not shared. Facility investment costs are also not capacity dependent. In addition to processing routes, we need *entry routes* and *delivery routes*. An entry route is the connection between a supply point and the *first* facility of a processing route. Entry routes can be used at a certain volume related cost, equivalent to the transportation cost between the two locations involved. Analogously, the customer is supplied with recovered products, components or materials via the delivery route. The 'delivery costs' are also volume related and equivalent to the transportation costs between the *last* facility of the processing route and the demand point. In the optimisation, an optimal configuration of entry, processing and delivery routes is determined, using a MILP-solver (LINGO). The model is single product and distinguishes no capacity levels or capacity constraints. The model formulation is relatively complex, resulting in long computation times for larger problem instances. This may be reduced by clustering of supply and demand points or reducing the set of possible routes by eliminating routes unlikely to be selected.

In our general approach, we design a reverse logistic network based on a recovery strategy. The models found in the literature do not fit our problem definition. Therefore we have developed a general model for this situation, which has been tailor made for this case. The general model and a discussion on differences and similarities with the models discussed above can be found in Krikke (1998a). In the remainder of this paper, we focus on the practical issues of applying our general model at Océ, concerning the re-design of the reverse logistic network for a copying machine.

The paper is built up as follows. In Sect. 1, we give an introduction to the problem and describe our approach. In Sect. 2, we analyse the problem in more detail, resulting in a description of the processing graphs, the transportation graph, locations and relevant processes as well as solutions and scenarios formulated by Océ management. In Sect. 3, we represent the problem by a network graph and formulate an optimisation model. We compare results for the various test situations defined. Finally, discussion and conclusions are the subject of Sect. 4.

1 Recovery of copiers at Océ

Océ is an international firm, which is active in 80 countries and has a yearly turnover of more than Dfl 4 billion. It holds headquarters in Venlo (The Netherlands) and employs about 16 500 people world-wide. Océ develops, produces and sells high quality copying, printing and plotting systems as well as consumables and imaging supplies for these systems. Also service and financing (leasing) are part of the product range of Océ. The main markets are the office market and the drawing table market and the company aims for a prominent position in these markets by producing advanced products of excellent quality, reliability, durability and environment friendliness. Here, the design and development of the products plays a key role. Therefore, Océ invests about 7% of its turnover in R&D. Moreover, one co-operates with co-developers and suppliers, thus strengthening the technology base. The technology base enables high level recovery of products and components, which is an important ingredient of the business strategy. Let us now describe the reverse chain for copying machines. We limit ourselves to the physical reverse chain, i.e., issues concerning planning/MRP, administration, controlling, quality management etc. are not addressed in this study.

The return process of machines can roughly be divided into two stages. In the *first stage*, customers return a machine to the local operating company. Océ machines are returned from the market due to the ending of lease contracts or active buy back because of market demand for recovered machines. The operating company is allowed to refurbish the machine and put it back into the market. If operating companies themselves are not interested in refurbishing, they return the machine to a recovery location of Océ, for which they receive a fee. The machine now enters the *second stage*, in which it can be recovered by three recovery-strategies:

– Revision-strategy. A returned machine is disassembled to a fixed level and cleaned. Subsequently, additional disassembly of non-functioning, suspicious or retrofitted parts is carried out. Released parts are reused, recycled or in some cases disposed of. New or repaired parts, that are the (sometimes improved) equivalent of the released parts, are built in. The new machine is thus an upgraded version of the returned machine. The machine is inspected and tested before sales and has an 'as good as new' quality status, but it is sold as a secondary machine.

– Factory Produced New Model-strategy. In FPNM, a returned machine is converted into a new model, that contains all features and functions of the old model and adds new ones. The return process is basically the same as for revision, except that not only released parts are replaced, but also entirely new parts and units are added to provide additional functionality. The resulting machine has a new quality status and is sold as a new machine.

– Scrap-strategy. The returned machine is disassembled to the part level. Some parts are cleaned and repaired for reuse in recovery, service or regular production of other machines. Other parts are recycled or disposed of.

The recovery strategies can, in principle, also be applied to returned components. Machines returned to Venlo are classified in four classes, expressing recovery feasibility. These classes are:

- A: excellent,
- B: reasonable,
- C1: poor,
- C2: extremely poor.

The higher the class of a returned machine, the better it is suited for revision and FPNM since fewer parts need to be replaced due to failure or retrofit. Machines in class C2 can only be scrapped.

In order to be flexible, a large stock of returned machines is built up in Venlo. The machines are actually 'pulled in' on market demand. The class A machines are used up first, then class B and finally class C1 machines. If the stock levels of A, B and C1 machines become too high, these machines are also scrapped, starting with C1, then B and finally A. Reusable parts are also cleaned and repaired on demand. Sometimes, they are stocked, before being repaired. If demand or quality is low, the parts are recycled.

Stage 2 activities are carried out in Europe and the USA. Until recently, the European recovery processes were centrally located in Venlo. For strategic reasons, it was decided to investigate instalment of activities in the Czech Republic, where a number of recovery processes should be located in Prague. The cause of this study is the desire of the management to evaluate whether the *strategically motivated* decision to move certain recovery activities to Prague is also economically beneficial. On the one hand, low wages and purchase prices may make (re-) manufacturing in the Czech Republic beneficial. On the other hand, increased flows incur extra transportation, inventory and handling costs. Moreover, overhead costs and non-quantifiable costs influence an optimal choice of locations. This study should reveal whether the (partial) assignment of recovery processes to Prague is economically justifiable and if so, which particular processes should be carried out in Prague and which in Venlo. The study is limited to one type of machine: the HV02-machine (this can be done, because processes and cost functions are product specific in this situation). This machine is the output machine resulting from applying an FPNM-strategy to a HV01-machine. Furthermore, we limit our study to Europe, hence we do not include recovery locations in the USA. Because of these limitations, the reverse logistic network can only be partially redesigned. To this end, we have to address two research questions:

- At which locations and at which capacity should recovery processes be installed?
- How should transportation flows run between the various locations?

Because optimality depends on many exogenous factors, such as recovery strategies applied, number and quality of returned machines, developments in wages, legislation, etc., we shall study various alternative situations. In addition, the management of Océ wishes to compare three network re-designs as intuitive managerial choices. We shall compare these three managerial solutions with our optimum.

Finally, we define some concepts used in the remainder of this paper.

Table 1. Methodology

Research step	Result	Method	Section
Problem analysis	Processing graphs of HV02 Set of potential locations Transportation graph Test situations	Documents Interviews	2
Modelling	Network graph Optimisation model	Theory (Krikke, 1998a)	3
Data-acquisition	Parameter settings	MRP-system Administrative system	3
Optimisation	Results per test situation	Solver	3
Discussion and conclusions	Evaluation of modelling Evaluation of results Directions for further research		4

- *Part* – smallest entity distinguished in the construction of a return product.
- *Unit* – a composition of parts.
- *Component* – a unit or a part.
- *Module* – a unit which can be sold as a separate product, e.g. a finishing module.
- *Transportation means* – reusable resources needed for transportation of parts, units and products, e.g. pallets, yokes, etc.
- *Main recovery processes* – most important processes in FPNM-recovery, being dismantling, preparation and re-assembly.
- *Supportive recovery processes* – all processes supporting the main recovery processes, being take back, stock keeping, recycling, repair and production of new modules.
- *Process supplies* – units, parts, packages and test material needed for recovery processes.

The research is carried out in five subsequent steps, which are described in Table 1.

2 Problem analysis

In order to make a detailed analysis of the problem, we discuss the following issues:

- the set-up of the processing graph for FPNM of the HV02-machine,
- potential locations for those processes that are subjected to locational decisions,
- locations of other processes and inventories serving as supply and demand points,
- the formulation of test situations.

We remark that the assumptions made in this study are based on recent management discussions. Although they are not entirely in line with current practice, we consider them as reasonable.

2.1 Processing graph for FPNM of the HV02-machine

In the reverse chain of the HV02-machine, the following processes take place:

– Return. This involves receipt of input HV01-machines returned from the market as well as classification (A/B/C) and stocking of these machines. This is done at an external inventory location in Venlo.

– Dismantling. Returned machines are disassembled to a certain fixed level. A predetermined standard set of parts, units and materials is removed. Released parts are inspected. Approved parts are restocked directly or after internal/external repair. Disapproved parts are recycled or disposed of at specialised firms. The removed finishing module is inspected and if approved recovered for reuse in other product lines. Disapproved units are broken down into parts, which follow the same procedure as the other parts. Also, toner is removed and recycled at the original plant after a cleaning process. Now, two big elements remain for the FPNM process: the engine (or process unit) and by-frame (or paper unit). Both are cleaned before entering the next process.

– Preparation. This process only applies to the two modules engine and by frame. Here, it is assessed whether the modules should be further disassembled for quality reasons. Hence, the degree of disassembly is not prefixed, but depends on the state of return of the incoming machines. Bad or suspicious parts are *replaced.* All these parts are supplied from the central stock. Released parts are recycled at specialised firms.

– Re-assembly. This process is similar to the assembly process in regular production. Here, a HV02-machine is assembled from the engine unit, the by-frame unit, a new finishing module delivered from the Czech Republic and parts supplied from parts inventory. The parts are pre-assembled into units before final assembly. We assume that this pre-assembly step is always carried out at the same location as the final assembly. The assembled machines are tested and packaged before being delivered to the finished machine inventory.

– Central stock keeping of supplies and transportation means. This involves new and secondary units, parts, packaging materials, test materials as well as pallets, yokes, etc. This inventory is located in at the Océ plant in Venlo.

– Buffer stocking. This involves clustering of parts, units, packages, modules and products for inbound and outbound transportation. This is only necessary for transportation between processes (incl. central stock keeping) installed at different locations. Transportation means are also buffered for transportation.

– Stock keeping of finished products. Finished products are collected at a central inventory in Venlo, before delivery.

– External repair. Internal repair is part of the dismantling process and therefore no separate process. External repair is often done by the original suppliers of the parts in the area around Venlo. Repaired parts are always returned to Venlo to be stocked.

– Recycling. Recycling of parts is done by specialised firms in the area of Venlo or Prague. Toner is recycled by the original plant near Venlo.

– Disposal. Parts can be disposed of to landfills in the area of Venlo or Prague.

– New production. New finishing modules are manufactured in the Czech Republic to supply the re-assembly process.

– Transportation. By this we mean the external transportation of parts, packages, units, modules, products and transportation means between processes (including stock keeping) at various locations.

– It should be noted that internal transport and working inventories are neglected in this study.

The processing graph of the HV02-machine is given in Fig. 1. The processes described above are part of this graph. Box A provides an explanation of the symbols and abbreviations in order to link the above process descriptions with Fig. 1.

2.2 Locations and transportation graph

Although we analyse the entire reverse chain in the processing graphs, only the main recovery processes *preparation* and *re-assembly* will be subjected to locational analysis. The process dismantling is not subjected to locational analysis, because its activities are closely linked to stock-keeping and forward logistic activities. Moving dismantling would require a complete re-design of the logistic system and this is beyond the scope of the project. Locations of inventories and supportive processes are assumed fixed, also because of their interaction with other (forward) processes in the company. Due to the limitations in the scope, we merely optimise a sub-network, concerning the main processes preparation and re-assembly as well as the transportation links connecting these processes to other locations, which serve as supply and demand points. The sub-system studied is reflected in Fig. 1 by printing the elements inside the sub-system bold and the elements outside the sub-system dotted. Three locations are available for the processes preparation and re-assembly:

- Venlo 1 (an Océ location in Venlo), feasible for preparation,
- Venlo 2 (another Océ location in Venlo), feasible for preparation and re-assembly,
- Prague (Czech Republic), feasible for preparation and re-assembly.

The motivation for distinguishing two locations in Venlo lies in the internal logistics. Discussion with the management reveals that transportation costs justify the distinction of two separate locations, because these costs may influence the network design seriously. These costs are thus seen as external transportation costs. In principle, a process can only be assigned to one location, hence at most $3 * 2 * 2 = 12$ assignments of the main processes preparation and re-assembly to locations are possible. Some additional constraints are discussed in 2.3.

We optimise over all possibilities and moreover compare three assignments or *cases*, described in paragraph 2.3, which are interesting from a managerial point of view. As we mentioned before, all other locations are fixed. The locations of all processes and inventories part of the processing graph are summarised in Table 2. We use artificial names for all non Océ-recovery locations for confidentiality. We

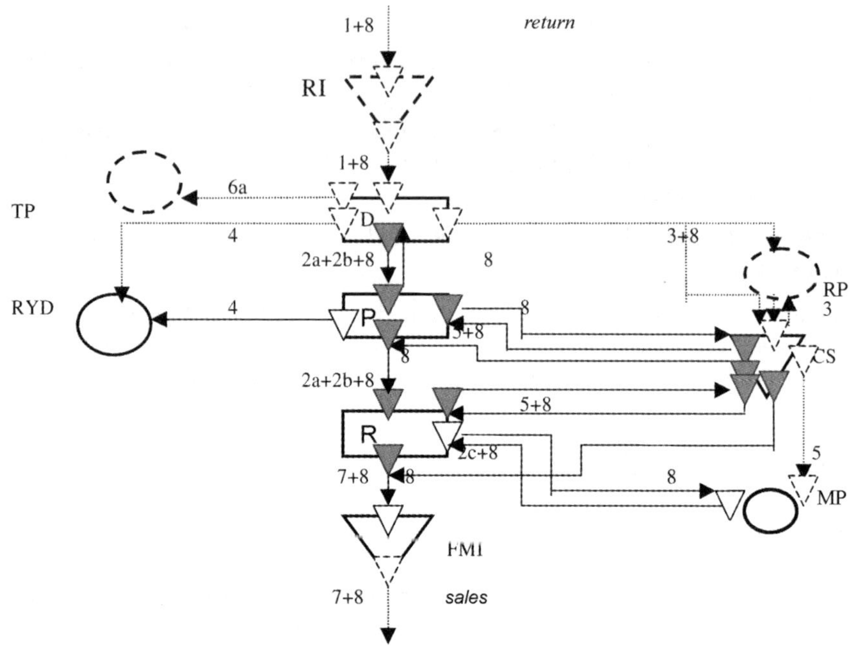

Material flows:

1. return machine: HV01
2. process supplies: modules
a. by-frame (paper unit)
b. engine (process unit)
c. twin finishing module
3. repairables (parts)
4. parts for recycling packages
5. process supplies: parts, testmaterials and
6. materials for recycling
6.a. toner
7. finished product
8. transportation means (pallets, yokes and part boxes)

Fig. 1. Processing graph for FPNM of HV02-machine

Box A Symbols and terminology of Fig. 1

Meaning of symbols:

▽ (large)	inventory	▼ (shaded)	optional buffer inventory for transportation needed in case subsequent process or inventory is at other location
○	supportive process	▭	main FPNM-process
▽ (small)	buffer inventory for transportation	→	transport flow

Note: bold: inside sub-system considered, dotted: outside sub-system considered

Meaning of abbreviations:

MP	Module Production	RI	Returned machine Inventory
RP	external RePair	CS	Central Stock (parts, packages, transportation means, test material)
TP	Toner Production	FMI	Finished Machine Inventory
RYD	specialised RecYcling firms and Disposal facilities	D	Dismantling
MR	Module Recovery	P	Preparation
		R	Re-assembly

Table 2. Specification of locations

Process/inventory	Firm	Location	Status	Inside or outside sub-system
TP	Océ	Venlo 3	fixed	outside
RP	suppliers	around Venlo	fixed	outside
MR	firm 1	NL1	fixed	outside
MP	Océ	CZ3 (in Czech Rep.)	fixed	inside
Disposal Venlo parts	firm 2	NL3 (near Venlo)	fixed	outside
Disposal Prague parts	local	CZ1 (near Prague)	fixed	outside
Recycling Venlo parts	firm 3	NL2 (near Venlo)	fixed	inside
Recycling Prague parts	local	CZ2 (near Prague)	fixed	inside
RI	firm 4	Venlo 4	fixed	outside
CS	Océ	Venlo 2	fixed	inside
FMI	Océ	Venlo 5	fixed	inside
Dismantling	Océ	Venlo 1	fixed	inside
Preparation	Océ	Venlo 1, 2 or Prague	variable	inside
Re-assembly	Océ	Venlo 2 or Prague	variable	inside

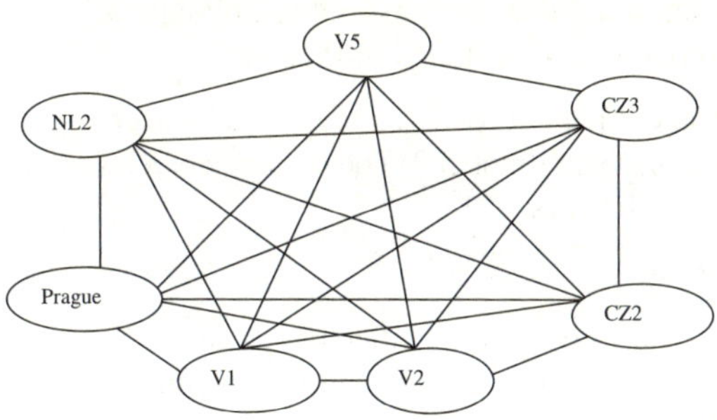

Fig. 2. Bi-directed transportation graph (only for locations inside sub-system)

note that buffer inventories are always located at the location of the processes and inventories involved.

Regarding transportation, there are no limitations in transportation links, hence the transportation graph is a full bi-directed graph with all locations of Table 2 that are within the sub-system considered as nodes. There are no capacity constraints set on the arcs of the transportation graph. Practical complications concerning customs etc. are incorporated in the cost price per transportation link per item to be transported.

We also assume that combined transport is used: products, modules, units, parts, transportation means and supplies can be transported together. In Fig. 2 the transportation graph is reflected.

2.3 Cases and problem parameters

The HV02-machine has been manufactured both in regular new production and by FPNM-production in Venlo 2 for a couple of years. Now, new production is phased out, FPNM production is extended. The current facilities in Venlo 2 can be used for both processes preparation and re-assembly. Facilities in Venlo 1 and Prague have to be established new, although some facilities may be taken over from Venlo 2.

As we mentioned, the processes preparation and re-assembly should be assigned to optimal locations. However, these assignments are restricted by the following managerial constraints:

- Preparation can be installed at all three locations, re-assembly only in Venlo 2 or Prague.
- Processes cannot be combined randomly. If preparation is assigned to Venlo and re-assembly to Prague, then preparation should be located in Venlo 1. If both processes are assigned to Venlo, they should both be assigned to Venlo 2.

The optimisation is divided in two parts. Firstly, we shall optimise the problem over all possibilities. Secondly, we shall compare the results of this optimisation with the intuitive management solutions: the management of Océ wishes to compare three network re-designs, here referred to as *cases*:

- case 1: both processes are installed in Venlo 2,
- case 2: preparation in Venlo 1, re-assembly in Prague,
- case 3: both processes are installed in Prague.

The optimisation is carried out for three scenarios, concerning exogenous factors. We come back to this in paragraph 3.2. The parameters relevant in this optimisation can be categorised as follows.

– Cost parameters

Investments costs. These costs are specified per product (!) per process per location. They are built up of costs for installing equipment, education of staff, the learning curve, translation of instructions and costs for planning and project management. The depreciation period is either 10 years or 1 year, depending on the kind of costs.

Constant costs. These costs concern the wages and taxes of overhead staff, costs for space, etc. They are specified per process per product per location per year.

Variable processing costs. These costs consist of labour and handling costs. These costs are also specified per process per product per location.

Recycling tariffs. These costs are specified per location per material type, hence we need to assign released parts to a material cluster. In the sub-system studied, only parts are released in preparation and these are all recycled.

Distribution costs. This concerns costs for transportation (incl. customs fee) and transportation related handling of all types of flows: products, parts, units, packages and transportation means. Costs are specified per transportation link per type of flow. Transportation frequencies are assumed to be fixed: which is daily between Venlo 1 and 2, daily between Prague and CZ3 and weekly between the Venlo locations and the Czech locations.

Inventory costs. If transport takes place between processes (incl. stock holding) at separate locations, an inbound and outbound buffer inventory must be installed between these processes, to facilitate batch transportation. The size of these inventories depends on the frequency of transportation, because buffer stocks increase if transport frequency goes down. Inventory costs are specified per type of flow (return product or process supplies) per year and calculated as a percentage of the capital value. For return products they are also defined per process, since their capital value increases while moving through the reverse chain. Holding costs for parts to be recycled or disposed of (after disassembly) are assumed to be zero, since no capital or risk is involved.

Overall costs. These involve costs of a Telecom contract concerning all phone, fax and email communications between Venlo and Prague. These costs are spread over all processes assigned to Prague. If no processes are installed in Prague, no telecommunications costs are incurred.

– Volumes.

Supply and demand. This involves the number of input and output machines.

Processing graphs. The processing graph concerns the successor relations of the recovery processes. Together with the supply and demand volumes, this determines the product flows in the system. Moreover, it determines the flows of transportation and flows of process supplies, since these are both linked to the product flows.

Volume data are specified per product.

- We assume that no capacity limitations are set on processes and transportation links.

In paragraph 3.2. we describe data acquisition with respect to these problem parameters.

3 Optimisation

Here, we describe the optimisation of the problem.

3.1 Modelling in terms of general reverse logistic network design

In this paragraph we discuss the modelling of the network design problem. Following our methodology, we construct a network graph and subsequently formulate an optimisation model based on this graph. Let us begin with the construction of the network graph of the HV02-machine. In Fig. 3, we visualise this network graph.

The dismantling process D in Venlo 1, the central stock CS in Venlo 2 and module production MP in CZ3 serve as supply points. The intermediate nodes are formed by the combinations of the processes preparation P and re-assembly R and the available locations Venlo 1, Venlo 2 and Prague. Demand points are the finished machine inventory FMI in Venlo 5 and the recycling locations RYD in NL2 and

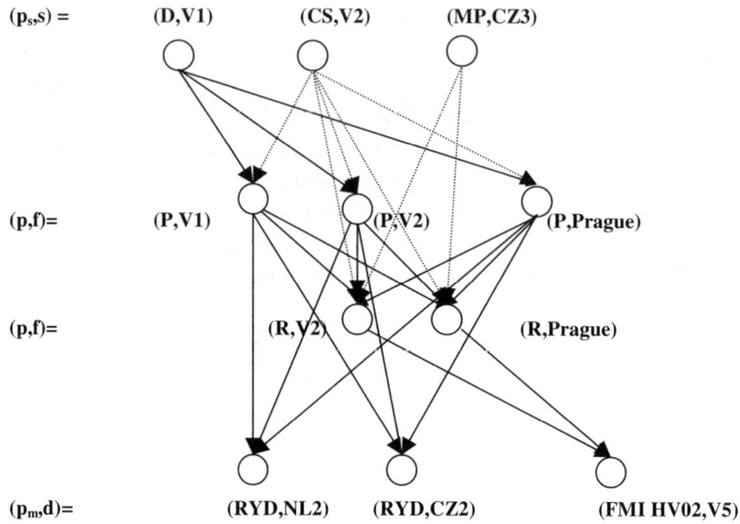

Fig. 3. Network graph for the HV02-machine

CZ2. Returned machines are identified as return products, finished machines and recyclables or disposables are defined by their end markets. Note that (p_s, s) stands for supply point, (p_m, d) for market m at demand location d (together demand point), p for process and f for facility location (together a facility). Also note that the main stream of input machine being transformed into an output machine or recyclables m is depicted as a bold line, while process supplies are dotted. This is a two-echelon problem. The optimisation (MILP-) model can be found in Appendix 1.

In the next paragraph it is described how data are collected. Subsequently, the results are discussed for each of the cases for several scenarios defined below. It turns out that the optimal solution obtained by the MILP-model coincides with one of the cases as suggested by the management.

3.2 Data acquisition and scenarios

Here, we describe data acquisition and scenarios analysed. The planning period chosen is December 1997–December 1998, because this matches the financial year. A common feature of reverse logistics is a high level of uncertainty in quantity, quality and timing of both demand and supply, i.e., secondary market volumes and collection volumes respectively. At Océ, a lot of experience is built up in managing this uncertainty. For example, the remanufacturing of copiers is largely independent of the input quality of return machines in the sense that A, B and C1 machines can all be remanufactured, of course with increasing effort. Moreover, the return of high quality machines is enhanced by using a differentiated return fee schedule. However, a few parameters -concerning volumes and processing costs- may change severely due to external circumstances. Therefore, Océ management has formulated three scenarios, in order to test the robustness of the solutions (cases) chosen.

The following scenarios are analysed. In scenario 0, the parameter settings, based on current forecasts of Océ for the financial year 1998, are as given in Appendix 2 and the problem is optimised for this year. In scenario 1, the problem is optimised for the financial years 1998 and 1999, assuming that variable cost in the Czech Republic (=labour+handling cost) increase by 20% in 1999. In scenario 2, the problem is again optimised for one year (1998) and the market volume is increased to from 875 to 1000 items. Thus, three scenarios are analysed for three cases, resulting in nine test situations. Data are supplied by officials of the company, who retrieve them from planning- and MRP-systems as well as the financial system. The units used are numbers for volumes and not given for costs (for confidentiality reasons). Data with respect to recycling had to be converted from kg to numbers.

3.3 Results

In the optimisation, we minimise the sum of yearly constant and variable processing costs, inventory costs and transportation costs for the HV02-machine with the MILP-model of Appendix 1. Investment costs are left out of the optimisation, because we do not wish that the 'optimality' of solutions depends on the depreciation method or depreciation period chosen. Note that the investment costs are relative: for convenience only differences in investments are calculated. The optimal solution for all three scenarios corresponds with case 3. For analytical purposes, it is interesting to compare this optimum with the other cases. We do so, by removing locations from the set of feasible locations for both processes such that only one network design can be obtained, namely the predefined one. Subsequently, we repeat the 'optimisation'. The yearly costs per case per scenario are set out against the investment costs in Table 3. Note that although case 3 is the optimum, it also requires the highest investments.

In cases 2 and 3, a part of the telecommunication costs (20 000 per year) is assigned to the HV02-machine. It is not known which part exactly. Optimisation was done with the solver LINDO. The computation times are only a couple of seconds, since problem instances involve only five boolean variables and around fifty continuous variables.

Table 3. Results of reverse logistic sub-network re-design (optimum bold)

	Yearly cost scenario 0 (1998)	Yearly cost scenario 1 (1998+1999)	Yearly cost scenario 2 (1998)	Investment costs
Case 1	2 184 698	2 184 698	2 394 190	0
Case 2	2 215 213	2 248 288	2 391 850	299 000
Case 3	**2 083 091**	**2 128 678.5**	**2 227 390**	655 000

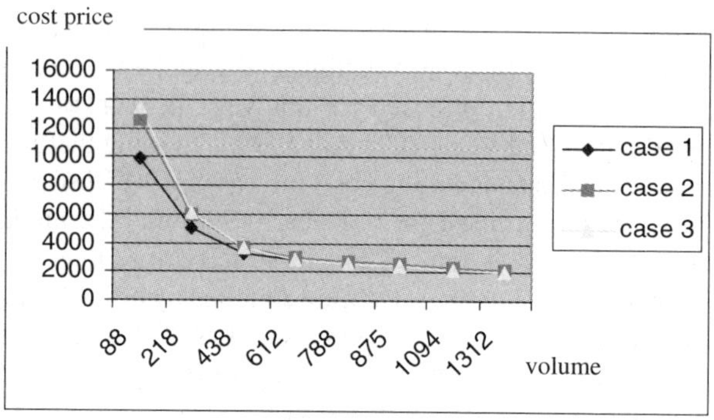

Fig. 4. Cost price per machine as function of volume

3.3.1 Some sensitivity analysis

To show the sensitivity of optimality of decisions, we depict the total cost price as a function of market volumes for all three cases in Fig. 4.

We can see that with increasing volumes (in a range from 88 to 1312 machines per year), case 1 loses its cost advantage over the other solutions, but also we see that the cost prices per machine converge with increasing volume. The cost price does not include yearly investment costs.

4 Discussion and conclusions

This section falls apart in a discussion on modelling, a discussion on results and some final conclusions.

4.1 Discussion of modelling

The discussion focuses on five subjects:

– Firstly, some specific modelling elements are applicable in this case, in particular process supplies, transportation means and buffer inventories. We shall discuss them one by one. In an FPNM-situation, process supplies are essential to the recovery strategy to convert an input machine into an output machine. They can have a strong impact on locational decisions because their supply points are fixed, magnitudes can be substantial and transportation costs severe. Transportation means have to be distinguished for transportation of products and process supplies. Partly, they follow the products and process supplies and partly they have their own routing. Although we distinguish this element, these flows are optimised as a decoupled problem after an optimal sub-network design for products and process supplies has been determined. Buffer inventories incur extra costs for external transportation. In our case, these costs are variable and they could be assigned to the transportation links. However, in other situations, fixed costs or capacity constraints may be related to installing a buffer inventory. Thus, buffer inventories should be distinguished as a separate entity, with fixed and/or variable costs and/or capacity constraints. As a constraint, the buffer capacity should be larger than the yearly volume divided by a delivery frequency augmented with a safety buffer.

– Secondly, Océ Controlling assumes that all processes and hence cost components (with the exception of the telecom costs) are specified per product per process per location. However, it can be argued that synergy exists between processing various products. In addition, it is reasonable to assume that economies of scale can be achieved *per location*. For example, it is reasonable to assume that the management capacity needed in the Czech Republic will not double if two processes are installed instead of one. Nevertheless, the Controlling Department treats cost functions as described in Sect. 3 and it is virtually impossible to determine the real cost behaviour within the scope of this study. Therefore, we have optimised the reverse logistic network for a single product.

– Thirdly, we have taken an approach in which the reverse chain is optimised separately from the forward chain. This can be done, because we only consider a sub-network concerning two main recovery processes, assuming that all other locations are fixed. One should realise that some of these locations fulfil a task in the forward as well as the reverse chain. For example, the central stock not only supplies preparation and re-assembly, but also forward chain processes. Also, transportation means and transportation links are used in forward and return flows. Moreover, for some machines (although not the HV02) the re-assembly process is integrated with regular assembly. Because of the strong interaction of processes, either directly or via inventories or transportation, an integral optimisation of forward and reverse chain is in our view preferable in a remanufacturing situation.

– Fourthly, if an input machine is processed, process supplies are added to it. As a result, an output machine is created step by step. During these steps, the weight of a machine increases. This can be handled in two ways. One way is to add variables X_{sk}^{tm}, X_{kj}^{tm} and X_{kd}^{tm} and serve each market m with the proper process supplies, analogously to the way products i serve markets m. The other way is to set the successor relations $v^i(p_k, p)$ and/or $w^i(p_k, p_m)$ to a value v $(v > 1)$, reflecting the gain in weight. The market demand V_m^d (also expressed in kg) should be expressed accordingly. We experienced no problems, because we used numbers instead of kg.

– Fifthly, the management of Océ had some clear solutions in mind for the reverse logistic network design and wished to compare three solutions. We suspect that this will often be the case and therefore a MILP-model may not always be needed (an LP-model is sufficient for the comparison of management solutions with prefixed locations).

4.2 Discussion of results

The results show clearly that there are only minor cost differences between the cases. Case 2 is more expensive in all scenarios, at most the yearly costs are about 7% higher than in the other two cases. Furthermore, it appears that installing both processes at one location is cheaper than installing them at different locations due to lower transportation and inventory costs. With the parameter settings chosen, case 3 is a little bit cheaper than case 1, but scenarios can be imagined in which case 1 is cheaper than case 3. Moreover, investments for case 3 are higher than for case 1.

Another observation concerns the choice of optimisation criteria. Since Océ is primarily a quality driven company one can question whether cost minimisation is the right optimisation criterion, instead of performance indicators concerning e.g. JIT delivery or the controllability of return flows. One should realise that the logistic costs part of this optimisation represent about 10% of the overall cost price. Also, constraints may be altered. In this study, we assume that processes can be installed at different locations, which implies that the processes dismantling, preparation and re-assembly are separable. However, it can be argued that separating these three processes has serious drawbacks in terms of the maintenance of know-how on machines, education of staff, instruction of staff, quality control, logistic control and administration. These arguments are in favour of keeping all three processes on one location, which changes the optimisation problem.

Finally, opportunities for improvement may also be found in internal logistics and the forward chain. Especially the possibility of an integral re-design of forward and reverse logistic network may be interesting because of the strong interactions between forward and reverse processes. This goes not only for the HV02-machine, but for a broader product range. Internal logistics may be addressed after external logistics has been dealt with.

4.3 Conclusions

We have optimised a sub-network for the HV02-machine for three scenarios and also we have compared three re-designs for the reverse logistic sub-network given by Océ-management. From the results we conclude that the decision to install activities in the Czech Republic for the HV02-machine must be well motivated from a strategic point of view, since there are no convincing economical reasons. Moreover, the underlying optimisation criteria (cost minimisation) can be criticised in relation to Océ's business strategy, hence it should be reviewed whether quality indicators should have a higher weight. Other future issues for Océ include the improvement of internal logistics and the advisability of an integral re-design of the reverse and forward logistic network.

With respect to modelling, we found that some specific elements we used, in order to deal with the Océ situation. Moreover, the assumption that processes and cost functions are product specific was questioned, but could not be adjusted within this study. The reverse chain was partly re-designed for one machine and the forward chain was also excluded. It may be worthwhile to develop models for integral optimisation of forward and reverse networks. Also, we found that an LP-model suffices to compare the management solutions, but that an MILP-model is necessary to determine the overall optimum.

Appendix 1
MILP-model for optimisation of HV02-machine network

Based on the network graph of Fig. 3, we construct a single product MILP-model. The following symbols are used to identify the modelling elements:

Indices:

p	process, $p = p_s, p_1, p_2, p_m$ (process at supply location: preparation, re-assembly and secondary markets respectively)
f	facility location, $f = f_1, f_2, f_3$ (Venlo 1, Venlo 2, Prague)
$F(p)$	set of feasible locations f for process p, with $F(p_1) = \{f_1, f_2, f_3\}$, $F(p_2) = \{f_2, f_3\}$
k	facility (p, f), $f \in F(p)$ with $k_1 = (p_1, f_1)$, $k_2 = (p_1, f_2)$, $k_3 = (p_1, f_3)$, $k_4 = (p_2, f_2)$, $k_5 = (p_2, f_3)$
p_k	process p of facility k
$K(p)$	all facilities k with $p_k = p$
s	supply locations, $s = s_1, s_2, s_3$ (Venlo 1, Venlo 2, CZ3)
d	demand locations, $d = d_1, d_2, d_3$ (Venlo 5, NL2, CZ2)
m	secondary market, $m = m_1, m_2$ (HV02, parts recycling)
t	process supplies, $t = t_1, t_2$ (process supplies, twin fin module)
$\mathrm{IN}(k)$	inbound buffer stock of return flow at facility k
$\mathrm{OUT}(k)$	outbound buffer stock of return flow at facility k
$\mathrm{IN}(t, k)$	inbound buffer stock of process supplies t at facility k
$\mathrm{OUT}(t, s)$	outbound buffer stock of process supplies t at supply location s
$\mathrm{OUT}(s)$	outbound buffer stock of return flow at supply location s
$\mathrm{IN}(p_m, d)$	inbound buffer stock of return flow at demand point (p_m, d)
FIX_k	yearly constant processing costs when facility k is opened, investment costs are excluded

Parameters:

c_k	unit cost of processing return product at facility k
c_{sk}	unit cost of transporting return product from supply location s to facility k
c_{kj}	unit cost of transporting return product from facility k to facility j
c_{kd}	unit cost of transporting return product from facility k to demand location d
c^t_{sk}	unit cost of transporting process supplies t from supply location s to facility k
$ch^{\mathrm{in}(k)}$	yearly unit holding costs of return product flow entering facility k
$ch^{\mathrm{out}(k)}$	yearly unit holding costs of return product flow leaving facility k
$ch^{\mathrm{out}(s)}$	yearly unit holding costs of return product flow leaving supply location s
$ch^{\mathrm{in}(m)}$	yearly unit holding costs of a return product flow ready to be used in market m

ch^t	yearly unit holding costs of process supplies t
λ_{sk}	av. no. of days between two transports from s to k divided by (2*365)
λ_{kj}	av. no. of days between two transports from k to j divided by (2*365)
λ_{kd}	av. no. of days between two transports from k to d divided by (2*365)
$u(p_s, p)$	fraction of return product flow at supply location s to be supplied to process p installed at any facility k, $k \in K(p)$
$v(p_k, p)$	fraction of return product flow processed by process p_k at facility k to be processed by process p at any facility j, $j \in K(p)$
$w(p_k, p_m)$	fraction of return product flow processed by process p_k at facility k to be reused in market m at any demand location d with $V_d^m > 0$
$u^t(p_s, p_k)$	number of units of process supplies t supplied from s, needed when one unit of return product flow is subjected to process p_k installed at k
V_s	yearly supply of return product flow at supply location s
V_d^m	yearly demand of market m at location d
V_s^t	maximal yearly supply possible for process supplies t at supply location s
M	a very large number

Decision variables:

X_k	total no. of return flow units processed at facility k per year
Y_k	boolean, value is 1 if facility k is opened else 0
X_{sk}	total no. of return flow units transported from supply location s to facility k per year
X_{kj}	total no. of return flow units transported from facility k to facility j per year
X_{kdm}	total no. of return flow units transported from facility k to market m at demand location d per year
X_{sk}^t	total no. of process supplies units t supplied from supply location s to facility k per year
$X^{\text{in}(k)}$	average no. of return flow units stored in inbound buffer at facility k
$X^{\text{in}(t,k)}$	average no. of process supplies units t stored in inbound buffer at facility k
$X^{\text{out}(k)}$	average no. of return flow units stored in outbound buffer at facility k
$X^{\text{out}(t,s)}$	average no. of process supplies units t stored in outbound buffer at supply location s
$X^{\text{out}(s)}$	average no. of return flow units stored in outbound buffer at supply location s
$X^{\text{in}(pm,d)}$	average no. of return flow units stored in inbound buffer at demand point (p_m, d)

Now the MILP-model becomes:

MINIMISE *(processing costs + transportation costs + inventory costs)* (1)

$$\sum_k c_k * X_k + \sum_k \mathrm{FIX}_k * Y_k$$
$$+ \sum_s \sum_k c_{sk} * X_{sk} + \sum_k \sum_j c_{kj} * X_{kj} + \sum_k \sum_d \sum_m c_{kd} * X_{kdm}$$
$$+ \sum_t \sum_s \sum_k c^t_{sk} * X^t_{sk} + \sum_k ch^{\mathrm{in}(k)} * X^{\mathrm{in}(k)} + \sum_k ch^{\mathrm{out}(k)} * X^{\mathrm{out}(k)}$$
$$+ \sum_t \sum_s ch^t * X^{\mathrm{out}(t,s)} + \sum_t \sum_k ch^t * X^{\mathrm{in}(t,k)}$$
$$+ \sum_s ch^{\mathrm{out}(s)} * X^{\mathrm{out}(s)} + \sum_d \sum_m ch^{\mathrm{in}(m)} * X^{\mathrm{in}(pm,d)}$$

subject to

$$V_d^m = \sum_{k, p_k = p_2} X_{kdm} \qquad \text{with } d = d_1, m = m_1 \tag{2}$$

$$V_d^m \geq \sum_{k, p_k = p_1} X_{kdm} \qquad \text{with } d = d_2, d_3, m = m_2 \tag{3}$$

$$X_k * w(p_k, p_m) = \sum_d X_{kdm} \qquad \forall k \, \forall m \tag{4}$$

$$X_k * v(p_k, p) = \sum_{j \in K(p)} X_{kj} \qquad \forall k \, \forall p \neq p_k \tag{5}$$

$$V_s * u(p_s, p) \geq \sum_{k \in K(p)} X_{sk} \qquad \forall s \, \forall p \tag{6}$$

$$X_k = \sum_s X_{sk} + \sum_{j \neq k} X_{jk} \qquad \forall k \tag{7}$$

$$V_s^t \geq \sum_k X^t_{sk} \qquad \forall t \, \forall s \tag{8}$$

$$X^t_{sk} = X_k * u^t(p_s, p_k) \qquad \forall t \, \forall s \, \forall k \tag{9}$$

$$X^{\mathrm{in}(k)} = \sum_s X_{sk} * \lambda_{sk} + \sum_{j \neq k} \lambda_{jk} * X_{jk} \qquad \forall k \tag{10}$$

$$X^{\mathrm{out}(k)} = \sum_{j \neq k} X_{kj} * \lambda_{kj} + \sum_d \sum_m X_{kdm} * \lambda_{kd} \; \forall k \tag{11}$$

$$X^{\mathrm{in}(t,k)} = \sum_s X^t_{sk} * \lambda_{sk} \qquad \forall t \, \forall k \tag{12}$$

$$X^{\mathrm{out}(t,s)} = \sum_k X^t_{sk} * \lambda_{sk} \qquad \forall t \, \forall s \tag{13}$$

$$X^{\text{out}(s)} = \sum_k X_{sk} * \lambda_{sk} \qquad \forall s \tag{14}$$

$$X^{\text{in}(p_m,d)} = \sum_k X_{kdm} * \lambda_{kd} \; \forall m \, \forall d \tag{15}$$

$$Y_k * M \geq X_k \qquad \forall k \tag{16}$$

$$\sum_{k, p_k = p} Y_k = 1 \qquad \forall p \tag{17}$$

$$Y_{k2} = Y_{k4} \tag{18}$$

$$Y_{k1} + Y_{k3} = Y_{k5} \tag{19}$$

$$Y_k = 0, 1 \tag{20}$$

all $X_k, X_{sk}, X_{kj}, X_{kdm}$ *are integer*
all $X^t_{sk}, X^{\text{in}(k)}, X^{\text{in}(t,k)}, X^{\text{out}(k)}, X^{\text{out}(t,s)}, X^{\text{out}(s)}, X^{\text{in}(p_m,d)} \geq 0$

Objective function (1) minimises overall costs. The balance equations (2) to (7) make sure that input and output of facilities are balanced in the system according to the recovery strategies implemented. Constraints (8) to (9) guarantee that sufficient process supplies are provided. Constraints (10) to (15) make sure that buffer inventories are installed correctly. Constraint (16) and (17) enforce that one facility is opened for each process. Constraints (18) and (19) reflect the limitations in combining processes at locations, such as described in paragraph 2.3. Finally, some logical constraints are formulated in (20) which speak for themselves. The parameters λ are used to estimate the average time an item is stored in a buffer, which is assumed to be 0.5*(mean time between transport). Our modelling assumptions include:

- Supply and demand are equally spread over the planning period.
- The capital value of return flows is the sum of the material values, added value through labour and transportation is zero. Parts to be recycled have zero capital value.
- All parts released in dismantling are re-assembled in re-assembly.
- All parts released in preparation are re-assembled in preparation.
- Pipeline inventories, i.e., items in trucks during transportation are ignored.
- Working inventories, i.e., items in process, are ignored.
- Transportation costs are calculated per single trip, assuming that trucks are also full on related return trips.
- Internal transport is neglected.
- The transportation matrix is symmetric.
- A week has 7 days, a year 365 days.

Finally, we note that a part of the telecommunications costs should be added to the objective function, in case processes are installed in Prague.

Appendix 2
Specification of parameter settings HV02-machine

For reasons of confidentiality, numbers in this appendix have been multiplied by certain factors and the monetary unit is not mentioned.

Table 4. Unit processing costs per facility

k	c_k
$k_1 = (p_1, f_1)$	429
$k_2 = (p_1, f_2)$	429
$k_3 = (p_1, f_3)$	143
$k_4 = (p_2, f_2)$	1116
$k_5 = (p_2, f_3)$	378

Table 5. Unit transportation costs between supply locations and facility locations

c_{sk}	$f_k = f_1$	$f_k = f_2$	$f_k = f_3$
s_1	0	2.5	132
s_2	2.5	0	132
s_3	132	132	∞

Table 6. Unit transportation costs between facility locations

c_{kj}	$f_j = f_1$	$f_j = f_2$	$f_j = f_3$
$f_k = f_1$	0	2.5	132
$f_k = f_2$	2.5	0	132
$f_k = f_3$	132	132	0

Table 7. Unit transportation costs between facility locations and demand points

c_{kd}	(p_{m1}, d_1)	(p_{m2}, d_2)	(p_{m2}, d_3)
$f_k = f_1$	2.5	-12	∞
$f_k = f_2$	2.5	-12	∞
$f_k = f_3$	132	∞	0

NB: other possible transportation links are not relevant in Table 7. For (p_{m2}, d_2), costs are negative due to revenues from recycling.

Table 8. Unit transportation costs from supply locations to facility locations for $t = t_1$

c_{sk}^{t1}	$f_k = f_1$	$f_k = f_2$	$f_k = f_3$
s_1	∞	∞	∞
s_2	2.5	0	35
s_3	∞	∞	∞

Table 9. Unit transportation costs from supply locations to facility locations for $t = t_2$

c_{sk}^{t2}	$f_k = f_1$	$f_k = f_2$	$f_k = f_3$
s_1	∞	∞	∞
s_2	∞	∞	∞
s_3	∞	93	12

Table 10. Unit holding costs per year per facility k for inbound buffer stock

p_k	$ch^{\mathrm{in}(k)}$
p_1	1259.25
p_2	1635.20

Table 11. Unit holding costs per year per facility k for outbound buffer stock

p_k	$ch^{\mathrm{out}(k)}$
p_1	1635.20
p_2	4234.00

Table 12. Unit holding costs per year for process supplies

t	ch^t
t_1	365
t_2	773.80

Table 13. (Number of days between transport between supply and facility locations)/(2*365)

λ_{sk}	$f_k = f_1$	$f_k = f_2$	$f_k = f_3$
s_1	0	2.74 10E-3	9.58 10E-3
s_2	2.74 10E-3	0	9.58 10E-3
s_3	9.58 10E-3	9.58 10E-3	2.74 10E-3

Table 14. (Number of days between transport between facility locations)/(2*365)

λ_{kj}	$f_j = f_1$	$f_j = f_2$	$f_j = f_3$
$f_k = f_1$	0	2.74 10E-3	9.58 10E-3
$f_k = f_2$	2.74 10E-3	0	9.58 10E-3
$f_k = f_3$	9.58 10E-3	9.58 10E-3	0

Table 15. (Number of days between transport between facility and customer locations)/(2*365)

λ_{kd}	d_1	d_2	d_3
$f_k = f_1$	2.74 10E-3	2.74 10E-3	∞
$f_k = f_2$	2.74 10E-3	2.74 10E-3	∞
$f_k = f_3$	9.58 10E-3	∞	2.74 10E-3

Table 16. Fraction of return flow transported from s to any k with $p_k = p$

$u(p_s, p)$	p_1	p_2
s_1	1	0
s_2	0	0
s_3	0	0

Table 17. Fraction of return flow transported from k to any j with $p_j = p$

$v(p_k, p)$	$p = p_1$	$p = p_2$
$p_k = p_1$	0	1
$p_k = p_2$	0	0

Table 18. Fraction of return flow transported from k to any d with demand for market m

$w(p_k, p_m)$	$m = m_1$	$m = m_2$
$p_k = p_1$	0	1
$p_k = p_2$	1	0

Table 19. Number of units of process supplies $t = t_1$ needed for processing one unit of return flow

$u^{t1}(p_s, p_k)$	$p_k = p_1$	$p_k = p_2$
$s = s_1$	0	0
$s = s_2$	0.6	5.3
$s = s_3$	0	0

Table 20. Number of units of process supplies $t = t_2$ needed for processing one unit of return flow

$u^{t2}(p_s, p_k)$	$p_k = p_1$	$p_k = p_2$
$s = s_1$	0	0
$s = s_2$	0	0
$s = s_3$	0	1

Table 21. Volumes

Flow:		t_1	t_2	m_1	m_2
Location:	V_s	V_s^t		V_d^m	
s_1	∞	0	0		
s_2	0	∞	0		
s_3	0	0	∞		
d_1				875	0
d_2				0	∞
d_3				0	∞

Table 22. Constant processing costs per facility

	Fixed costs
Facility	
k_1	228,875
k_2	228,875
k_3	323,125
k_4	489,375
k_5	749,875

References

1. Barros L, Dekker R, Scholten V (1998) A two-level network for recycling sand: a case study. European Journal of Operational Research 110: 199–214
2. Caruso C, Colorni A, Paruccini M (1993) The regional urban solid waste management system: a modelling approach. European Journal of Operational Research 70: 16–30
3. Domschke W, Krispin G (1997) Location and layout planning, a survey. OR-Spektrum 19: 181–194
4. Gottinger HW (1988) A computational model for solid waste management with application. European Journal of Operational Research 35: 350–364
5. Krikke HR (1998a) Recovery strategies and reverse logistic network design. PhD thesis, University of Twente, The Netherlands
6. Krikke HR, van Harten A, Schuur PC (1998b) On a medium term product recovery and disposal policy for durable assembly products. International Journal of Production Research 36(1): 111–139

7. Kooi EJ, Krikke HR, Schuur PC (1996) Physical design of a reverse logistic network: a multi-echelon model. Proceedings of First International Working Seminar on Reuse, Eindhoven, The Netherlands
8. Linders H (1996) Remanufacturing manual. Océ, Venlo
9. Marks DH (1969) Facility location and routing models on solid waste collection systems. PhD thesis, Johns Hopkins University, Baltimore (USA)
10. Ossenbruggen PJ, Ossenbruggen PM (1992) SWAP, a computer package for solid waste management. Computers, Environment and Urban Systems 16: 83–99
11. Pugh MP (1993) The use of mathematical models in evaluating resource recovery options. Resources, Conservation and Recycling 8: 91–101
12. Spengler T, Püchert H, Penkuhn T, Rentz O (1997) Environmental integrated production and recycling management. European Journal of Operational Research 97: 308–326
13. Thierry M, Salomon M, van Nunen J, van Wassenhove L (1995) Strategic issues in product recovery management. California Management Review 37(2): 114–135
14. Thierry M (1997) An analysis of the impact of product recovery management on manufacturing companies. PhD thesis, Erasmus University Rotterdam
15. van Dijk R (1997) MQ Coach M&L-OS2 HV, private communication
16. van de Ven J (1997) MQ Coach AR, private communication
17. Hendriks R (1997) Unit Coach Logistics M&L-OS2, private communication
18. Wouters P (1997) Head of Asset Recovery Department, private communication
19. Gelderblom A (1997) Head of Business Unit OS2, private communication
20. Linders H (1997) Project Manager M&L-OS2, private communication
21. Janssen-Joosten M (1997) Planner inventory control recovery and return flows. Private communication
22. van der Vliet RH (1997) Logistics engineer AR. Private communication
23. Schreurs A (1997) MQ-engineer AR. Private communication
24. Coopmans F (1997) Controller. Private communication
25. Michels M (1997) Logistics engineer AR. Private communication

Models for recycling electronics end-of-life products*

Modelle zum Recycling von Altprodukten aus dem Elektronikbereich

Manbir S. Sodhi and Bryan Reimer

Industrial and Manufacturing Engineering, University of Rhode Island,
Kingston, RI 02881, USA

Abstract. Increasing environmental concerns about the disposal of mass produced products have resulted in efforts to take back end-of-life consumer products. Legislation aimed at forcing manufacturers to take back electronics products at the end of their useful lives has either been adopted or is impending in many countries. This, along with shrinking landfill capacity and the reluctance of communities to open new waste sinks underscores the importance of developing methods and models for the management of end-of-life materials and products.

This paper reports a study of the reverse channels for recycling of electronics products. The economics of electronics recycling are modeled from the viewpoints of the generators, recyclers, and material processors separately. A variety of mathematical programming models, representative of the many ways in which the recycling industry currently operates, have been proposed along with numerical illustrations. Models integrating disassembly and material recovery decisions are also presented. These models can be used by recyclers and processors for optimizing recycling operations and thus contribute towards the economic sustainability of electronics recycling.

Zusammenfassung. Zunehmende Berücksichtigung von Umweltgesichtspunkten bei der Abfallbehandlung hat zu einer Verstärkung der Bedeutung der Rücknahme von Altprodukten nach Ablauf ihrer Gebrauchsphase geführt. So wird in vielen Ländern durch gesetzgeberische Maßnahmen ein immer größerer Druck auf die Hersteller von Elektro- und Elektronikgeräten ausgeübt, ihre Erzeugnisse nach Ende ihrer Nutzungsdauer wieder zurück zu nehmen. Zugleich nehmen vorhandene Deponiekapazitäten zur Beseitigung solcher Produkte ab und es verringert

* This research is supported in part grant BES9727248 from the National Science Foundation.

Correspondence to: M.S. Sodhi (e-mail: sodhi@egr.uri.edu)

sich die Neigung seitens der Gebietskörperschaften, neue Deponien zu eröffnen. Damit nimmt die Bedeutung der Entwicklung von Methoden und Modellen zum Management der Verwertung von Altprodukten in diesem Bereich immer mehr zu.

Der vorliegende Beitrag befaßt ich mit der Untersuchung entsorgungslogistischer Aktivitäten im Rahmen des Recyclings von Elektronikschrott. Dabei werden die Planungsprobleme für ein ökonomisch vorteilhaftes Recycling getrennt aus der Sicht der Erzeuger, Entsorger und Verwerter der Altprodukte modelliert. Zur Darstellung der Entscheidungsprobleme im Recyclingbereich wird eine Reihe von Modellen der Mathematischen Optimierung vorgestellt und mit numerischen Beispielen untermalt. Hierbei werden auch Modelle zur Integration von Demontage- und Verwertungsplanung präsentiert. Solche Modelle können ebenso von Entsorgern wie von Verwertern zur Optimierung ihrer Recyclingaktivitäten genutzt werden.

Key words: Electronics recycling – End-of-life models – Material recovery – Reverse logistics

Schlüsselwörter: Elektronikschrott – Altprodukt-Recycling – Recyclingmodelle – Entsorgungslogistik

Introduction

Producers of electronics goods are increasingly facing pressure from users and by government mandates to recycle end-of-life products. One recent example of a government mandate to force electronics recycling is in Massachusetts, where a ban on CRT disposal in landfills has been enacted in 1998 to reduce the amount of lead contamination in runoff from landfills [1]. The law, which focuses only on banning the landfilling of Cathode Ray Tubes (CRTs), does not specify who has the responsibility for recycling. While this is a ground breaking law in the United States, many European countries have already gone one step further and started focusing on a system of Extended Producer Responsibility (EPR). EPR is "the extension of the responsibility of producers for the environmental impacts of their products to the entire product life cycle"[2]. Consequently producers are now responsible for the take-back, recycling and disposal of their products and packaging, shifting the end-of-life responsibilities to private industries and away form the public sector. EPR originated in the "German Packaging Ordinance of 1991," which was the first law requiring producer responsibility for taking back and recycling sales packaging ([2] and [3]). After Germany, 27 other European countries have started producer responsibility systems for packaging as of May 1998, and 16 have started programs for batteries [4]. In April of 1998 the first draft of a directive on EPR for end-of-life electronics was being looked at by the EU, while Switzerland, the Netherlands, Italy, and Norway have already adopted some form of EPR for end-of-life electronics [2]. While a number of studies on the collection and disposal of packaging in the form of household recyclables have been reported in the literature ([5], [6] and [7]), recycling end-of-life electronic products and waste differs from household recycling in two significant ways. First, household recyclables are usually composed of single material packaging, whereas

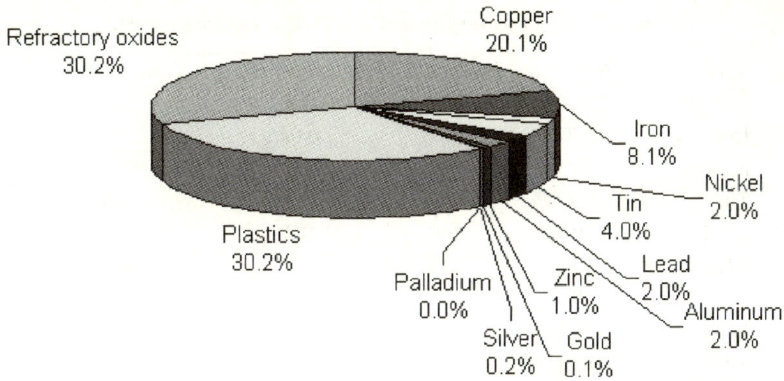

Fig. 1. Typical material composition of electronic scrap

electronic waste is a composite of several materials, some of which are considered hazardous. Second, household recyclables are usually made of low value materials, thus limiting the amount of additional processing that can be done to separate "clean" streams economically. On the other hand electronics waste and end-of-life goods have small quantities of both precious and non-precious metals, as well as easily re-usable components, which sometimes makes it feasible to collect and process small quantities from widely distributed sources.

A certain amount of material recovery is possible from most electronics waste. An optimal level can be determined by balancing the cost involved in material recovery against the revenue that can be gained from the removed components/parts [8], and the metals [9]. For example when parts such as memory chips, CPUs, and hard drives are removed from old 386 and 486 computers, the profit from component recovery is low because of the high cost of manual labor for disassembly and the low value of these components. However, the profit generated from the sale of precious metals concentrated in electronics components can sometimes exceed the cost of collection and processing. Consequently, there may be a financial advantage in collecting and recycling high value electronics. However, some electronics waste contains hazardous materials, such as batteries and CRTs, that require complex and expensive processing for elimination. In such cases there may not be any recoverable profit from recycling, and an additional fee is usually charged for collection.

Electronics scrap is composed of a ratio of approximately 40:30:30 of metal, plastics and refractory oxides respectively. As shown in Figure 1, the typical metal scrap consists of copper (20%), iron (8%), tin (4%), nickel (2%), lead (2%), Zinc (1%), silver (0.2%), gold (0.1%), and palladium (0.005%) [10]. Polyethylene, polypropylene, polyesters, polycarbonates, and phenolformaldehyde are the typical plastic components. Using this material composition, one ton of electronic waste in June 1999 , processed efficiently, could yield upto U.S. $9193.46 if the metals were sold at market prices obtained from (www.recycle.net) and (www.grn.com). Table 1 shows the detailed calculation of this total value, which is diagrammed in Figure 2.

The principal methods for recycling electronics scrap efficiently are disassembly, bulk recycling and smelting. Disassembly involves the removal of components

Table 1. Typical recovery values from 1 T of electronic waste

Material	Percentage	Quantity	Value ($ per lb)	Total Value ($)
Copper	20.000%	400	0.98	392
Iron	8.000%	160	0.045	7.2
Nickel	2.000%	40	2.23	89.2
Tin	4.000%	80	2.35	188
Lead	2.000%	40	0.21	8.4
Aluminum	2.000%	40	0.71	28.4
Zinc	1.000%	20	0.48	9.6
Gold	0.100%	2	3885.57	7771.14
Silver	0.200%	4	34.4	137.6
Palladium	0.005%	0.1	5019.16	501.916
Plastics	30%	600	0.1	60

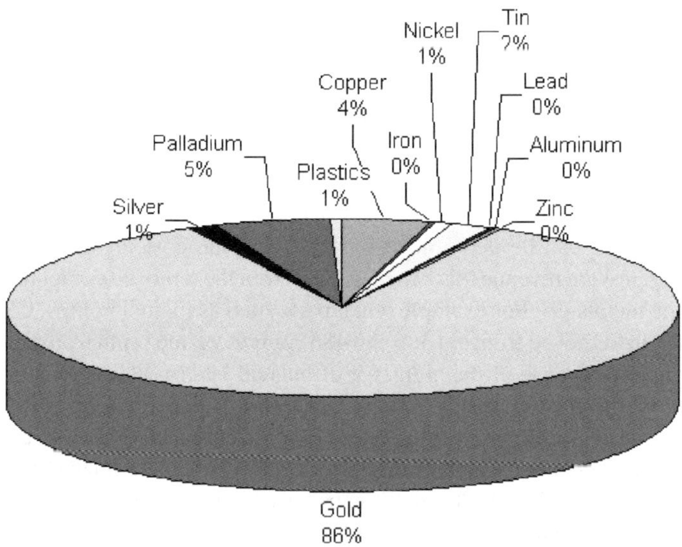

Fig. 2. Revenue breakdown by material from electronic waste

and specific materials from products. A product can be disassembled either completely or partially to remove some targeted materials [11]. Dissembly is a labor intensive task, and the cost of disassembly is proportional to the effort that must be expended to remove components. The optimal disassembly decision balances the cost of disassembling a product against the value of components removed, environmental liabilities, and the residual material value in the product ([11], [12], [13], [14] and [15]). ([16]) gives an overview of disassembly models as well. Environmental consequences of disposal are usually computed by methods such as Life Cycle Analysis [17] or single figure environmental indicators [18]. The residual value of a product which has been disassembled can be computed from the value of the material content and the weight of the item. In most situations bulk recycling is used in conjunction with disassembly – after a certain subset of hazardous and valuable parts are removed, the remaining parts can be efficiently bulk recycled.

Bulk recycling

Bulk recycling is the wholesale processing of recyclables, mostly for material recovery. In most cases some disassembly is done, primarily to remove hazardous materials or valuable parts. The remainder of the product is then shredded into flakes. Separation methods based on physical properties of materials can be used to separate these flakes into four categories: ferrous metals, nonferrous metals, plastics and a composite residual mixture [19]. The first three groups can then be further refined into pure streams for resale. Metals are removed from the composite mixture by magnets and eddy currents. Plastic separation is not as simple. Plastics can be further separated based upon different physical properties such as mass, density, or particle size. Techniques such as sink-float separation, air classification [20], and ultrasonic methods [21] take advantage of physical properties in separating plastics into pure streams. Methods for optimizing sink float separation sequences can be found in [22]. However, high value metals that are layered onto ceramics and plastics cannot be removed by physical separation methods. These are usually found in the composite residual mixture, and material recovery is typically effected by smelting and refining.

Various smelting methods can be used to separate lower value materials such as Copper, Aluminum, Zinc, Lead, Tin, Titanium and high value precious metals such as Gold, Silver, and Platinum group metals [23,24]. Precious metals are processed using a pyrometallurgical recovery process, where products containing the desired metals are first smelted in a blast furnace with litharge, coke, and pyrites. The precious metals are mostly collected in a lead bullion created by the litharge [24]. Silver can then be recovered from the bullion using a cupelation process, in which the Pb-Ag bullion is melted in a hearth furnace where Pb and other impurities are removed by preferential oxidation effected by air blowing. This results in a semi-pure silver that is cast into anodes, and this is further processed by electrorefining. Gold can be easily recovered using a chlorination process. Platinum group metals however, can be complicated to refine. While a chemical treatment plant can process out the six metals in the platinum group (PT, Rh, Ru, Ir, Os, and Pd), separating these metals requires a significantly greater degree of processing than for gold or silver since all six metals are usually present simultaneously [24].

In this paper the reverse channels for the recycling of electronics products are represented as a network of flows between generators, recyclers and material processors. While this is similar to the approach presented in [25], the models presented do not integrate the entire reverse production system (RPS). Rather, mathematical programming models, representative of the many ways in which the recycling industry currently operates in the U.S., detail the interactions between different components of the network. Since it is at the recyclers' level that most product disposal options are considered, models integrating material recovery and disassembly decisions for electronics recycling by recyclers have been developed.

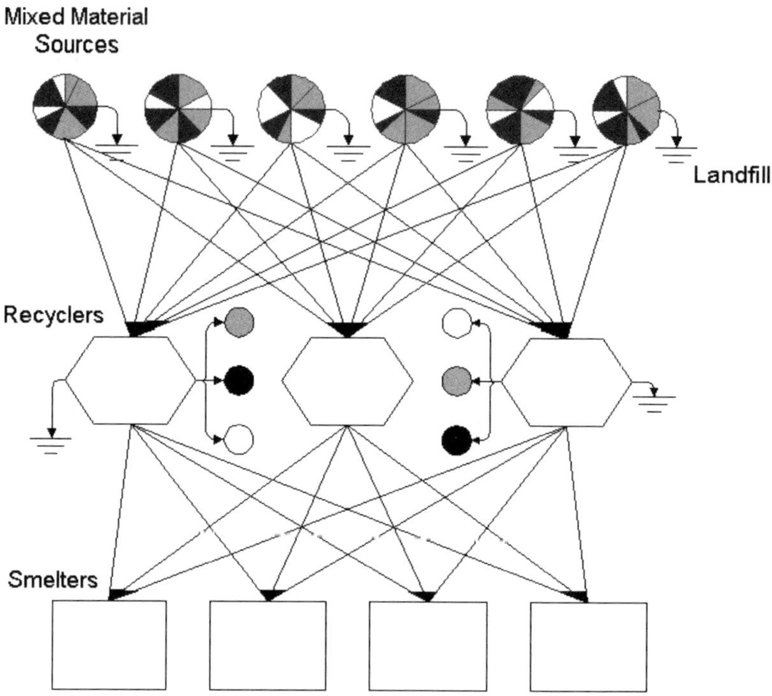

Fig. 3. The flow of a electronics recycling system

The electronics recycling network

A typical electronics recycling network is shown in Figure 3. Recyclers collect electronics waste from sources such as, cities, schools, office buildings, and manufacturers. The recyclers may then add some value by disassembly and processing to separate components and materials as previously discussed. These are then forwarded either to various processers/smelters for further processing into pure metal streams or for plastics recovery, or both. The entire system functions as a complex integrated economic entity, making monolithic representations large and intractable. In the discussion that follows, individual portions of the system have been modeled separately in relation to their specific goals. Although this decomposition may lead to suboptimal solutions for the total system, it facilitates the modeling effort and a detailed representation of the scope of decisions faced by individual components of the recycling system.

Source models

Sources of electronics waste can be organized into three categories. These are: (i) mixed groups of items in one location, (ii) a single item type spread over many locations, and, (iii) and various groups of items spread over many locations [26]. Examples of the first category include individual consumers, universities, offices, collection centers that serve individual consumers etc. The second category is representative of manufacturers or intermediate agencies such as municipalities or collection centers, who collect a particular product type at different locations for disposal. The last category may represent one or more manufacturers cooperating to share economies of scale for product disposal. The decision makers at this level do not involve themselves in detailed issues such as product disassembly, bulk recycling or material recovery from product disposal. In some cases they may participate fiscally through cost/profit sharing agreements with recyclers. In general, they are merely interested in the disposal of potentially hazardous materials, and may also pay for the convenience of having the material disposed with no residual liability. Other motivations for recycling may be environmental preservation, waste reduction or product end-of-life recycling mandates. The source problem is then identified as the decision problem for the generators of end-of-life products. It focuses on a waste generator who is selecting one or more recyclers to collect electronics waste from different locations, where each recycler charges different rates for collecting different materials/items. With an increasing number of recyclers willing to collect material, the goal of any source becomes one of minimizing the cost or, if possible, maximizing profit from recycling. While recycling at the source level is also motivated by environmental concerns, these are hard to measure, and this effect is not quantified by recycling service providers. Any negotiations are usually limited to the cost of the service. An additional option is landfilling - however this is limited only to non-hazardous materials and has an associated tipping fee (typically \$40–\$70 /T in the U.S.). A simple formulation of the source problem is:

$$Min\ z = \sum_{i=1}^{I}\sum_{p=1}^{P}\sum_{j=0}^{J}(SR_{ij} + TC_{ipj})V_{ipj} \tag{1}$$

$$s.t. \sum_{j=0}^{J} V_{ipj} = W_{ip}\ \forall\ (i,p) \tag{2}$$

$$V_{ip0} = 0\ \forall\ i\ \in h \tag{3}$$

$$V_{ipj} \in R^{+}\ \forall\ (i,p,j) \tag{4}$$

This formulation focuses on minimizing each source's net cost for material removal. The revenue from recycling each item i at a location p can be determined by adding the transportation cost to recycler j, TC_{ipj} to the cost a source pays the recycler to recycle an item (SR_{ij}). As indicated earlier, recyclers sometimes pay to remove items such as PC's and must be paid to remove items such as CRTs. Thus, this cost is negative if the recycler pays the source for the item and is positive if the source pays the recycler. Constraints (2) are material balance enforcements that ensures the aggregate quantity of each item at each location, W_{ip}, is sent either to a recycler or a landfill. V_{ipj}, the volume of item i that is transferred from source, p to recycler, j is the decision variable used to calculate the total disposal cost for all items. Landfilling too is easily accommodated by considering the landfill as a recycler, with appropriate tipping costs. Constraints, (3), are added to ensure only benign items are sent to the landfill ($j = 0$).

In this form, the solution of the source problem is trivially determined by assigning each item to the recycler who gives the highest return. Complications such as quantity discount, multi-item discounts and periodic pickup schedules can be easily included in extensions of this model.

Recycler models

Given that there are many different sources of electronics waste, electronics scrap recyclers have to make critical decisions on how to acquire material for optimizing operating costs, which include transportation, disassembly and material recovery costs. To acquire high value material, a recycler will usually need to pay some compensation to the source. This amount is based directly upon the composition of the material. As an example, printed circuit boards are known to have small amounts of precious metals and recyclers sometimes pay to pick up such items. In contrast, situations where the source will pay for collection and processing of electronics waste to comply with regulations also exist. One such example is that of computer monitors, where the sources pay the recycler for disposal because of the high processing cost to eliminate hazardous coated CRT glass.

A limited survey conducted with ten of the larger electronics scrap recyclers in the U.S. has found that a wide variety of schemes are currently being used for payment and compensation for collection and processing of electronic waste. In general, recyclers will pay for computer scrap without monitors. This payment is not always set in advance, although dealers of used computers do offer fixed prices for newer, working computers. The recycler collects the material, processes it and then returns some fraction of the profit to the supplier. Small electronics items such as telephones do not usually yield positive returns to the sources. The recycler may take these products for nothing, or charge a small fee for the service. As mentioned earlier monitors present a completely different situation. All recyclers surveyed charged a fee for picking up monitors. These charges range from \$ 5.50 to \$ 15.00 per unit in addition to any transportation cost.

In addition to deciding what materials to collect, the recycler also needs to decide where to send material for further processing for metal extraction. Different refiners/smelters specialize in the recovery of specific materials. To maximize return, the recycler needs to choose the smelter that returns the highest revenue for a given batch of materials. A related decision is that of determining an appropriate level of disassembly. It is often possible to disassemble products so that the concentration of a particular set of materials is increased, while either discarding materials or subassemblies that have little recovery value, or can be sold directly in an untreated form. This issue is dealt with in more detail later in this paper

Metal separation is usually done by a process where the smelter melts out all but the target materials. As stated earlier, end-of-life electronics products typically contain a mixture of precious metals in low concentrations. Since even small amounts of precious metal can be very valuable, different choices on how best to process a batch of electronics waste can yeild significantly dissimilar returns. Different processors have different setup fees, processing fees, transportation fees, minimum metal recovery stipulations and are capable of recovering different fractions of material from the metal mix. Smelters charge for material processing in many different ways. In some cases, they can charge a variable cost for each ton of dry input. A variable fee per ton for each material recovered is also typical. Additionally, setup fees for each metal recovered may also be levied. Another method of compensation is for smelters to retain a minimum amount of each metal they process. When long term relationships are established, contracts can be set up between the recycler and smelter for fixed quantities of material to be processed. These contracts stipulate minimum metal contents in the raw materials and aggregate quantities to be supplied over a certain time period. In most cases, penalties can be levied if toxic or undesirable elements are present in the mixture since these must be isolated prior to any recovery. Smelters can also generate profits for themselves if the recovery from materials processed exceeds the assayed values agreed upon with the suppliers. Given the combination of issues involved, selecting the best set of smelters for a given input material stream is not always an obvious decision.

Each smelter (l) can be characterized by a technological coefficient, T_{kl}, which is a fraction representing the amount of input metal (k) that the processor is capable of recovering. This recovery fraction is less than unity because material recovery processes are not perfect. Some small amounts of material may be economically impossible to recover. The variable cost charged for each unit processed can be incorporated into this recovery fraction. The additional setup cost is a fixed charge (S_{kl}), applied for each metal type recovered by the processor. Costs can also include transportation fees that are proportional to the distance between the recycler and processor (CS_l). Extra costs that should be considered relate to costs the collector will incur if the processor requires some special delivery needs. These might include items such as special bins, irregular labeling, extra unloading costs, or special trucking needs. Furthermore, acquisition costs can be assessed for each unit of a material acquired. Different models can be developed for different combinations of costs.

In many situations electronics recyclers collect pre-separated or partially manufactured electronics waste that does not require further disassembly. In these situations recyclers accrue a profit in two ways. First they can charge a fee for their services as described earlier, and second, they can make a profit from selling the materials they collect to different smelters. To generate revenue, the recycler is faced with two operating decisions. The first is a decision on what materials to collect, and the second is to decide on which smelter to send the material to. A model that handles both issues concurrently, thus maximizing the recyclers total revenue can be formulated as:

$$Max\ z = \sum_{k=1}^{K}\sum_{l=1}^{L}((G_{kl}\,R_k) - (Y_{kl}\,S_{kl})) - \sum_{i=1}^{I}\sum_{l=1}^{L}(C_i + CS_l)X_{il} \tag{5}$$

$$\text{s.t.} \sum_{i=1}^{I} X_{il}\,A_{ik}\,T_{kl} \geq G_{kl} \quad \forall\,(k,l) \tag{6}$$

$$M\,Y_{kl} \geq G_{kl} \quad \forall\,(k,l) \tag{7}$$

$$\frac{\sum_{i=1}^{I} X_{il}\, A_{ik}}{\sum_{i=1}^{I} X_{il}} \geq B_{kl}\, Y_{kl} \quad \forall\, (k,l) \tag{8}$$

$$\sum_{l=1}^{L} X_{il} \leq W_i \quad \forall i \tag{9}$$

$$Y_{kl} \in \; [0,1] \; \forall (k,l); \; X_{il}, G_{kl} \in R^{+} \; \forall\, (i,k,l) \tag{10}$$

In the model (5–10), the objective, (5), maximizes the recyclers profit after subtracting the cost of acquiring item i, C_i, as well as the cost of transporting the item to smelter l, CS_l. The recycler's revenue is calculated by subtracting the cost of setups from the profit received for recovered metals. This profit is calculated by multiplying G_{kl}, the amount of metal k recovered at smelter l with the unit selling price of metal k, R_k. In (6), G_{kl} is calculated as the amount of metal k that can be recovered from all the items sent to smelter l. In this constraint, A_{ik} is the coefficient representing the amount of metal k in item i. Constraints (7) trigger the binary decision variable Y_{kl} if metal k is recovered at smelter l. Constraints (8) are blending constraints that produce a mixture with the minimum concentration requirement (B_{kl}) for metal k at smelter l. However, this constraint is active only if metal k is recovered at smelter l. Finally, (9) requires the amount of item i shipped to smelter l, represented by the decision variable X_{il}, to be less than the weight of item i available for shipment W_i.

The formulation (5–10) is a non-linear, mixed integer formulation. While general approaches based on Benders decomposition [27] and outer approximations [28] may be used to solve this model, a branch and bound scheme is preferred since since fixing Y_{kl} gives a linear formulation which is easy to solve. This is further developed in [29].

As an illustration of this model, assume three items (computers, Video Casssette Recorders (VCRs), and televisions) containing different amounts of Gold, Silver, and other grouped metals, are available to a recycler for collection. Furthermore, two smelters are available to process the items. The cost of transportation to the smelters are 0.025 and 0.015 ($ per lb.) respectivly, and the value of the metals is: gold 3885.57, silver 34.4 and other metals 60.0 ($ per lb.). The example is solved by enumerating on Y_{kl} and solving the remaining linar model using CPLEX and GAMS. These data and the solution are shown in Figure 4 below.

Recycler models with product disassembly

In many cases recyclers recover value by performing some disassembly operations on end-of-life electronics items prior to processing for material recovery. The disassembly step can enhance value in several ways. First, removed components might be worth more sold directly than as contributions to the material recovery revenue. Second, additional materials not recoverable from the original product because of concentration limitations might now be recoverable from disassembled subassemblies. Lastly, separated subassemblies can be sent to different smelters for a more profitable material recovery than the entire product being sent to a single smelter.

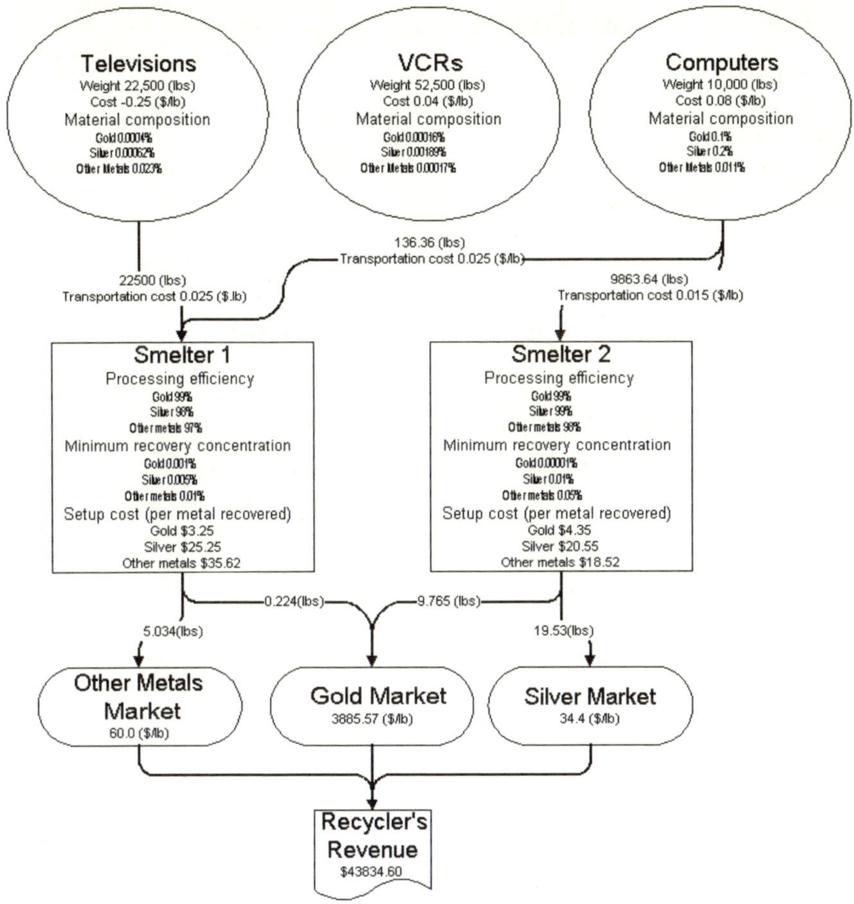

Fig. 4. Illustration of recycler's model (5)–(10)

Although several different models of the disassembly problem have been reported in the literature, ([13, 14] and [30]), a process for integrated consideration of material and component/subassembly recovery is difficult to develop because of the large number of component/subassembly subsets possible by disassembly. This is also consistent with the discussion in [15]. However, it is possible to limit consideration to a smaller set of possible disassembly subsets with associated disassembly costs, by investigating candidate subsets for material recovery. These can then be included in a material recovery formulation, prescribing the best combined disassembly and material recovery decision.

As an illustration, consider the data for a personal computer in Table 2. If the entire P.C. is processed for metal recovery, the gold content is 0.005%. However, the motherboard has a gold concentration ten times higher. Using these data, subsets of components can be grouped and proposed as candidates for metal recovery evaluation. This is illustrated in Table 3.

Table 2. Metal composition of items in a P.C.

Component	Weight (lbs)	Percent of metal in component					Wt Fraction
		Gold	Silver	Copper	Iron	Pt. Group	
Motherboard	1.430	0.05	0.1	0.8	0	0.001	0.06
Daughter boards	0.95	0.05	0.1	1	0	0.005	0.04
Case	14	0	0	0	95	0	0.61
Floppy drive	1.12	0.0001	0	5	12	0	0.05
Hard drive	1.03	0.0005	0.001	2	9	0.0001	0.05
Power supply	4.3	0	0	35	25	0	0.19
Total	22.83	0.005	0.01047	7.019	63.960	0.0003	1

Table 3. Metal composition of items grouped together

Row	Items grouped together	Weight (lbs)	Percent of metal in component					Fraction of P.C. wt.
			Gold	Silver	Copper	Iron	Pt. group	
1	PC	22.83	0.005	0.01047	7.019	63.960	0.0003	100
2	Mother board, Daughter board	2.38	0.05	0.1	0.88	0	0.0026	0.104
3	Case, Power supply	18.3	0	0	8.22	78.552	0	0.802
4	Mother-board, Daughter-board, Case	16.38	0.0073	0.0145	0.13	81.197	0.0004	0.717
5	Mother-board, Daughter-board, Case, Power supply	20.68	0.0058	0.0115	7.38	69.512	0.0003	0.906
6	Mother-board, Daughter board, Hard drive, Floppy drive	4.53	0.0264	0.0528	2.15	5.013	0.0014	0.198

The problem of simultaneous determination of optimal component disassembly and material recovery can now be formulated by using the disassembly cost for obtaining each proposed subset of materials of item i, D_{is}, which is the net cost or revenue after factoring in any value recovered from components sold directly. This will be negative if there is a profit - i.e. the cost of disassembly is less than the revenue returned from the sale of components. However, if the levels of disassembly described by each subset of parts, is predetermined and therefore, not optimized. Thus, model (5) - (10) can be extended as:

$$\max z = \sum_{k=1}^{K}\sum_{l=1}^{L}((G_{kl}R_k) - (Y_{kl}S_{kl})) - \sum_{l=1}^{L} CS_l \sum_{i=1}^{I}\sum_{s\in S_i} X_{ils} - \sum_{i=1}^{I}\sum_{s\in S_i}\frac{D_{is}}{f_{is}}\sum_{l=1}^{L} X_{ils} \quad (11)$$

$$\text{s.t.} \sum_{i=1}^{I}\sum_{s\in S_i} X_{ils}\, A_{iks}\, T_{kl} \quad \geq \quad G_{kl} \quad \forall\,(k,l) \quad (12)$$

$$M\, Y_{kl} \quad \geq \quad G_{kl} \quad \forall\,(k,l) \quad (13)$$

$$\frac{\sum_{i=1}^{I}\sum_{s\in S_i} X_{ils}\, A_{iks}}{\sum_{i=1}^{I}\sum_{s\in S_i} X_{ils}} \quad \geq \quad B_{kl}\, Y_{kl} \quad \forall\,(k,l) \quad (14)$$

$$\sum_{s\in S_i}\frac{1}{f_{is}}\sum_{l=1}^{L} X_{ils} \quad = \quad W_i \quad \forall i \quad (15)$$

$$Y_{kl} \in\ [0,1]\ \forall(k,l);\ X_{ils}, G_{kl} \in R^+\ \forall\,(i,k,s,l) \quad (16)$$

This model is similar to (5)–(10), except for the disassembly cost of selected subsets which is subtracted from the objective, and the fractional weight material balance in (15). The decision variables in this model are X_{ils}, which represents the amount of subset s of item i sent to smelter l. $B_{kl}, CS_l, G_{kl}, R_k, S_{kl}, Y_{kl}$ all have the same definition as in (5) – (10). The coefficient A_{ik} is replaced here by A_{iks}, and this can now be interpreted as the amount of metal k in subset s of item i. The cost of acquiring item i, C_i is not factored into the objective function because it is assumed that this model is solved by the recycler who has a certain quantity of items in inventory, and has to determine optimal disassembly and dispersal strategies. However, acquisition costs can be easily accommodated if necessary. In this model, several candidate plans for disassembly of the product are first generated. Corresponding to each disassembly plan, there is a set of disassembled components that can be sold directly, and a residual set intended for material recovery. There is also a cost, D_{is}, associated with each plan, which represents the revenue from the disassembled compoments reduced by the cost of disassembly. Because of the direct disposal of components, a reduction factor, f_{is} is computed as:

$$f_{is} = \frac{\text{Weight of residual set in plan}s}{\text{Original weight of assembly}}.$$

As an example, the data in Table 3 shows the approximate material content and several candidate disassembly plans for a personal computer. Although the PC consists of several smaller components, such as connectors and wires in addition to those shown, for the discussion that follows we limit our attention to the major components only. Thus, if the computer is sold without any processing (disassembly), the fraction of gold recoverable from the shredded mixture is 0.005%. However, if the power supply, case and drives are removed by disassembly (as shown in Row 2), and the remaining fraction is processed for metal recovery, the concentration of gold increases to 0.05%, almost a ten fold increase. This is approximately true for silver and platinum as well. Thus, the cost of separating the case and power supply can now be compared with the savings in processing the higher concentration motherboard and daughter-board scrap, and an economically optimal decision can be made. Thus, several candidate disassembly sets can be generated a-priori for each product type in the recycler's inventory, and the model selects the optimal level of each candidate plan that must be used. Several disassembly plans may be selected for the same product type, because of the varying contributions of different plans to different metal concentrations. The requirement that the disassembly plans be generated manually is not considered to be excessively limiting since there will usually be only a few dominant disassembly plans for electronics products. Also, although a bipartite paprtition of each disassembled subset has been considered here, a multipartite separation of each subset can also be accommodated without any change in the structure of the model. Landfilling options can also be included by representing the landfill as a (dummy) smelter, from whom no metal is recovered and a charge is applied for any amount shipped. Plastics recovery can also be addressed by designating a smelter as specializing in the recovery of plastics. However this model does not optimize pastics separation and related issues explicitly.

As an illustration of the application of this model, a hypothetical scenario is represented in Figure 5 below, where two products are being evaluated for recycling. Two smelters are available, both capable of processing all metals (metal 1 and 2) in these products. Product 1 can be processed either as-is, or can be further disassembled (at a cost of $0.40/item). The quantities and other model specific parameters are shown in the figure. Solving (11)–(16) with this data results in an optimal solution where 5200 pounds of product 1 is sent to smelter 1 for processing after disassembly, and the rest of product 1 is sent to smelter 2 for metal recovery. All of product 2 is sent to smelter 2 for processing. The split in Product 1 allows metal 2 to be recovered from product 2 by raising the composite concentration to a level above the minimum required.

This example helps illustrate the complexity of decisions that arise when recycling electronics for material recovery, and the utility of models such as (11) – (16) in facilitating the economic sustainability of electronics recycling.

Smelter model

Smelters receive items or parts from recyclers for processing. In some cases, smelters are required by contract to accept many of the items sent for metal recovery, but non-contract materials may also be processed to fill the smelters remaining capacity. When the smelter acquires items to recover metals for profit, a variable cost per unit of recovered metal and a fixed cost for extracting metal is incurred.

A model that allows the smelter to maximize profit (17) subject to a material balance constraint (20), a minimum required material content for recovery (21), and a capacity limitation (22) is:

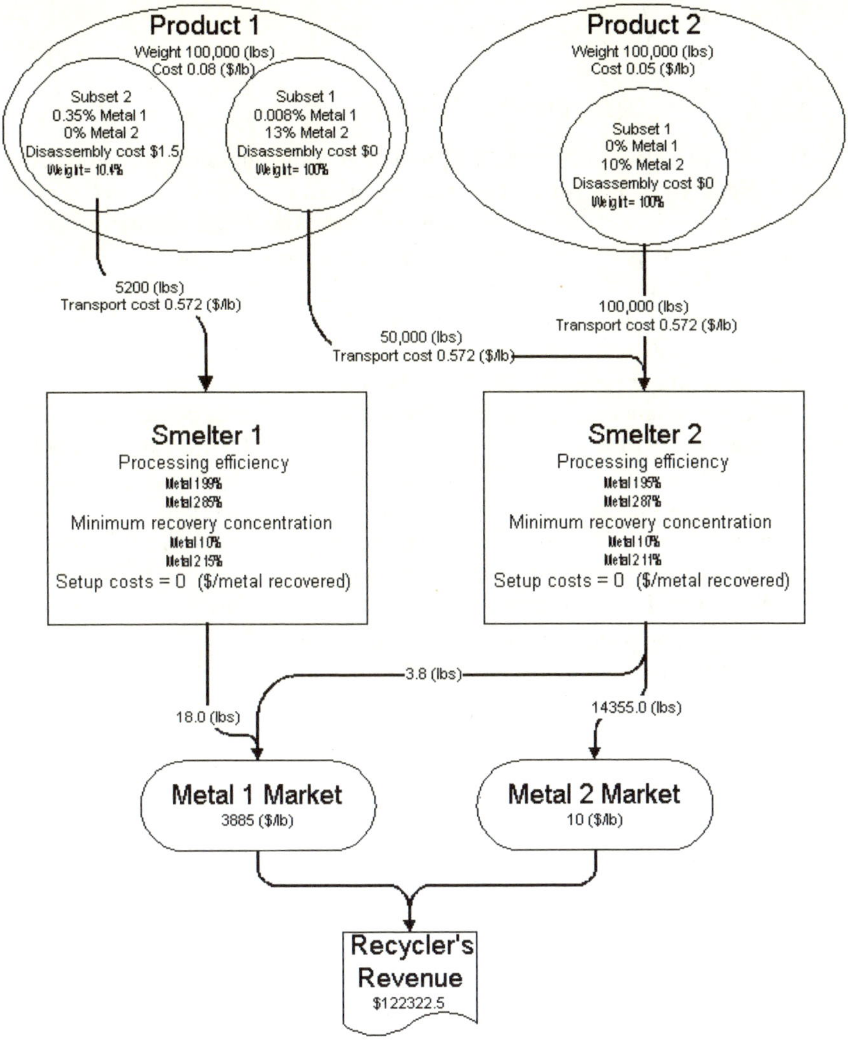

Fig. 5. Illustration of recycler's model (11)–(16)

$$Max\ z = \sum_{k=1}^{K} \sum_{j=1}^{J} (R_k G_{kj} - F_k\, Y_{kj} - \pi \sum_{i=1}^{I} X_{ij}) \tag{17}$$

s.t.

$$G_{kj} \leq \sum_{i=1}^{I} X_{ij}\, A_{ik}\, T_k \quad \forall\, (k,j) \tag{18}$$

$$M\, Y_{kj} \geq G_{kj} \quad \forall\, (k,j) \tag{19}$$

$$X_{ij} \leq W_{ij} \quad \forall\, (i,j) \tag{20}$$

$$\frac{\sum_{i=1}^{I} X_{ij}\, A_{ik}}{\sum_{i=1}^{I} X_{ij}} \geq Y_{kj}\, B_k \quad \forall\, (k,j) \tag{21}$$

$$\sum_{i=1}^{I}\sum_{j=1}^{J} X_{ij} \leq Cap \tag{22}$$

$$X_{ij}, G_{kj} \in R^{+} \; \forall\,(i,j,k);\; Y_{kj} \in [0,1] \; \forall\,(k,j) \tag{23}$$

The objective (17), maximizes the profit after deducting setup and processing charges. Here, R_k is the per unit revenue from metal k. F_k is the fixed cost of processing metal k. π is the per unit cost of processing at the smelter, which is assumed to be independent of the item type processed. However, if this is not the case, the corresponding term would be $\sum_{j=1}^{J}\sum_{i=1}^{I}\pi_i X_{ij}$. The first constraint (18), calculates G_{kj}, the amount of metal k recovered from recycler j and limits this quantity to be less than the amount of metal recoverable from the material sent from recycler j. Constraint (19), requires Y_{kj} to be one if metal k is recovered from the material sent by smelter j, i.e. G_{kj} is greater than zero. The quantity constraint, (20) requires the amount of item i sent from recycler j, X_{ij} to be less than the amount of item i available for processing from recycler j, W_{ij}. The blending constraint (21) requires that the concentration of metal k, in the items sent for processing from recycler j, $(\sum_{i=1}^{I} X_{ij} A_{ik})/(\sum_{i=1}^{I} X_{ij})$ to be greater than B_k, the minimum concentration requirement for metal k, if metal k is to be recovered from the material sent from recycler j, if the variable Y_{kj} is set to one. The capacity limitation (22), requires the total number of items sent for processing $\sum_{i=1}^{I}\sum_{j=1}^{J} X_{ij}$ to be less than the smelters capacity Cap. Finally, (23), requires Y_{kj} to be binary and G_{kj} and X_{ij} to be positive.

This model is also non-linear, because of constraints (21). Here too a branch-and-bound scheme by fixing the binary variables may be preferred to general approaches such as benders decomposition [27], outer approximations [28] because of the linear nature of the problem once the binary variables are fixed.

Conclusion

In this paper, a study of reverse channels for end-of-life electronics products is presented in a framework similar to that developed by [15]. The economics of electronics recycling have been modeled from the viewpoints of the generators, recyclers, and material processors separately. A variety of mathematical programming models, representative of the many ways in which the recycling industry currently operates, have been proposed, along with a discussion of solution techniques. A numerical illustration has been included for the recycler model. In addition, an integrated consideration of the disassembly and material recovery problem has been formulated at the recycler's level. Although an intergrated model representing the entire system would result in lower costs than those obtained using the models that separately address the decisions at each level, it is unlikely that the actual execution of decisions would be guided towards such an idealized goal. The models proposed here depict the generators, the recyclers and the processors/smelters of end-of-life electronics products each acting in their best interests, which is perhaps a more accurate representation of the actual behavior of these players in reality. The models set up in this paper can be used by recyclers and processors for optimizing recycling operations, and can also be used to study decisions such as pooling recycling efforts at the generators level, subsidizing takeback for recyclers etc.

Nomenclature
Source model (1–4)

i = item
j = recycler
p = location
SR_{ij} = per unit cost of sending item i to recycler j
TC_{ipj} = per unit transportation cost to send item i from location p to recycler j
V_{ipj} = volume of item i sent from location p to recycler j
W_{ip} = weight of item i available for pickup at location p

Recycler model (5–10)

A_{ik} = amount of metal k in item i
B_{kl} = minimum concentration of metal k for processing at smelter l
C_i = per unit acquisition cost of item i
CS_l = per unit transportation cost to smelter l
G_{kl} = amount of metal k processed at smelter l
i = item
k = metal
l = smelter
M = a large number
R_k = per unit revenue from metal k
S_i = subsets (disassembly) plans for part i
S_{kl} = setup cost for metal k at smelter l
T_{kl} = transformation coefficient of metal k at smelter l
W_i = weight of item i available for processing
X_{il} = amount of item i sent to smelter l
Y_{kl} = assignment variable for metal k at smelter l

Recycler model with product disassembly (11–16)

A_{iks} = amount of metal k in subset s of item i
B_{kl} = minimum concentration of metal k for processing at smelter l
CS_l = per unit transportation cost to smelter l
D_{is} = cost to disassemble subset s from item i
f_{is} = fraction of total weight of item i in subset s
G_{kl} = amount of metal k processed at smelter l
i = item
k = metal
l = smelter
M = a large number

R_k = per unit revenue from metal k
s = subset of parts
S_{kl} = setup cost for metal k at smelter l
T_{kl} = transformation coefficient of metal k at smelter l
W_i = weight of item i available for processing
X_{ils} = amount of subsets s from item i sent to smelter l
Y_{kl} = assignment variable for metal k at smelter l

Smelter model (17–23)

A_{ik} = amount of metal k in item i
B_k = minimum concentration of metal k for processing
Cap = capacity of smelter
F_k = fixed charge for processing metal k
G_{kj} = amount of metal k processed from items sent from recycler j
i = item
j = recycler
k = metal
M = a large number
π = per unit cost of processing an item
R_k = per unit revenue from metal k
T_k = transformation coefficient of metal k
W_{ij} = amount of item i available for processing from recycler j
X_{ij} = amount of item i processed from recycler j
Y_{kj} = assignment variable for processing metal k from recycler j

References

1. Business Publishers (1998) Slants and trends. Solid Waste Reports, vol 29
2. Fishbein BK (1998) Epr: What docs it mean? Where is it headed? Pollution Prevention Review 8: 43–55
3. Fleischmann M, Bloemhof-Ruwaard J, Dekker D, van der Laan E, van Nunnen JA, Van Wassenhove LN (1997) Quantitative models for reverse logistics: A review. European Journal of Operational Research 103: 1–17
4. Raymond R (1998) Will europe's producer responsibility system work? Resource Recycling Journal, http://www.raymond.com/RRJART.htm
5. Jahre M (1995) Logistics systems for recycling – efficient collection of household waste. PhD thesis, Chalmers University of Technology Sweeden
6. Beullens B, Van Oudheusden D, Cattrysse D (1999) Bi-directional waste collection: Impact of vehicle type and operations on transportation costs. In: Proceedings of the Twenty-fifth International Conference on Computers and Industrial Engineering, New Orleans, Louisiana, pp 5–14
7. Bullock D, Burk D (1989) A different scenario commingled vs. curbside sort. Biocycle 30(6): 35–36
8. Jayaraman V, Guide Jr DR, Srivastava R (1997) A reverse logistics supply chain model within a remanufacturing environment. In: Management of Engineering and Technology 1997 Portland International Conference on Management of Engineering and Technology

9. Ferrer G (1997) The economics of personal computer remanufacturing. Resource, Conservation and Recycling 21: 79–108
10. Sum EYL (1991) The recovery of metals from electronics scrap. JOM 43(4): 53–61
11. Sodhi M, Knight W (1998) Product design for disassembly and bulk recycling. Annals of the CIRP 47(1): 115–118
12. Lambert AJD (1999) Linear programming in disassembly / clustering sequence generation. Computers and Industrial Engineering 36: 723–738
13. Penev KD, de Ron AJ (1996) Determination of a disassembly strategy. International Journal of Production Research 34(2): 495–506
14. Johnson MR, Wang MH (1995) Planning product disassembly for material recovery opportunities. International Journal of Production Research 33(11): 3119–3142
15. Krikke HR, van Harten A, Schuur PC (1998) On a medium term product recovery and disposal strategy for durable assembly products. International Journal of Production Research 36: 111–139
16. Gungor A, Gupta SM (1999) Issues in environmentally conscious manufacturing and product recovery: A survey. Computers and Industrial Engineering 36: 811–853
17. Bloemhof-Ruwaard J, Salomon M, VanWassenhove LN (1996) The capacitated distribution and waste disposal problem. European Journal of Operational Research 88: 490–503
18. Kalisvaart S, Remmerswal J (1994) Proc. integrating impact assessment into lca. SETAC 55–61
19. Lauder P (1998) Batch decisions for bulk recycling. Master's thesis, The University of Rhode Island
20. Stessel RI, Pelz S (1994) Air classification of mixed plastics. In: National Waste Processing Conference Procedings ASME, pp 333–339
21. Hull JB, Langton CM, Jones AR (1994) An ultrasonic method for identifying polymeric materials/components for reclamation and recycling. Macromolecular Reports 31: 1191–1199
22. Sodhi MS, Young J, Knight WA (1999) Modeling material separation processes in bulk recycling. International Journal of Production Research 37(10): 2239–2252
23. Kaplan RS, Ness H (1987) Recycling of metals. Conservation and Recycling 10(1): 1–13
24. Moore JJ (1978) Recycling of non-ferrous metals. International Metals Reviews 23(5): 241–264
25. Ammons JC, Realff MJ, Newton DJ, Ceray S (1999) Determining reverse production systems infrastructure for electronic assemblies. In: 1999 IEEE International Symposium on Electronics and The Environment: Life-Cycle Environmental Stewardship for Electronic Products, Danvers, Mass
26. Reimer B (1999) Collection logistics for recycling. Master's thesis, The University of Rhode Island
27. Geoffrion A (1972) Generalized benders decomposition. JOTA 10(4): 237–260
28. Duran M, Grossmann I (1986) An outer-approximation algorithm for a class of mixed-integer nonlinear programs. Mathematical Programming 36: 307–339
29. Reimer B, Sodhi M (1999) Solving fixed charge blending models. Working Paper Industrial and Manufacturing Engineering Department, The University of Rhode Island
30. Stuart J, Ammons J, Turbini L (1999) A product and process selection model with multidisciplinary environmental considerations. Operations Research 47(2): 221–234

Integrated planning of acquisition, disassembly and bulk recycling: a case study on electronic scrap recovery*

Thomas Spengler, Martin Ploog, and Marcus Schröter

Department of Production Management, Technical University of Braunschweig, Katharinenstraße 3, 38106 Braunschweig, Germany (e-mail: t.spengler@tu-bs.de)

Abstract. Due to national and supranational legislation activities, the recovery of discarded products will attain an increasing momentum. Electronic equipment consists of many different parts and materials. Therefore, the related recovery process is often divided into disassembly to remove harmful substances or reusable parts and into bulk recycling to recover ferrous and non-ferrous metals. In order to consider the interactions between choice of scrap to be recovered (acquisition problem), disassembly and bulk recycling, a mixed-integer linear programming model for integrated planning of these stages is presented in this case study. It is applied to determine the daily allocation of products to processes for a major electronic scrap recovery centre that faces limited processing capacities and market restrictions. The optimization calculations covering typical discarded electronic products to be recycled in the related centre lead to a relevant improvement of the economic success.

Key words: Reverse logistics – Electronics recovery – Recovery planning

1 Introduction

It is estimated that in the European Union more than 6 million tons of "Waste on Electrical and Electronic Equipment" (WEEE) were generated in 1998. This quantity is predicted to increase by 3–5 % per year (European Commission, 2000). These discarded products should be recovered to prevent leachate pollution from

* The authors would like to thank the German "Bundesministerium für Bildung und Forschung" (Federal Ministry of Education and Research) for supporting the research project "Substance Flow Oriented Closed Loop Supply Chain Management in the Electrical and Electronic Equipment Industry (STREAM)".

Correspondence to: T. Spengler

landfills caused by substances like heavy metals. The recovery and the reuse of the valuable materials and parts may be another valid reason.

The recovery of electronic scrap is a multistage process. Logistic issues concern collection, grading, transport and allocation of discarded products as well as of recovered materials. Disassembly in order to remove hazardous substances and reusable parts is often followed by bulk recycling to recover separated material fractions that are sent to metal recycling facilities or other recycling specialists. In order to consider the complex interactions between these stages appropriately, the application of mathematical decision support appears promising (Fleischmann, 2000). In this case study, the determination of the daily recovery program concerning processed items and executed processes is examined for a major recovery centre.

The paper is structured as follows. In Section 2 the case is defined. Hence a review of technical processes and material flows, the planning problem and data as well as the related literature are outlined. Based on this, a mixed-integer linear programming model for the considered planning problem is developed and solved in Section 3. The results for the basic data set and the discussion of different solution procedures are presented in subsequent Section 4. Section 5 consists of a detailed interpretation of the numerical results for different scenarios. Conclusions and recommendations for future research are derived in the final Section 6.

2 Problem definition

2.1 Case description

This case study refers to the acquisition, disassembly and bulk recycling planning of the Electrocycling GmbH that recovers industrial and consumer electronics in an integrated recovery centre in Goslar (Germany). The fixed costs represent about a half of the total treatment costs due to the related investments of 15 MEuro. The operating costs result especially from disposal fees, maintenance and labour wages for more than 100 workers. A detailed description of the recovery centre is given by Koch and Kasper (1996).

In the considered planning problem, the system boundary is limited to the recovery works profit centre, not including the input storage. Other stages like transport, smelting or marketing of reusable parts are excluded since they are accomplished by external services. The material flow throughout the enterprise is shown in Figure 1. Long-term contracts for scrap collected by public and industrial collectors contribute less than 10% of the recovered mass. Thus electronic scrap is actively acquired by the input storage of the enterprise. This is a separate profit centre that may sell scrap to external customers as well as to the internal customer recovery works. After the recovery works profit centre has ordered the daily range of scrap types to be processed from the input storage, the delivery to the adjacent recovery works is accomplished immediately. The scrap is assessed with a given transfer price that is dependant on the scrap market. Therefore, the recovery works manager faces the internal "acquisition" decision as to which products have to be taken

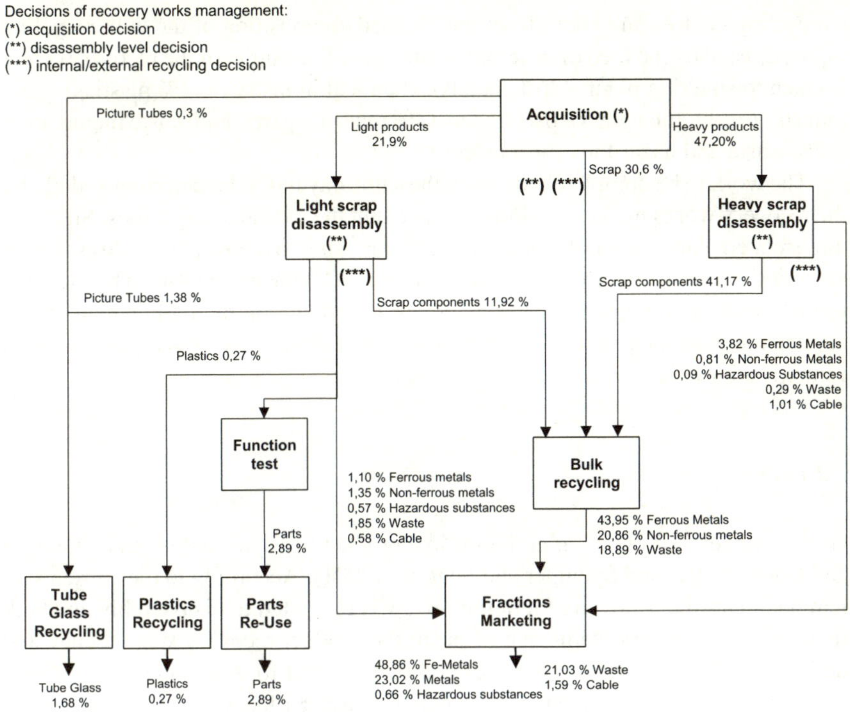

Fig. 1. Material flow of the recovery centre (Mass-%)

from storage and recovered in the considered daily planning period as seen at point (*) in Figure 1.

The first recovery step "disassembly" is composed of manual or partly-automated processes. Mechanical assistance addresses only transport and lifting tasks. In planning the disassembly step, the recovery works manager has to determine the disassembly level at point (**) in Figure 1, taking into account that some products do not have to be disassembled at all, that the reusable part demand as well as prices change daily and that some disassembly activities are mandatory in order to eliminate hazardous substances.

The second step "bulk recycling" is designed to recover precious fractions such as ferrous and non-ferrous metals from mixed electronic scrap, using unit operations like crushing and separation. When planning the bulk recycling step, the recovery works manager has to decide which scrap types are to be recycled internally or marketed externally as seen at point (***) in Figure 1. This is a complex problem due to the fact that bottlenecks can appear in different units. Based on a specific input composition and the expected separation of material fractions, the required capacity for every unit had been calculated in the process design phase. Thus the unit capacities have been fixed for the long term planning horizon. Deviations from the expected composition can lead to bottlenecks in the subsequent units if the components that should be separated in the preceding units of the facility, e.g. ferrous metals, are underrepresented in the feed. In order to balance these deviations, a recycling centre can benefit from the blended composition of the bulk recycling input adjusted by the feed of different scrap types. The output of the recovery centre is taken to smelting plants (70% metals), chemical industry ($< 1\%$ plastics), glass industry ($< 2\%$ tube glass), part traders (3% reusable parts) and waste incineration (22% waste and hazardous substances).

The work to be completely done on the same day has to be determined daily by the recovery works manager, taking into account the available capacities. Since the three related planning problems choice of "scrap types to recover" (*), "disassembly activities" (**) and "internal/external treatment" (***) have to be solved by the same person, show a complex interaction and are dependant on the daily situation, the application of an integrated short term planning model for the recovery works profit centre is advisable.

2.2 Literature review

General reviews of planning problems in a product life cycle are given by Gungor and Gupta (1999) and Spengler and Schröter (2001). According to the chosen system boundary the following literature review is restricted to disassembly and bulk recycling planning problems that occur in the product recovery stage. Penev and de Ron (1996) describe a static cost comparison tool to determine an economic disassembly level and sequence of a single product. Krikke et al. (1998) develop a method to determine a good product recovery and disposal policy of one product type taking into account technical, economic and ecological criteria. For this purpose a two-phased dynamic programming algorithm is presented. Meacham et al. (1999) provide an optimal disassembly configuration algorithm for single as well as for multiple products. Spengler et al. (1997) develop a mixed-integer linear programming model based on linear activity analysis for integrated dismantling and recycling of buildings. The purpose of the model is to obtain optimal dismantling strategies but also an optimal assignment of components and parts to recycling techniques. A planning tool for integrated planning of recovery and remanufacturing is given by Uzsoy and Venkatachalam (1998). Sodhi and Reimer (2001) describe an integrated disassembly and material recovery model for discarded electronic products. In addition to other works the presented model includes the process stage of metal smelting.

Only a few papers refer to operation research models for the economic assessment of bulk recycling operations. Lund et al. (1994) use linear programming to analyse the design and the operation of municipal waste material-recovery-facilities. Sodhi et al. (1999) present a dynamic programming model for the determination of the unit operations sequence for float-sink materials separation by density. Alternative disposal strategies for cars are assessed using goal programming by Isaacs and Gupta (1998), taking into account the profit for disassembler and shredder. Stuart and Lu (2000a,b) offer decision support for processing and reprocessing options in electronic scrap bulk recycling centres. Rudolph (1999) provides an integrated model for disassembly, remanufacturing and bulk recycling planning without giving a practicable processing description.

The planning situation presented in the preceding subsection differs from the assumptions of these papers concerning two major facets. On the one hand, in this case study scrap has to be acquired actively. The focus in the recent publications is laid rather on the question of how to treat given discarded products (Penev and de Ron, 1996; Krikke et al., 1998; Spengler et al., 1997; Uzsoy and Venkatachalam, 1998), e.g. on the determination of the disassembly level. On the other hand, the recovery centre referred to consists of disassembly and multi-level bulk recycling. The capacity restrictions are given in each processing step and can not be subsumed due to the complex interactions of disassembly activities, unit operations, material composition and chosen products. Since the recycling centre can benefit from the calculated composition of bulk recycling input, the consideration of the impact of input composition in combination with the basic engineering foundations of the processes in individual separation units is advisable. Few models take into account the interaction of subsequent recovery processes (Sodhi and Reimer, 1999; Rudolph, 1999; Isaacs and Gupta, 1998), but the unit capacities are not considered separately. The detailed bulk recycling models (Lund et al., 1994; Stuart and Lu, 1998a,b) neglect the interactions with disassembly. Thus the recent publications that deal with the application of operations research in recovery planning can not be directly applied in the considered case study. The development of an integrated model for the daily planning of acquisition and recovery processes is necessary.

2.3 Planning data

The application of recovery planning models is often obstructed by the lack of data. In this case study, information is provided by a database that contains product and processing data. Since the composition of discarded electronic products shows significant variations, the products have to be classified into different types. Because of the variations within these types, a representative product has to be chosen for each type, and a detailed range of possible disassembly activities as well as an inventory of material composition have to be prepared. The delivery of recovery-relevant information is a complex problem that should be solved by production enterprises, referring to the forthcoming directive of the EU on Waste on Electric and Electronic Equipment (European Commission, 2000). A practical concept seems to be the "recycling passport" of the Belgian-German equipment producer AGFA-Gevaert AG that contains data about hazardous substances, composition and reusable parts (Verein Deutscher Ingenieure, 2000). It can be used to plan recovery processes and to estimate disassembly times. Based on this, the development of a database with recovery data provides necessary information for planning at recycling facilities.

2.4 Recovery process analysis

The input-output-relations throughout the execution of the possible disassembly activities are given in Table 1. The index i ($i = 1, \ldots, 29$) describes the scrap types (products, parts, materials). The range of discarded products ($i = 1, \ldots, 6$) that is

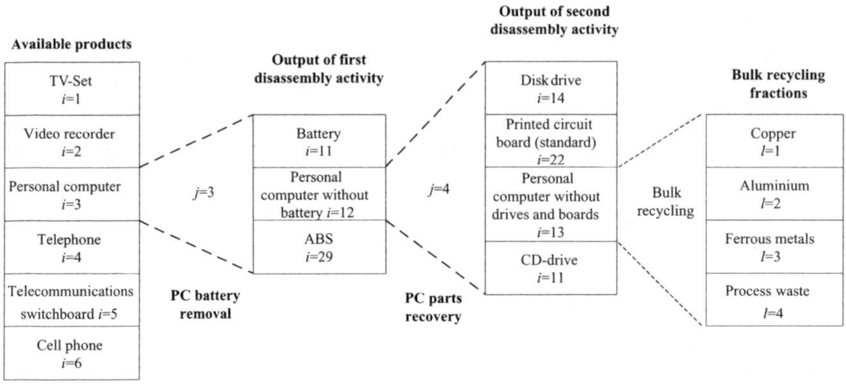

Fig. 2. Model of discarded product recovery (example PC)

available in input storage as well as of subsequent disassembly output materials ($i = 29$) and parts ($i = 7, \ldots, 28$) are shown in Table 1. The column index j ($j = 1, \ldots, 10$) specifies the possible disassembly activities. Each row i describes the result of one execution of a certain disassembly activity j that represents the disassembly of exactly one piece of a defined scrap type. Negative values of the given coefficient v_{ij} represent the mass input of the scrap type to the activity, positive values the mass output. A time factor for the execution of one activity j is presented in the bottom row.

The structure of the product recovery including disassembly activities and bulk recycling is depicted in Figure 2. A "personal computer" chosen from the discarded products range must be opened and the harmful part "battery" has to be removed (disassembly activity $j = 3$). The option to recover spare parts ($j = 4$) can be exercised if there is a market for these parts. Residues can be treated in bulk recycling in order to recover metals or can be marketed externally. As seen in Table 1, one "personal computer" ($i = 3$) of 10 kg is dismantled into one "battery" ($i = 7$) of 0,05 kg, into "ABS" ($i = 29$) from casings of 0,5 kg and into one "personal computer without battery" ($i = 12$) of 9,45 kg. This "personal computer without battery" ($i = 12$) can be disassembled by the activity "personal computer spare parts recovery" ($j = 4$) leading to one "CD-drive" ($i = 11$), one "disk-drive" ($i = 14$), one "printed circuit board (standard)" ($i = 22$) and one "personal computer without drives and boards" ($i = 13$) with the given masses. In Table 1 typical disassembly activities like the required removal of hazardous substances or the recovery of reusable parts are shown. Since these activities are manual work, the relevant cost-driver is labour. Disassembly output has to be directed towards external or internal treatment. In case of external treatment, the output is collected in boxes and marketed externally, e.g. as spare parts or to be treated by recycling specialists. Otherwise, the disassembled parts are discharged by a belt or by gravitation to the input of the bulk recycling facility where they are mixed with other products and parts.

The flowsheet of the bulk recycling step is depicted in Figure 3. The multi-level bulk recycling consists of size-reduction and separation units. The size-reduction to isolate and concentrate the valuable materials is accomplished by several

Table 1. Scrap types and disassembly activities Input-Output-Matrix

	Activity	TV dismantling	Video dismantling	PC battery removal	PC parts recovery	Telephone dismantling	Switch-board opening	Switch-board board removal	Switch-board parts removal	Cellular phone accu removal	Cellular phone board removal
	v_{ij} $j=$	1	2	3	4	5	6	7	8	9	10
		[kg]	[kg]	[kg]	[kg]	[kg]	[kg]	[kg]	[kg]	[kg]	[kg]
	Products										
$i=1$	TV-set	−25	0	0	0	0	0	0	0	0	0
2	Video recorder	0	−4	0	0	0	0	0	0	0	0
3	Personal computer	0	0	−10	0	0	0	0	0	0	0
4	Telephone	0	0	0	0	−1.5	0	0	0	0	0
5	Telecommunication switchboard	0	0	0	0	0	−100	0	0	0	0
6	Cellular phone	0	0	0	0	0	0	0	0	−0.2	0
	Parts										
7	Battery	0	0	0.05	0	0	0	0	0	0	0
8	Coverings	0	0	0	0	0	5	0	0	0	0
9	Accumulator	0	0	0	0	0	0	0	0	0.1	0
10	Cathode ray tube	13.75	0	0	0	0	0	0	0	0	0
11	CD-Drive	0	0	0	0.2	0	0	0	0	0	0
12	Personal computer without battery	0	0	9.45	−9.45	0	0	0	0	0	0
13	Personal computer without drives	0	0	0	8.09	0	0	0	0	0	0
14	Disk-drive and boards	0	0	0	0.2	0	0	0	0	0	0

Table 1 (continued)

	Activity	TV dismantling	Video dismantling	PC battery removal	PC parts recovery	Telephone dismantling	Switch-board opening	Switch-board board removal	Switch-board parts removal	Cellular phone accu removal	Cellular phone board removal
	v_{ij} $j =$	1	2	3	4	5	6	7	8	9	10
		[kg]	[kg]	[kg]	[kg]	[kg]	[kg]	[kg]	[kg]	[kg]	[kg]
	Parts										
15	Mixed entertainment electronics parts	6.25	3.2	0	0	0	0	0	0	0	0
16	Housing parts	3.75	0.4	0	0	0	0	0	0	0	0
17	Telephone back	0	0	0	0	0.6	0	0	0	0	0
18	Cellular phone body	0	0	0	0	0	0	0	0	0.1	−0.1
19	Cellular phone waste	0	0	0	0	0	0	0	0	0	0.05
20	Printed circuit board (precious)	0	0	0	0	0	0	4.75	0	0	0
21	Printed circuit board (cheap)	1.25	0.4	0	0	0	0	0	0	0	0
22	Printed circuit board (standard)	0	0	0	0.96	0	0	0	0	0	0.05
23	Switch board parts	0	0	0	0	0	0	0	9.025	0	0
24	Telephone front	0	0	0	0	0.45	0	0	0	0	0
25	Telecommunication switchboard without boards	0	0	0	0	0	0	90.25	−90.25	0	0
26	Telecommunication switchboard (open)	0	0	0	0	0	95	−95	0	0	0

Table 1 (continued)

	Activity		TV dismantling	Video dismantling	PC battery removal	PC parts recovery	Telephone dismantling	Switch-board opening	Switch-board board removal	Switch-board parts removal	Cellular phone accu removal	Cellular phone board removal
	v_{ij}	$j =$	1	2	3	4	5	6	7	8	9	10
			[kg]	[kg]	[kg]	[kg]	[kg]	[kg]	[kg]	[kg]	[kg]	[kg]
	Parts											
27	Telecommunication switchboard (frame)		0	0	0	0	0	0	0	81.225	0	0
28	Telephone receiver		0	0	0	0	0.45	0	0	0	0	0
	Materials											
29	ABS		0	0	0.5	0	0	0	0	0	0	0
	Time Factor t_j^Z	[h/act]	0.1	0.03	0.03	0.02	0.006	0.01	0.05	0.05	0.005	0.015

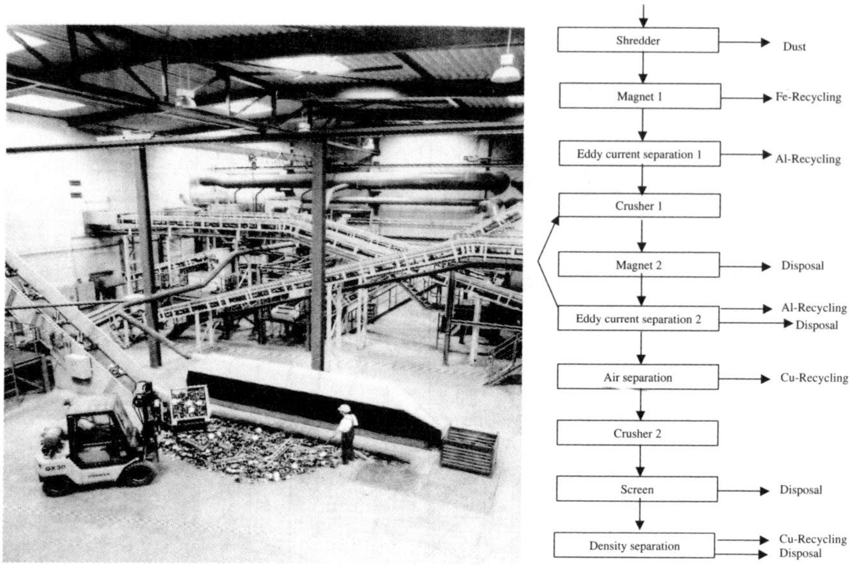

Fig. 3. Bulk recycling picture and flowsheet

totally automated shredding and crushing units and has decisive importance for the success of the separation. Ferrous metals represent a major share of the bulk recycling input. Their removal by magnetic separators takes place after the first shredding step since the size-reduction of ferrous metals is very energy- and wear-intensive. Other valuable materials like aluminium and copper are recovered by eddy-current separators in which metals are isolated due to their electric conductivity (Schubert, 1994), by density-separation processes and by classifying stages. The remaining non-valuable materials like paper or mixed plastics can be found in the process waste that is sent to external incineration. The range of material components in the input of the units as well as of separated material fractions in the output of the units is indicated with the index l as shown in Table 2. Relevant cost drivers in bulk recycling are maintenance, energy consumption and labour (Meier-Staude and Mersmann, 1997).

Table 2. Materials in bulk recycling

l	Material
1	Copper
2	Aluminium
3	Ferrous metals
4	Process waste

2.5 *Planning horizon*

The acquisition planning in the enterprise is divided into two levels. First, the scrap has to be acquired at the market by the input storage profit centre. For the related tasks like supplier negotiations, forecasting or logistics, the planning horizon ranges from one week to six months. In this case study, however, due to the "recovery works" as the given system boundary, only the daily internal acquisition is considered. Thus acquisition and delivery from the adjacent storage is executed within few minutes, and acquisition capacity constraints are given by the masses available from the input storage of the recovery enterprise. The available disassembly capacity is determined by the maximal number of operators who can be delegated to disassembly that day. In bulk recycling, each unit may treat a defined mass per period. The related time period is influenced by the number of shifts per day and by expected downtimes due to maintenance operations. In the case examined, the daily available unit capacities result from one shift without downtimes. The total time required for internal acquisition and transport as well as for disassembly and bulk recycling is exceeded significantly by the given capacity constraint horizon, and thus the time needed for acquisition and processing can be taken into account using the given capacity restrictions. Based on the analysis of recovery processes, the daily planning problem of determining the short term recovery program becomes apparent: Taking into consideration the structure and composition of these scrap types, the different disassembly activities and unit operations, market prices and costs, the possible reuse options, achievable proceeds as well as capacity and market constraints, the recovery works manager has to determine daily the optimal choice of scrap types to recover, the optimal allocation of disassembly activities and the optimal choice of scrap types input to bulk recycling.

3 Model formulation and solution procedure

In this section, an integrated short term planning model based on the material flows throughout the recovery centre is provided. It leads to an optimal solution of the presented decision problem maximising the total achievable marginal income. The chosen model formulation is based on linear activity analysis (Koopmans, 1951) that permits a very simple and appropriate model formulation of recovery planning problems (Souren, 1996; Spengler, 1998). In the model the depiction of the disassembly sector corresponds to the mixed-integer disassembly optimization model that is given by Spengler et al. (1997). The structure of the material flows is shown in Figure 4.

It is assumed that a number of different discarded products, parts and materials i are available in the input storage at a given negative or positive transfer price. These I scrap types can be disassembled by the execution of J different disassembly activities and can be processed by a bulk recycling plant using K process units. The following notation is used:

Indices

- i Scrap types: products ($i \in \{1, \ldots, I_1\}$), parts ($i \in \{I_1 + 1, \ldots, I_2\}$), materials ($i \in \{I_2 + 1, \ldots, I\}$)
- j Disassembly activity $j \in \{1, \ldots, J\}$
- k Unit operation $k \in \{1, \ldots, K\}$
- l Material $l \in \{1, \ldots, L\}$

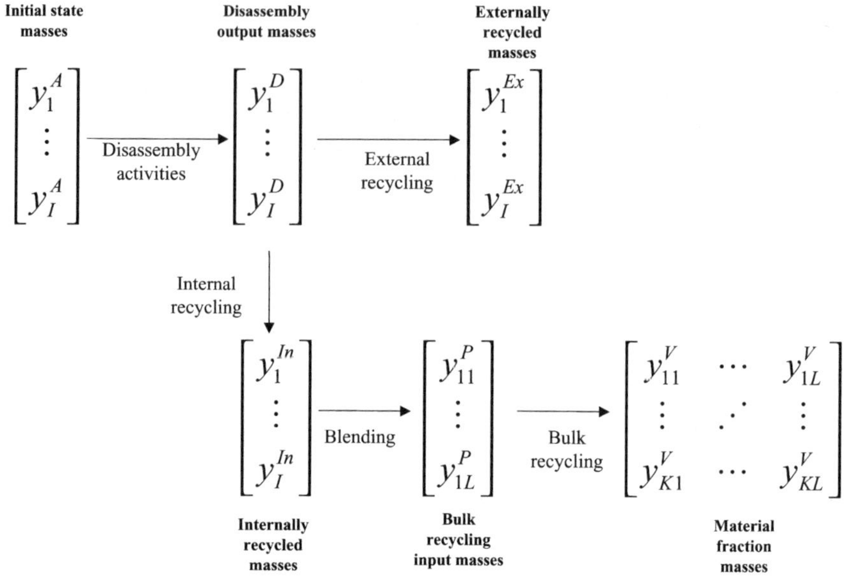

Fig. 4. Model of material flows

Parameters, coefficients and limits

a_{il}	Composition factor that assigns scrap type i composition to material component l [kg/kg] $(0 \leq a_{il} \leq 1)$
c_k^P	Bulk recycling cost factor for unit operation k [Euro/kg]
c^Z	Disassembly labour cost factor [Euro/h]
e_i^{Ex}	Cost (−) or price (+) factor for scrap type i to external recycling [Euro/kg]
e_i^A	Acceptance cost $(-)$ or price $(+)$ factor for scrap type i [Euro/kg]
e_{kl}^V	Recycling material sale cost $(-)$ or price $(+)$ factor of isolated material fraction l separated by separation unit k [Euro/kg]
m_i	Mass of one piece of scrap type i [kg] (m_i is only defined for products and parts if reuse has to be considered)
$T^{\max}$	Limit for disassembly labour time [h]
t_j^Z	Disassembly time needed for one execution of activity j [h/act]
u_{kl}^P	Fraction of material l extracted at unit operation k [kg/kg] $(0 \leq u_{kl}^P \leq 1)$
v_{ij}	Disassembly activity coefficient for the input $(-)$ or output $(+)$ masses of scrap type i caused by one execution of activity j [kg/act]
$y_i^{A,\max}$	Limit for the maximum mass of scrap type i to be obtained from storage [kg]
$y_i^{Ex,\max}$	Limit for sale capacity of scrap type i to external recycling [kg]
$y_k^{P,\max}$	Limit for equipment capacity of separation unit k [kg]
$y_{kl}^{V,\max}$	Limit for sale capacity of isolated material fraction l separated by unit k [kg]

Variables

$x_j \in N_0$	Number of executions of disassembly activity j [act]
$y_i^A \in \left[0; y_i^{A,\max}\right]$	Mass of scrap type i to be taken from storage for processing [kg] with $y_i^A / m_i \in N_0$ if m_i is defined
$y_i^D \in [0; \infty)$	Mass of scrap type i after disassembly [kg]
$y_i^{Ex} \in \left[0; y_i^{Ex,\max}\right]$	Mass of scrap type i to external recycling [kg] with $y_i^{Ex} / m_i \in N_0$ if m_i is defined
$y_i^{In} \in [0; \infty)$	Mass of scrap type i to internal recycling [kg]; $y_i^{In} = 0$ if i contains hazardous substances
$y_{kl}^P \in [0; \infty)$	Mass of material component l in the mixture that is treated in unit operation k [kg]
$y_{kl}^V \in \left[0; y_{kl}^{V,\max}\right]$	Mass of isolated material fraction l separated by unit operation k [kg]

Based on these notations the decision problem can be formulated as a mixed-integer linear programming (MILP) model.

$$MAX \sum_{i=1}^{I} (y_i^A \cdot e_i^A + y_i^{Ex} \cdot e_i^{Ex}) + \sum_{k=1}^{K} \sum_{l=1}^{L} y_{kl}^V \cdot e_{kl}^V - \sum_{j=1}^{J} x_j \cdot t_j^Z \cdot c^Z - \sum_{k=1}^{K} c_k^P \cdot \left(\sum_{l=1}^{L} y_{kl}^P \right) \tag{1}$$

subject to:

$$y_i^D = y_i^A + \sum_{j=1}^{J} x_j \cdot v_{ij} \qquad i = 1, \ldots, I \tag{2}$$

$$y_i^D = y_i^{Ex} + y_i^{In} \qquad i = 1, \ldots, I \tag{3}$$

$$y_{kl}^P = \begin{cases} \sum_{i=1}^{I} a_{il} \cdot y_i^{In} & \text{for } k = 1 \\ y_{(k-1)l}^P - y_{(k-1)l}^V & \text{for } k = 2, \ldots, K \end{cases} \qquad l = 1, \ldots, L \tag{4}$$

$$y_{kl}^V = u_{kl}^P \cdot y_{kl}^P \qquad k = 1, \ldots, K \; l = 1, \ldots, L \tag{5}$$

$$\sum_{l=1}^{L} y_{kl}^P \leq y_k^{P,\max} \qquad k = 1, \ldots, K \tag{6}$$

$$\sum_{j=1}^{J} x_j \cdot t_j^Z \leq T^{\max} \tag{7}$$

The objective function (1) maximizes the total achievable marginal income resulting from acceptance revenues/costs, disassembly output revenues/costs, bulk recycling output revenues/costs, variable disassembly costs and variable process costs. The three short term decisions are depicted by the decision variables:

- the mass of scrap type i to be taken (y_i^A),
- the number of executions of disassembly activity j (x_j),
- the mass of scrap type i directed to internal recycling (y_i^{In}).

The other variables y_i^D, y_i^{Ex}, y_{kl}^P and y_{kl}^V are not essential due to the fact that they could be expressed by means of the decision variables. However, the explicit consideration of these material flows is a precondition for the detailed interpretation of the solution. Thus the values of y_i^D, y_i^{Ex}, y_{kl}^P and y_{kl}^V are determined by the constraints (2)-(5). Disassembly activities are modelled with the number of executions of the disassembly of this activity x_j and input-output coefficients v_{ij} that represent the input-output-relationships of the disassembly activities seen in Table 1 (2). The obtained disassembly output y_i^D has to be directed either to external or to internal treatment (3). The disassembly output to internal treatment y_i^{In} is completely processed in the bulk recycling units.

It is assumed that the product structure is completely destroyed in the first unit shredder. This simplification is based on the scrap refining studies of Koch and Kasper (1996) who state a liberation degree for ferrous metals of about 90% after shredding and for copper of more than 96% after crushing. At the moment of destruction, the composition coefficients a_{il} assign the scrap types i to a material component l. The input masses of the other units can be calculated by the mass balance equations for each unit (4). The separation units are described by linear coefficients u_{kl}^P representing the share of the available material l in the input of the unit k that is directed into material fraction l of this unit (5). As seen in Figure 5, a share of 0,75 of available "ferrous metals" ($l = 3$) is separated in unit "magnet 1" ($k = 2$). The values of u_{kl}^P have been determined from empirical data and are shown in Table 3. Bottlenecks in bulk recycling can appear at every unit due to variations in feed composition. Thus capacity constraints must be depicted by limits for every unit (6). Disassembly capacity refers to a maximum of labour time (7).

Input supply capacity and output sales capacity are given by the domains of the variables y_i^A; y_i^{Ex}; y_{kl}^V. The mandatory removal of hazardous substances before the treatment in bulk recycling is forced by the setting $y_i^{In} = 0$ initiating external treatment for scrap types i that contain hazardous substances. Disassembly activities x_j as well as the number of discarded products and parts y_i^A/m_i; y_i^{Ex}/m_i are modelled as integer variables. This might not be essential if the values of these variables are high enough. In that case the LP-relaxation of the MILP will lead to an acceptable solution quality.

With the product range that is shown in Table 1, the mixed-integer linear optimization model consists of 68 integer variables, 224 non-integer variables and 131 constraints. Standard optimization software packages using Branch-and-Bound techniques provide a quick solution in case of the given problem complexity. Therefore, the presented integrated recovery planning model has been implemented on a personal computer (Pentium III / 600 MHz) referring to EXCEL-spreadsheets that contain the data sets and using the commercial solver LINGO for the optimization calculations. Using these standard software tools, one takes advantage of their public availability and the absence of additional implementation costs. The solution time for the given problem is less than 2s.

Table 3. Linear coefficients u_{kl}^P and prices e_{km}^V in bulk recycling units (other $u_{kl}^P = 0$)

			Dust from shredder	Ferrous metals from magnet 1	Aluminium from Eddy current 1	Aluminium from Eddy current 2	Ferrous metals from magnet 2	Waste from Eddy current 2	Copper from air separation	Waste from Screen	Copper from density separation	Waste from density separation
Unit	$k =$		1	2	3	4	4	4	5	7	8	8
Material	$l =$		4	3	2	2	3	4	1	4	1	4
Separation coefficient	u_{kl}^P	[kg/kg]	0.3	0.75	0.75	1	1	0.45	0.7	0.22	1	1
Unit	$k =$		1	2	3	4	4	4	5	7	8	8
Material	$l =$		4	3	2	2	3	4	1	4	1	4
Material price	e_{kl}^V	[Euro/kg]	−0.15	0.05	1	1	0.04	−0.15	1.5	−0.15	1.5	−0.1

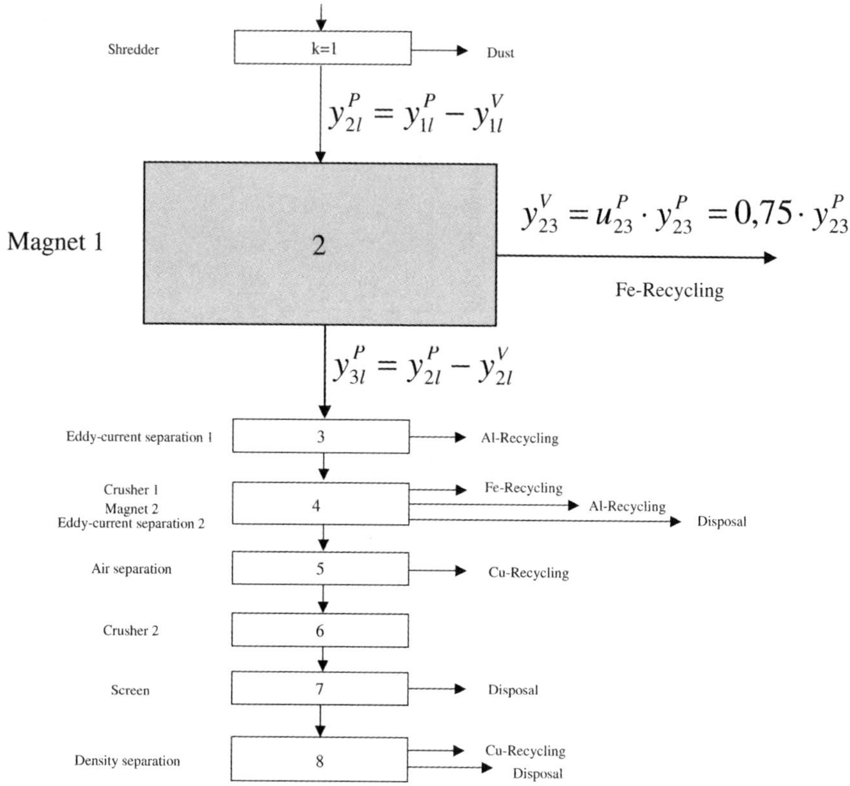

Fig. 5. Bulk recycling model

4 Numerical results

In this section, the complete solution data for one optimization calculation are presented. The impact of different solution procedures for the given planning problem like a non-integer model or intuitive strategies is discussed. The solution to the short term recovery program is shown in Tables 4 and 5. The masses of treated scrap types are given by the values of decision variables y_i^A and y_i^{In} in Table 4. In Table 5 the number of disassembly activities to be executed is stated by the value of decision variable x_j.

All available "TV-sets" ($i = 1$) are taken and disassembly ($j = 1$) is enforced. The subsequent parts like "cathode ray tubes" ($i = 10$), "housing parts" ($i = 16$) and "cheap circuit boards" ($i = 21$) are marketed externally. A share of the obtained "mixed entertainment electronic parts" ($i = 15$) is treated internally, another share of them is treated externally due to the capacity restrictions in the bulk recycling units. "Video recorders" ($i = 2$), which are all accepted, are sent to bulk recycling directly. Only a fraction of available "personal computers" ($i = 3$) is taken. This quantity is limited by the market restriction for "CD-drives" ($i = 11$). Mandatory disassembly ($j = 3$) is done and the "battery" ($i = 7$) is sent to a recycling specialist. The resulting parts from the operation spare part recovery ($j =$

Table 4. Scrap type data and decision variable y_i^A and y_i^{In} results

			Acceptance price	External recycling price	Taken mass	Maximal mass to be processed	Inter-nally recycled mass	Exter-nally recycled mass	Maximum of external recycled mass	Mass per piece	Composition factor Cu	Composition factor Al	Composition factor Fe	Composition factor waste	Contains hazardous substances
i	Name		e_i^A	e_i^{Ex}	y_i^A	$y_i^{A,\max}$	y_i^{In}	y_i^{Ex}	$y_i^{Ex,\max}$	m_i	a_{i1}	a_{i2}	a_{i3}	a_{i4}	
	Unit		[Euro/kg]	[Euro/kg]	[kg]	[kg]	[kg]	[kg]	[kg]	[kg/piece]	[–]	[–]	[–]	[–]	
1	TV-set	Product	0.625	–	3,000	3,000	0	–	0	25	0,03	0,00	0,13	0,85	yes
2	Video recorder	Product	0.3	–	2,000	2,000	2.000	–	0	4	0,09	0,00	0,40	0,51	no
3	Personal computer	Product	0.1	–	300	2,000	0	–	0	10	0,15	0,00	0,47	0,38	yes
4	Telephone	Product	0.25	–	600	600	600	–	0	1.5	0,09	0,00	0,08	0,84	no
5	Telecommuni-cation switchboard	Product	0	–	2,000	2,000	0	–	0	100	0,11	0,10	0,62	0,17	no
6	Cellular phone	Product	1	–	412	500	0	–	0	0.2	0,05	0,00	0,00	0,95	yes
7	Battery	Part	–	−0.1	–	0	0	1.5	–	–	0,00	0,00	0,00	1,00	yes
8	Coverings	Part	–	−0.1	–	0	0	100	–	–	0,00	0,00	0,90	0,10	no
9	Accumulator	Part	–	−0.25	–	0	0	206	–	–	0,00	0,00	0,00	1,00	yes
10	Cathode ray tube	Part	–	−0.05	–	0	0	1650	–	–	0,00	0,00	0,00	1,00	yes
11	CD–Drive	Part	–	5	–	0	0	6	6	0.2	0,20	0,00	0,30	0,50	no
12	Personal computer without battery	Part	–	−0.1	–	0	0	0	–	–	0,16	0,00	0,44	0,40	no

Table 4 (continued)

i	Name		Acceptance price	External recycling price	Taken mass	Maximal mass to be processed	Inter-nally recycled mass	Exter-nally recycled mass	Maximum of external recycled mass	Mass per piece	Composition factor Cu	Composition factor Al	Composition factor Fe	Composition factor waste	Contains hazardous substances
			e_i^A	e_i^{Ex}	y_i^A	$y_i^{A,\max}$	y_i^{In}	y_i^{Ex}	$y_i^{Ex,\max}$	m_i	a_{i1}	a_{i2}	a_{i3}	a_{i4}	
	Unit		[Euro/kg]	[Euro/kg]	[kg]	[kg]	[kg]	[kg]	[kg]	[kg/piece]	[–]	[–]	[–]	[–]	
13	Personal computer without drives and boards	Part	–	–0.1	–	0	242	0	–	–	0,15	0,00	0,50	0,35	no
14	Disk-drive	Part	–	1	–	0	1	5	5	0.6	0,20	0,00	0,30	0,50	no
15	Mixed entertainment electronics parts	Part	–	–0.1	–	0	351.3	398.7	–	–	0,10	0,00	0,50	0,40	no
16	Housing parts	Part	–	–0.1	–	0	0	450	–	–	0,00	0,00	0,00	1,00	no
17	Telephone back	Part	–	–0.1	–	0	0	0	–	–	0,10	0,00	0,20	0,70	no
18	Cellular phone body	Part	–	–0.1	–	0	0	206	–	–	0,10	0,00	0,00	0,90	no
19	Cellular phone waste	Part	–	–0.1	–	0	0	0	–	–	0,00	0,00	0,00	1,00	no
20	Printed circuit board (precious)	Part	–	1	–	0	0	95	–	–	0,30	0,00	0,00	0,70	no
21	Printed circuit board (cheap)	Part	–	0	–	0	0	150	–	–	0,10	0,00	0,00	0,90	no

Table 4 (continued)

i	Name		Acceptance price	External recycling price	Taken mass	Maximal mass to be processed	Inter-nally recycled mass	Exter-nally recycled mass	Maximum of external recycled mass	Mass per piece	Composition factor Cu	Composition factor Al	Composition factor Fe	Composition factor waste	Contains hazardous substances
			e_i^A	e_i^{Ex}	y_i^A	$y_i^{A,\max}$	y_i^{In}	y_i^{Ex}	$y_i^{Ex,\max}$	m_i	a_{i1}	a_{i2}	a_{i3}	a_{i4}	
	Unit		[Euro/kg]	[Euro/kg]	[kg]	[kg]	[kg]	[kg]	[kg]	[kg/piece]	[−]	[−]	[−]	[−]	
22	Printed circuit board (standard)	Part	–	0.5	–	0	0	28.8	–	–	0,20	0,00	0,00	0,80	no
23	Switch board parts	Part	–	0	–	0	0	0	–	–	0,20	0,20	0,50	0,10	no
24	Telephone front	Part	–	−0.1	–	0	0	0	–	–	0,10	0,00	0,00	0,90	no
25	Telecommunication switchboard without boards	Part	–	0	–	0	1805	0	–	–	0,11	0,11	0,64	0,15	no
26	Telecommunication switchboard (open)	Part	–	0	–	0	0	0	–	–	0,12	0,10	0,60	0,17	no
27	Telecommunication switchboard (frame)	Part	–	0	–	0	0	0	–	–	0,10	0,10	0,65	0,15	no
28	Telephone receiver	Part	–	−0.1	–	0	0	0	–	–	0,05	0,00	0,00	0,95	no
29	ABS	Material	–	0.05	–	0	0	15	–	–	0,00	0,00	0,00	1,00	no

Table 5. Number of executions of disassembly activities – decision variable x_j results

Disassembly activity j	Input scrap type	Number of executions of disassembly activity (=disassembled pieces)
		x_j
1	TV-set	120
2	Video recorder	0
3	Personal computer	30
4	Personal computer without battery	30
5	Telephone	0
6	Telecommunication switchboard	20
7	Telecommunication switchboard (open)	20
8	Telecommunication switchboard without boards	0
9	Cellular phone	2060
10	Cellular phone body	0

4) – "disk-drives" ($i = 14$), "CD-drives" ($i = 11$) and "standard circuit boards" ($i = 22$) – are marketed externally. The entire amount of "telephones" ($i = 4$) is also accepted and sent immediately to bulk recycling ($j = 5$). "Telecommunication switchboards" ($i = 5$) are all taken and disassembled ($j = 6, 7$) to recover "precious circuit boards" ($i = 20$) for sale. The "accumulators" ($i = 9$) of taken "cellular phones" ($i = 6$) need to be removed ($j = 9$) and treated by specialists. Further disassembly ($j = 10$) of cellular phones is not recommended. The disassembly time capacity constraint and the capacity restriction of the first unit shredder are reached as seen in Table 6. As a result of bulk recycling capacity restrictions, the feed composition of the bulk recycling plant is adjusted consciously through the choice of taken products and marketed parts. The consequence of mono-treatment would have been a decrease in capacity of up to 40% compared to the unit 1's capacity for some scrap types since bottlenecks can appear in the units 1, 4, 5 or 6 depending on the composition of the scrap types. Thus the blending of different scrap types in the feed of bulk recycling represents an economic value for the recycling centre. The objective function represents a value of 2,906 Euro per daily planning period as seen in Table 7. The main contribution to the objective value results from acceptance fee revenues.

In Table 7 the results of different solution procedures for the basic data set are shown. The integer variables cause exponentially increasing solution time due to the complexity of the optimization problem. When the basic data set has been solved by an optimization model without integer variables, the calculated short term recovery program has not changed at all. Concerning other data sets, there were no changes at all in most of the cases, in some cases there were small deviations. However, the use of integer variables concerning the given problem complexity causes negligible extra solution time and can be justified by the more precise model formulation. In

Table 6. Unit cost factors and capacities

		Process cost factor	Maximum available capacity	Used capacity in basic scenario
		c_k^P	$y_k^{P,\max}$	$\sum_{l=1}^{L} y_{kl}^P$
Unit	k	[Euro/kg]	[kg]	[kg]
Shredder	1	0.02	5,000	5,000
Magnet 1	2	0.005	5,000	4,397
Eddy-current separation 1	3	0.005	4,000	2,697
Crusher 1, Magnet 2, Eddy-current separation 2	4	0.02	3,000	2,530
Air separtion	5	0.005	2,000	1,275
Crusher 2	6	0.02	1,500	924
Screen	7	0.005	2,000	924
Density separation	8	0.005	1,500	754

cases of growing problem complexity, the elimination of integer variables may be advisable to provide a quick solution time.

In order to point out the benefits of the optimization model, the calculated result should be compared with basic concepts to determine short term recovery programs that have been used in the enterprise as seen in Table 7. The first concept, "Re-Use", aims at high disassembly output revenues and demands the choice and disassembly of products that contain parts with positive market value, e.g. computers with CD-drives. Disassembly is continued until all parts with a positive value are liberated. Parts that do not contain any hazardous substances and that have a negative value are recycled internally. If disassembly and equipment capacity constraints are not reached this way, the products with the highest acceptance prices are treated. As shown in Table 7, this approach leads to an objective value of 2,411 Euro. The lower value is caused mainly by lower acceptance fees due to smaller TV-set ($i = 1$) quantities.

The second concept "Acceptance Fees" determines the choice of discarded products in sequence of acceptance prices. Only mandatory disassembly is done until the constraints are met. Parts that do not contain any hazardous substances and have a negative value are recycled internally. This concept is assessed with an objective value of 2,557 Euro. The success of this concept is impaired by lower bulk recycling output revenues caused by the treatment of less valuable scrap from TV-sets ($i = 1$) and telephones ($i = 4$) in the bulk recycling units. These intuitive strategies have been compared to the optimization concerning other data sets. Basically, a benefit of at least 10% can be obtained using the optimization model proposed in Section 3.

Table 7. Effect of application of different solution procedures for the short term planning problem

Solution procedures		"Optimization"	"Non-integer model"	"Re-use"	"Acceptance fees"
Note		Optimization based on mixed-integer linear programming	Optimization based on linear programming	Intuitive: Choose scrap types that contain parts with positive value!	Intuitive: Choose scrap types that have high acceptance fees!
Acceptance fees		3,067 Euro	3,067 Euro	2,511 Euro	3,141 Euro
Disassembly output revenues		−105 Euro	−105 Euro	15 Euro	−145 Euro
Bulk recycling output revenues		788 Euro	788 Euro	729 Euro	419 Euro
Disassembly costs		625 Euro	625 Euro	625 Euro	625 Euro
Process costs		219 Euro	219 Euro	220 Euro	233 Euro
Objective function value		2,906 Euro	2,906 Euro	2,411 Euro	2,557 Euro
Chosen scrap types		[kg]	[kg]	[kg]	[kg]
TV-set	i=1	3,000	3,000	2,450	3,000
Video recorder	2	2,000	2,000	1,496	2,000
Personal computer	3	300	300	300	160
Telephone	4	600	600	2	600
Telecommunication switchboard	5	2,000	2,000	2,000	600
Cellular phone	6	412	412	500	500
Disassembly activities notes		None	No changes	Disassembly is done until all parts with positive value are gained.	Only mandatory disassembly
Recycling direction notes		None	No changes	Internal recycling of parts with negative value	Internal recycling of parts with negative value

5 Scenario analysis

The interpretation of the solution for the considered short term planning model reveals bottlenecks and risks concerning the presented recovery system. Recommendations for future planning can be derived, including design improvement hints and promising strategic positions for the recycling enterprise. For that purpose, the system behaviour under changing conditions has been analysed by the assessment of different scenarios based on the sensitivity analysis (review of dual prices and reduced costs) for the non-integer solution of the basic data set.

According to the basic scenario in Table 7, the acceptance fees and the bulk recycling output revenues represent a major share of the objective value. Therefore, price variations have been examined as seen in Table 8. With regard to the acquisition of scrap, all acceptance fees are changed to zero and to double value of the basic scenario respectively. The volatility of material markets is considered by changing the prices for copper and aluminium to half and to double value. Changes in cost coefficients have been analysed as well, but the related impacts are very small. The impact of different capacity constraints can be seen in Table 9. Labour capacity is increased to double value due to the option of hiring temporary workers and decreased to half value due to possible sick leaves of operators. Unit capacity may be harmed by failure downtimes. The extension of acceptance constraints leads to the outranking devices "TV-set" and "video recorder".

The review of dual prices concerning the capacity restrictions of the non-integer solution leads to two bottlenecks and thus to potential improvements for the design of the present recovery system.

On the one hand, one should consider the idea of hiring more workers in the long run. As a result of the sensitivity analysis, 1 Euro that is additionally spent for human work in disassembly would lead to an increasing marginal income of 0.32 Euro. However, this way represents only a small contribution to improve the result. Even if the disassembly capacity is increased up to 150% in the scenario "More Labour Capacity" in Table 9, the benefit for the objective function value amounts to only 1% due to the limited range of attractive disassembly activities. The reduced costs concerning the disassembly activities x_j for $j = 2, 8, 10$ in the basic scenario exceed the value of the objective coefficients $t_j^Z \cdot c^Z$. Therefore, the accomplishment of these activities would not be recommendable even if there were no disassembly capacity restrictions. The main explanation is to be found in the very restricted market for reusable parts, which is also shown through the poor results of the basic "Re-Use" approach. A decrease in disassembly capacity of 50% in the scenario "Less Labour Capacity" leads only to a deviation of the objective function of about 5%.

On the other hand, the sensitivity analysis reveals that every 1 kg additional daily capacity in the shredder ($k = 1$) would improve the objective value by 0.17 Euro. A decreasing unit capacity up to the half in the scenario "Less Unit Capacity" in Table 9 results in a strong reduction of about 17% concerning the objective function value. More scrap types with low material value, such as "mixed entertainment electronic parts" ($i = 15$), are recycled externally. Scrap types that require disproportionate unit capacity, like telecommunication switchboards, are not taken for treatment.

Table 8. Effect of different scenarios concerning prices

Scenarios		Basic scenario	"Metal prices double"	"Metal prices half"	"Acceptance prices double"	"Acceptance prices half"	"Acceptance prices zero"
Assumptions		Basic scenario	The prices for copper and aluminium fractions grow to double value	The prices for copper and aluminium fractions sink to half value	The acceptance prices for all scrap types grow to double value	The acceptance prices for all scrap types sink to half value	The acceptance prices for all scrap types sink to zero
Acceptance fees		3,067 Euro	2,949 Euro	3,067 Euro	6,194 Euro	1,263 Euro	0 Euro
Disassembly output revenues		−105 Euro	−158 Euro	−105 Euro	−140 Euro	11 Euro	149 Euro
Bulk recycling output revenues		788 Euro	1,973 Euro	313 Euro	768 Euro	866 Euro	998 Euro
Disassembly costs		625 Euro	625 Euro	625 Euro	625 Euro	387 Euro	195 Euro
Process costs		219 Euro	215 Euro	219 Euro	219 Euro	214 Euro	214 Euro
Objective function value		2,906 Euro	3,924 Euro	2,431 Euro	5,977 Euro	1,539 Euro	738 Euro
Chosen scrap types	y_i^A	[kg]	[kg]	[kg]	[kg]	[kg]	[kg]
TV−set	i=1	3,000	3,000	3,000	3,000	3,000	–
Video recorder	2	2,000	1,996	2,000	2,000	2,000	1,244
Personal computer	3	300	1,310	300	–	520	2,000
Telephone	4	600	600	600	600	–	–
Telecommuication switchboard	5	2,000	2,000	2,000	2,000	2,000	2,000
Cellular phone	6	412	195	412	472	–	–
Disassembly activities notes		None	Telephones are not shredded due to low material value	None	None	PC disassembly is performed more often	PC disassembly is performed more often
Recycling direction notes		None	Scrap types with low material value are recycled externally	None	None	Complete internal recycling of mixed entertainment electronic parts	None

Table 9. Effect of different scenarios concerning restrictions

Scenarios		Basic scenario	"Less labour capacity"	"More labour capacity"	"Less unit capacity"	"Bigger scrap market 1"	"Bigger scrap market 2"
Assumptions		Basic scenario	The disassembly capacity sinks to half value	The disassembly capacity grows to double value	The unit capacity of all units sinks to half value	All acceptance restrictions are extended to 3/2 value	There are no acceptance restrictions
Acceptance fees		3,067 Euro	2,625 Euro	3,267 Euro	3,005 Euro	3,991 Euro	5,406 Euro
Disassembly output revenues		−105 Euro	−126 Euro	−160 Euro	−247 Euro	−277 Euro	−421 Euro
Bulk recycling output revenues		788 Euro	782 Euro	929 Euro	395 Euro	825 Euro	425 Euro
Disassembly costs		625 Euro	313 Euro	880 Euro	625 Euro	625 Euro	625 Euro
Process costs		219 Euro	219 Euro	214 Euro	126 Euro	215 Euro	229 Euro
Objective function value		2,906 Euro	2,749 Euro	2,942 Euro	2,402 Euro	3,699 Euro	4,556 Euro
Chosen scrap types	y_i^A	[kg]	[kg]	[kg]	[kg]	[kg]	[kg]
TV-set	i=1	3,000	3,000	3,000	3,000	4,500	6,250
Video recorder	2	2,000	2,000	2,000	2,000	3,000	5,000
Personal computer	3	300	–	1,420	280	300	–
Telephone	4	600	600	600	600	900	–
Telecommunication switchboard	5	2,000	2,000	2,000	700	2,200	–
Cellular phone	6	412	–	500	352	24	–
Disassembly activities notes		None	Mandatory TV-set disassembly and switchboard parts removal	Disassembly of telephone and pc; capacity not reached	None	Disassembly of telephones is done due to low material value	Only mandatory TV-set disassembly
Recycling direction notes		None	None	None	Scrap types with low material value are recycled externally	Scrap types with low material value are recycled externally	Internal video recorder recycling, TV-set parts are marketed externally

Thus a valuable improvement hint for future process design modifications may be raising the capacity of the bottleneck unit 1 in bulk recycling in the long run.

The decision concerning internal or external recycling of scrap types is influenced strongly by the market situation. The variation of prices as seen in Table 8 directs the program depending on the amount of valuable materials. With the growing relevance of bulk recycling output revenues, scrap types with big metal shares should be recycled internally. Higher (lower) acceptance prices, lower (higher) metal prices and bigger (smaller) scrap markets lead to a choice of more (less) low-value scrap types like cellular phones or telephones and of less (more) scrap types with valuable parts like personal computers. Scrap types with low material value, for instance telephone parts, are recycled externally in order to reduce the amount of plastics in bulk recycling. As a result more unit capacity for the more valuable inputs like parts of personal computers is available.

The reduced costs concerning the internally recycled masses y_i^{In} give relevant hints to decide in which cases one should think about external recycling of scrap types. Those parts that contain very high ($i = 20; 22$) shares of valuable metals have the highest reduced costs. As a conclusion, it is difficult to obtain any benefit from trying to improve the purity of nearly pure scrap types by mixing it with other scrap in the feed of the bulk recycling plant.

As a result of the optimization calculations for the present system, disassembly is only advisable if hazardous or very precious parts are removed. Disassembly can not compete with bulk recycling concerning the recovery of valuable materials due to the high variable costs. However, the long-term allocation between manual activities and unit operations may be completely different because of the high impact of fixed cost in bulk recycling that have not to be taken into account in this short term planning calculation.

The data set of this case study refers to a typical, but small range of electronic devices. Classifying the considered products into several groups and taking into account the outcomes of the scenario analyses, one is enabled to generalize a recycling characterization for other products.

- Some devices like "TV-sets" ($i = 1$) or "cellular phones" ($i = 6$) contain harmful substances that require mandatory disassembly activities. The reuse possibilities and the material values are low. This leads to a decreasing share in the short term recovery program when acceptance prices fall, which is shown in the scenarios "Acceptance Prices Zero" and "Acceptance Prices Half" as seen in Table 8. Thus the only reason to treat this group is the acceptance fee revenue. Other devices in that group may be refrigerators or monitors.
- "Video recorders" ($i = 2$) and "telephones" ($i = 4$) can be shredded directly since they do not contain any harmful substances. The motive to treat this group is mainly the acceptance fee, but the material value represents a small contribution as well. Hi-fi devices and some domestic appliances may be characterized in the same way.
- "Personal computers" ($i = 3$) and "telecommunication switchboards" ($i = 5$) as well as lot of other investment goods contain reuse parts and valuable materials. The recovery of these represents a considerable economic value. The

relevance of treating this group increases with sinking acceptance prices as well as growing material and parts prices and market extent.

It becomes obvious that the acceptance fees have the highest impact on the success of the company. The share of the acceptance fees in the overall marginal income is nearly 106% in the basic scenario as seen in Table 4. Hence the "disassembly" and "bulk recycling" steps of the optimal recovery program have a negative value. Furthermore, the objective function value of the basic strategy "Acceptance Fees" touches almost the level of the basic scenario (88%). The scenarios with lower or no acceptance prices reduce the objective value of the basic scenario to 53% and 25% respectively. In comparison, the scenario with lower prices for copper and aluminium leads only to a reduced objective value of about 16%. Basically, it would be very profitable at the moment to try to take more scrap types with high acceptance fees.

Considering these results and ongoing trends in the legal and economic environment such as the forthcoming European directive concerning Waste on Electrical and Electronic Equipment (European Commission, 2000), one can also derive recommendations for strategic planning. Since recycling business depends on two markets that may contribute to success, two strategic positions determining a market focus seem to be promising. Both of them may derive benefit by long-term cooperation with producers of equipment in order to avoid deviations in supply. An important aspect of both strategies is to assure the delivery of the necessary input of discarded products (Guide and van Wassenhove, 2001).

Concentrating on the acceptance fee revenues, one should secure the delivery of scrap through supplier contracts. An important change concerning future scrap markets is the extended producer responsibility in the electronics industry, which will lead to charging producers for collecting, sorting and recovery of WEEE from private households. Producers have to provide systems to assure the treatment of collected WEEE. As a result, producers or consortia of producers will obtain greater importance as customers of the recycling industry. In order to meet fixed recycling quotas, e.g. through the isolation of pure plastic fractions, the high-end disassembly of "low-valued" scrap will obtain an increasing status. The main challenge for the recycling company is to enter these systems by signing long-term-contracts. This will lead to economy-of-scale benefits due to greater quantities of WEEE from private households (TV-sets, video recorders, personal computers) and long-term fixed acceptance fee revenues. The other side of this strategy is characterized by marketing incentives for the incurring recycling output.

A different strategic position focuses on the output side of recycling. Variations in the metal markets can not be avoided by fixed-price contracts because of the small market relevance of recycling enterprises in comparison to smelting plants. However, the delivery of reusable parts for spare parts management and equivalent-to-new-parts for the use in production processes are a practical way to secure output proceeds, especially concerning investment goods. The scrap supply may be accomplished by financial incentives or by cooperation with producers. In the latter case, recycling companies should be integrated into the producers supply chain, e.g. as a supplier of spare parts. In order to support this cooperation, producers and recycling enterprises should use joint information and planning tools.

6 Conclusions and recommendations for future research

In this paper, an integrated short term recovery planning problem for electronic scrap has been analysed and formulated as a mixed-integer linear programming model, based on the linear activity analysis. The purpose of this daily task is to obtain an optimal choice of recovered products for disassembly and bulk recycling as well as an optimal allocation of disassembly activities. Compared with the models found in the literature, the presented model has some specific characteristics, especially the description of material flows throughout the bulk recycling units, the consideration of the choice of taken products in combination with capacity constraints and the explicit modelling of interactions between disassembly and bulk recycling. Though the planning tool refers to electronic scrap recovery, it can also be applied to other recovery planning problems with bulk recycling processes.

The integer constraints in the present model formulation are not essential, but justified by the precise depiction of the reality and the short solution time for the given complexity. The presented model provides practicable decision support concerning the amount and types of discarded products as well as disassembly and bulk recycling activities for recovery facilities. Compared with basic short term recovery programs used in the examined recycling enterprise, a benefit of at least 10% can be obtained. When we model a multistage electronic scrap recovery facility including disassembly and bulk recycling, our results show that disassembly is only advisable if hazardous or very precious parts are removed. Therefore, rough data are sufficient for a model of scrap recycling if there is no functional value in the recycled parts.

Useful insights concerning operation of the recycling system can be given by the analysis of reduced costs, dual prices and scenarios. On the one hand, the share of valuable scrap types treated in the bulk recycling unit should grow with sinking acceptance prices and unit capacities as well as with increasing material prices. On the other hand, using bulk recycling for nearly pure scrap types seems to be questionable.

The data set of this case study consists of a typical range of products that represent major groups of electronic scrap. This classification is not based on standardized collection groups because these groups depend on the collector at present. In the future, municipal collection systems with specific scrap categories are expected to be realized to comply with regulations like the forthcoming directive on WEEE (European Commission, 2000). The scrap market itself and therefore the decisions of recovery enterprises will be affected by this classification. Hence the adaptation of these scrap categories in the planning model will be necessary.

Long-term cooperation between producers of equipment and recycling enterprises will attain an increasing relevance due to the expected benefits for both sides. Two basic strategic positions for recycling enterprises have been presented: on the one hand concentrating on acceptance fee revenues within the scope of implementing take back and recycling-systems and on the other hand intensifying efforts to recover reusable parts which will find a ready market through producers and suppliers in the traditional supply chain. Especially the last strategy will lead to closed loop supply chains. Future research is necessary to balance demand and supply of the reusable parts and to coordinate producers or suppliers demand planning and the recovery program planning of the recycling enterprise.

Since the presented model formulation is limited to one planning period, an enhancement to regard dynamic factors may be promising in order to take inventory decisions into account. Impurities affect the compositions and the prices of externally marketed material fractions. Though the assumption that only target materials are separated by the units is justified by the fact that the separated fractions are nearly pure, the consideration of these impurities in the separated fractions may be beneficial. In addition, the assessment of the derived hints for process improvements should be analysed in the context of a long-term planning period.

References

European Commission (2000) Proposal for a directive of the European Parliament and of the Council on Waste Electrical and Electronic Equipment. Brussels, Commission proposals: Document 500PC0347(01)

Fleischmann M (2000) Quantitative models for reverse logistics. Springer, Berlin Heidelberg New York

Guide VDR, Van Wassenhove LN (2001) Managing product returns for remanufacturing. Production and Operations Management 10: 142–155

Gungor A, Gupta SM (1999) Issues in environmentally conscious manufacturing and product recovery: a survey. Computers & Industrial Engineering 36: 811–853

Koch P, Kasper R (1996) Dismantling and process technology for electronic scrap and discarded electric appliances. Aufbereitungstechnik 37: 211–219

Koopmans TC (1951) Efficient allocation of resources. Econometrica 19: 455–465

Krikke HR, Van Harten A, Schuur PC (1998) On a medium term product recovery and disposal strategy for durably assembly products. International Journal of Production Research 36: 111–139

Isaacs J, Gupta SM (1998) Economic consequences of increasing polymer content for the U.S. automobile recycling infrastructure. Journal of Industrial Ecology 1: 19–33

Lund JR, Tchobanoglous G, Anex RP, Lawyer RA (1994) Linear programming for analysis of material recycling facilities. Journal of Environmental Engineering 5: 1082–1094

Meacham A, Uzsoy R, Venkatadri U (1999) Optimal disassembly configurations for single and multiple products. Journal of Manufacturing Systems 18: 311–322

Meier-Staude R, Mersmann A (1997) Cost-structure and cost factors for mechanical processing of electro-scrap. Aufbereitungstechnik 38: 287–296

Penev KD, De Ron AJ (1996) Determination of a disassembly strategy. International Journal of Production Research 34: 495–506

Rudolph A (1999) Altproduktentsorgung aus betriebswirtschaftlicher Sicht. Physica-Verlag, Heidelberg

Schubert H (1994) Eddy-current separation – Fundamentals, separators, application. Aufbereitungstechnik 35: 553–562

Sodhi MS, Reimers B (2001) Models for recycling electronics end-of-life products. OR Spektrum 23: 97–115

Sodhi MS, Young J, Knight WA (1999) Modelling material separation processes in bulk recycling. International Journal of Production Research 37: 2239–2252

Souren R (1996) Theorie betrieblicher Reduktion: Grundlagen, Modellierung und Optimierungsansätze stofflicher Entsorgungsprozesse. Physica-Verlag, Heidelberg

Spengler T (1998) Industrielles Stoffstrommanagement, Betriebswirtschaftliche Planung und Steuerung von Stoff- und Energieströmen in Produktionsunternehmen. Erich-Schmidt-Verlag, Berlin

Spengler T, Püchert H, Penkuhn T, Rentz O (1997) Environmental integrated production and recycling management. European Journal of Operational Research 97: 308–326

Spengler T, Schröter M (2001) Einsatz von Operations Research im produktbezogenen Umweltschutz – Stand und Perspektiven. Betriebswirtschaftliche Forschung und Praxis (BfuP) 53: 227–244

Stuart JA, Lu Q (2000a) A model for discrete processing decisions for bulk recycling of electronics equipment. IEEE Transactions on Electronics Packaging Manufacturing 4: 314–320

Stuart JA, Lu Q (2000b) A refine-or-sell decision model for a station with continuous reprocessing options in an electronics recycling center. IEEE Transactions on Electronics Packaging Manufacturing 4: 321–327

Uzsoy R, Venkatachalam G (1998) Supply chain management for companies with product recovery and remanufacturing capability. International Journal of Environmentally Conscious Design & Manufacturing 7: 59–72

Verein Deutscher Ingenieure (2000) Recyclingorientierte Produktentwicklung. VDI-Richtlinie 2243 Blatt 1 Entwurf. VDI, Düsseldorf